JONATHAN DRUMMOND

# Structural dynamics for engineers

# Structural dynamics for engineers

H. A. Buchholdt

Thomas Telford

Published by Thomas Telford Publications, Thomas Telford Services Ltd, 1 Heron Quay, London E14 4JD

First published 1997

Distributors for Thomas Telford books are
*USA:* American Society of Civil Engineers, Publications Sales Department, 345 East 47th Street, New York, NY 10017-2398
*Japan:* Maruzen Co. Ltd, Book Department, 3–10 Nihonbashi 2-chome, Chuo-ku, Tokyo 103
*Australia:* DA Books and Journals, 648 Whitehorse Road, Mitcham 3132, Victoria

A catalogue record for this book is available from the British Library

**Classification**
*Availability:* Unrestricted
*Content:* Original analysis
*Status:* Author's invited opinion
*User:* Structural and Civil engineers

ISBN: 0 7277 2559 9



Typeset in Great Britain by Alden Bookset, Didcot.
Printed in Great Britain by Redwood Books, Trowbridge, Wiltshire.

# Contents

# Preface

This book is intended as an introduction to the dynamics of civil engineering structures. It has evolved from lectures given to industrially based MSc students in order to improve their understanding and implementation of modern design codes, which increasingly require a greater knowledge and understanding of vibration caused by either man or the environment. It is also intended to give practising engineers a better understanding of the dynamic theories that form the basis of computer packages. Examples of such packages, which include dynamics analysis and soil–structure interaction, are:

- ANSYS (US)—general purpose finite element program
- LUSAS (UK)—civil engineering and design program
- DYNA 3 (UK/Japanese/US)—dynamics and soil–structure interaction program.

Experience has shown that it is all too easy to make mistakes in the input data and still accept the results obtained. It is hoped that the methods presented will aid the practising engineer to judge the validity of the dynamic response calculations obtained using the above programs. This book is not intended to be an advanced course in theoretical structural dynamics: for this the author recommends *Dynamics of structures* by R. W. Clough and J. Penzien. However, some topics have been developed further than in most textbooks, namely the evaluation of damping values and the use of spectral and cross-spectral density functions, or power spectra, to predict response to wind and earthquakes and to generate correlated wind and earthquake histories needed when analysing nonlinear structures.

As this book is intended for the practising engineer, certain older techniques have been omitted, such as the Duhamels integral used in time domain analysis. In the author's experience, the Newark $\beta$-equations or the Wilson $\theta$-equations are easier to understand and equally effective. Also, no reference has been made to computer methods for solving large eigenvalue problems. For these the reader should consult mathematical textbooks.

This book has two major omissions. The first is that no reference is made to wave loading such as experienced by dams and offshore structures. For this the reader is referred to *Dynamic analysis of offshore structures* by C. A. Brebbia, which gives a very good introduction to the subject. Another omission, in this case a partial one, is the subject of soil–

structure interaction, which is important as it modifies the dynamic behaviour of structures. The interaction between structure and ground can be taken into account either by representing the stiffness and damping properties of soil as equivalent springs and dampers respectively, or by modelling the soil by finite elements. The former is at best an approximate method, which requires some experience to use. For the latter a great deal of experience is necessary, as this is a highly specialized field, and it is therefore considered to be outside the scope of this book. For this reason only the concept of numerical modelling of the soil by springs and dampers is presented. For more detailed information the reader is, in the first instance, referred to *Earthquake design practice for buildings* by D. E. Key.

There are a number of other topics that it has not been possible to include, as the main purpose of this work is to give the reader an introduction to the vibration of structures and, it is hoped, to make other more advanced or specialized texts easier to follow. Most of the omitted topics can be found in a handbook on vibration entitled *Shock vibration* by C. M. Harris. Methods for dynamic response analysis of cable and cable-stayed structures are given in *Introduction to cable roof structures* by the author.

In order to follow the theoretical work presented the reader will need to have some knowledge of differential and integral calculus, first and second order differential equations, determinants, matrices, and matrix formulation of structural problems. These topics are included in the teaching of mathematics and the theory of structures in undergraduate engineering courses. A knowledge of the concept of eigenvalues and eigenvectors will be useful, but is not essential.

Chapter 1 gives a number of reasons why the modern structural engineer needs to have a knowledge of vibration. Many civil engineering structures vibrate predominantly in the first mode with a simple harmonic motion, and may therefore be reduced to mass–spring systems with only one degree of freedom. Also covered in this chapter is the concept that wind and earthquake histories may be considered to consist of a summation of harmonic components, and that most structures which possess a dominant frequency that falls within the frequency band of either history will tend to vibrate at resonance.

Chapter 2 shows how to make an initial estimate of the dominant first natural frequencies of loaded beam elements, continuous beams and multistorey structures by equating the maximum kinetic energy to the maximum strain energy at resonance.

The theories of free damped linear and torsional vibration of one degree of freedom systems are presented in chapter 3.

Chapter 4 provides closed-form solutions to the response of damped one degree of freedom systems subjected to rectilinear and torsional harmonic excitation caused by the rotation of unbalanced motors, and to harmonic support excitation.

The evaluation of structural damping is considered in chapter 5. Measurements of damping by the two classical methods described in all dynamic textbooks, namely the measurement of logarithmic damping from records of decaying vibrations and the measurement of damping ratios from amplitude–frequency curves, the so-called *bandwidth method*, usually leads to inaccurate results. The two main reasons are that the level of damping varies with the amplitude of vibration and structural damping is at best only approximately viscous. The latter method is also difficult to implement as it is usually difficult to obtain a set of satisfactory values near the peak of the curve on either side of resonance. The author has therefore included a few methods, not found in most other textbooks, by which the accuracy of these methods can be studied and improved on.

Chapter 6 is devoted to the formulation of step-by-step methods for calculating time histories of response of one degree and multi-degrees of freedom systems, when subjected to impulse loading, and time histories for wind, earthquakes and explosions.

Matrix formulation of the equations of motion of free vibration, and the calculation of natural frequencies and modeshapes for multi-degrees of freedom systems, are presented in chapter 7, together with methods for reducing the number of degrees of freedom of structures when this is required. Methods for determining the natural frequencies and mode-shapes of simplified structures are included.

Chapter 8 presents the classical method of mode superposition in which the dynamic response of an $N$ degrees of freedom structure is sought by transforming the global equations of motion into the equations of motion for $N$ one degree of freedom system. This transformation is made possible by the orthogonal properties of the eigenvectors or modeshape vectors for the structure and assumptions made with respect to the properties of the damping matrix.

The construction of damping matrices is considered in chapter 9. It has been pointed out that in practice such matrices need to be assembled only in the case of dynamic response analysis of nonlinear systems such as cable and cable stayed structures. These structures may respond in a number of closely spaced modes for which the method of mode superposition presented in chapter 8 is not appropriate. In this connection it should be mentioned that the use of inadequate damping matrices in the case of a cable-stayed bridge model resulted in calculated amplitudes of strains in the stays that differed considerably from those measured.

Chapters 10 and 11 deal with the nature and statistical properties of wind and the response to buffeting and vortex shedding. Chapter 11, dealing with dynamic response, is mainly concerned with frequency domain analysis using power spectra and the method of mode superposition developed in chapter 8. A considerable amount of space is devoted to the use and importance of cross-spectral density functions, which are not normally found in detail elsewhere.

Chapters 12 and 13 deal with the nature of, and dynamic response to, earthquakes. Again the emphasis is on frequency domain methods using the method of mode superposition, response and power spectra. Examples of both rectilinear and torsional response analyses are given. The new Eurocodes require that rocking motion caused by earthquakes should be taken into account in future designs. For this reason the author has constructed a power spectrum for rocking in order to demonstrate its use in dynamic analysis. The author wishes to emphasize that such spectra are introduced only in order to be able to demonstrate their use if and when they become available, as they appear not to be covered in the current literature. The spectrum in this text should not be used for design purposes.

Chapter 14 presents methods for generating spatially correlated wind histories, and families of correlated earthquake histories, as such histories need to be available in order to use the step-by-step methods given in chapters 6, 11 and 13. Earthquake histories may be generated either with the statistical properties of recorded earthquakes and the dominant ground frequency of the site. References to research work behind the development of these methods are given.

Throughout the text, solved examples are given in order to illustrate and demonstrate the use of theories presented; it is hoped that these will prove useful to the reader.

The author has learnt a great deal while writing this book and hopes that others also will benefit from his work. From those who read the book, he should be pleased to receive comments and suggestions for a possible revised edition, and to have his attention drawn to any errors that must inevitably exist and for which he alone is responsible.

Finally, the author should like to thank the following: Professor P. Regan, head of the School of Architecture and Engineering at the University of Westminster, for making the book possible by giving the author the opportunity to lecture on the School's MSc course for industrially based students; Dr S. Moossavinejad for his continued help and support; the author's old friends Øistein Bertheau and Harald Bjerke for suggestions and help with chapter 1; and his wife Reiko Buchholdt for encouraging the work and drawing a large number of the diagrams.

# 1. Causes and effects of structural vibration

## Introduction

An understanding of structural vibration and the ability to undertake dynamic analysis are becoming increasingly important. The reasons for this are obvious. Advances in material and computational technology have made it possible to design and construct taller masts, buildings with ever more slender frames and with skins that contribute little to the overall stiffness, and roofs and bridges with increasingly larger spans. In addition, masts, towers and new forms of construction such as offshore structures are being built in more hostile environments than previously. This, together with increasing vehicle weights and traffic volumes, requires that designers take vibration of structures into account at the design stage to a much greater extent than they have done in the past.

Sometimes the trouble caused by vibration is merely the nuisance resulting from sound transmission, or the feeling of insecurity arising from the swaying of tall buildings and light structures such as certain types of footbridge. Occasionally, however, vibration can lead to dynamic instability, fatigue cracking, or incremental plastic deformations. The first two types of problem may lead to a reduction in utilization of a structure, and the last type to costly repairs if discovered in time or, if not, to complete failure with possible loss of human life.

Unfortunately, there are numerous examples of structural vibrational failures, many of which have resulted in the loss of life. The disastrous effects of the numerous large earthquakes that have occurred during the twentieth century are obvious examples, but wind and waves have also taken their toll. Examples are the collapse of the Tacoma Narrows Bridge in the USA, the *Alexander Kjelland* platform in the North Sea, the cooling towers at Ferry Bridge in the UK, and numerous large cable-stayed masts. Many failures of such masts have occurred in Arctic environments, but a number have occurred in countries with more clement climates such as Britain, Germany and the USA. Frequently the cause of failure has been the development of fatigue cracks in the attachments of the guys to the tower, caused by the vibration of the guys.

Until quite recently is was assumed that the response of guyed masts to earthquakes need not be considered. Recent research has shown, however, that an earthquake can be as severe as any storm if the dominant frequency of the ground coincides with one of the main natural frequencies of the mast, and can cause local buckling of structural elements, which again can lead to a complete collapse.

Long-term vibration induced by traffic can lead to fatigue in structural elements and should not be underestimated. A number of old railway bridges have started to develop fatigue cracks in the gusset plates, which have had to be replaced. Costly repairs and modifications have also had to be undertaken on relatively new suspension and cable-stayed bridges, because the possibility of fatigue caused by traffic-induced vibration had not been sufficiently investigated at the design stages. The suspension bridge across the River Severn near Bristol in the UK is one example among many.

The effect of traffic is not confined to bridges—it must also be taken into account when the foundations of buildings situated next to railway lines or roads carrying heavy traffic are being designed. It is interesting that many ancient buildings such as cathedrals, which have been built next to main roads, tend to lean towards the road and in many cases also show signs of cracks as a result of centuries of minute amplitude vibrations caused by carts passing on the cobbled road surfaces.

In factories, rotating machinery can lead to large amplitude vibrations that can cause fatigue problems if not considered early enough. Such problems apparently occur more frequently than has been generally appreciated in the past, and some countries such as Sweden have produced design guides in an effort to overcome them. Other causes of vibration are currents in air and water, explosions, impact loading, and the rupture of members in tension. Currents can give rise to vibration as a result of vortex shedding. Explosions such as are used in demolitions will transmit pressure waves both through the ground and through the air and can, if insufficient precautions are taken, cause damage to nearby buildings and sensitive electronic instruments. The dynamic shocks set up by the ruptures of highly tensioned members such as steel cables in tension can be devastating. A number of mast failures, where one of the guys or attachments has ruptured because of the development of fatigue cracks, can be attributed to the magnitude of the bending moments caused by the resulting dynamic shock—these moments will have been several times greater than those the towers would have experienced if the guys had been removed statically. There are also examples of kilometres of electric transmission lines with towers collapsing because of the rupture of a single cable, and a number of hangers in a suspension bridge have snapped because a single hanger was broken when hit by a lorry.

The sudden release of forces restraining the movements of an element may also lead to structural failures. Thus, after a very heavy snowfall the steel box space ring containing the pretensioned cable net roof over the Palasport in Milan buckled. The space ring was supported on roller bearings on the top of inclined columns. A number of explanations for the failure were suggested; the most likely one is that the rollers, which were completely locked at the time the roof was subjected to resonance testing, suddenly moved under the exceptionally heavy snow load, and the resulting dynamic shock induced bending moments much larger than those for which the ring had been designed.

From the above it ought to be evident that it is important for engineers not only to develop an understanding of structural vibration, but also to be able to investigate the effects of dynamic response at the design stage, when a structure can readily be modified, rather than having to make possibly costly alterations later on.

Not every designer needs to be an expert in dynamic analysis, but all ought to have an understanding of the ways in which structures are likely to respond to different types of dynamic excitation, and of the fact that some types of structure are more dynamically sensitive than others. In particular, designers should know that all structures possess not one but a number of *natural frequencies*, each of which is associated with a particular *modeshape of vibration*, and should be aware of the fact that pulsating forces or pulsating force components with the same frequencies as the structural ones will cause the structure to vibrate with amplitudes much greater than those caused by pulsating forces with frequencies different from the structural ones. Therefore, the designer needs to be able to calculate the natural frequencies of a structure, identify and formulate the characteristics of different types of man-made and environmental forces, and to calculate the total response to these forces in the modes in which the structure will vibrate. An understanding of the importance of damping and the principles and methods to control and reduce the amplitudes of vibrations is also required.

The following are some types of structure and structural element that, experience has shown, can be dynamically sensitive:

- tall buildings and tall chimneys
- suspension and cable-stayed bridges
- steel-framed railway bridges
- free-standing towers and guyed masts
- cable net roofs and membrane structures
- cable-stayed cantilever roofs
- cooling towers
- floors with large spans and floors supporting machines
- foundations subjected to vibration
- structures during erection and structural renovation
- offshore structures
- electrical transmission lines.

This list is not intended to be exhaustive, but it indicates the range and variety of civil engineering structures whose dynamic responses need to be considered before they are constructed. Of the above, only the membrane and cable and cable-stayed structures are likely to respond in a relatively large number of modes, because their dominant frequencies are closely spaced within their respective frequency spectra.

The relationship between the dominant frequency of a structure and its degree of static structural stability also deserves attention. Both are functions of stiffness and mass. The criterion for instability is that the stiffness during any time of a load history becomes zero. If this happens the dominant frequency will also be zero, and the mode of collapse will be similar to that of the mode shape of vibration. Thus frequency analysis is a useful tool for investigating the stability of a structure, and the amount of load a structure can support before it becomes unstable.

Finally, engineers concerned with the design, operation and maintenance of nuclear installations need to have a thorough understanding of the effects of vibrations caused by any possible source of excitation, because of the very serious consequences of any failure.

## Vibration of structures: simple harmonic motion

The motion of any point of a structure when vibrating in one of its natural modes closely resembles *simple harmonic motion* (SHM). An example of simple harmonic motion is the type of motion obtained when projecting the movement of a point on a flywheel, rotating with a constant angular velocity, on to a vertical or horizontal axis. The motion of any point of a structure vibrating in one of its natural modes can therefore be described by

$$x(t) = x_0 \sin(\omega_n t) \tag{1.1}$$

$$\dot{x}(t) = x_0 \omega_n \cos(\omega_n t) \tag{1.2}$$

$$\ddot{x}(t) = -x_0 \omega_n^2 \sin(\omega_n t) \tag{1.3}$$

where $x(t)$ is the amplitude of motion at time $t$, $\dot{x}(t)$ is the velocity of the motion at time $t$, $\ddot{x}(t)$ is the acceleration of motion at time $t$, $x_0$ is the maximum amplitude of response, and $\omega_n$ is the natural angular frequency of the structure in rad/s.

As it happens, eq. (1.1) also represents the motion of a lumped mass suspended by a linear elastic spring, when the mass is displaced from its position of equilibrium and then released to vibrate. It is thus possible to model the vibration of a structure in a given mode as an equivalent mass–spring system where the lumped mass and spring stiffnesses are associated with a given mode shape. From Newton's law of motion

$$M\ddot{x} = -Kx \tag{1.4}$$

Substitution of the expressions for $x(t)$ and $\ddot{x}(t)$ given by eqs (1.1) and (1.2) yields

$$M\omega_n^2 = K \tag{1.5}$$

Hence

$$\omega_n = \sqrt{(K/M)} \tag{1.6}$$

$$f = \frac{1}{2\pi}\sqrt{(K/M)} \tag{1.7}$$

where $f$ is the frequency in cycles per second (Hz).

It should be noted that in eq. (1.7) the stiffness must be in N/m, and the mass must be in kg. If the weight of the vibrating mass is used then, since

$M = W/g$, the unit of weight must be N, and that of the gravitational acceleration $g$ must be m/s$^2$. When the weight rather than the mass is used, eq. (1.7) must be written as

$$f = \frac{1}{2\pi}\sqrt{Kg/W} \tag{1.8}$$

Single mass–spring systems are referred to as *one degree of freedom systems*, or one DOF systems. They are particularly useful as initial numerical models when one is first trying to ascertain the possible dynamic response of a structure, as most civil engineering structures mainly respond in the first mode. From eq. (1.7) it follows that to calculate the dominant natural frequency of a structure one need only calculate or obtain the equivalent spring stiffness and the magnitude of the corresponding vibrating mass. The stiffness may be obtained by elastic calculations using equations derived in linear elastic theory; from a computer program that calculates the deflection for a specified force; or from static testing of a model or a real structure.

Alternatively eq. (1.7) permits the calculation of the equivalent lumped vibrating masses of structures, if the stiffnesses and the frequencies of the structures are known. The frequencies in such cases must be found by dynamic testing or by the use of a computer using a standard eigenvalue program.

The determination of a first natural frequency and an equivalent vibrating mass is demonstrated in the following two examples.

**Example 1.1** Determine the frequency of a bridge with a 10 t lorry stationed at midspan. The bridge itself may be considered as a simply supported beam of uniform section having a total weight of 200 t. From a static analysis of the bridge it was found that the deflection at midspan due to a force of 1.0 kN applied at midspan is 1.5 mm.

The stiffness of the bridge at midspan is given by

$$K = 1000\,\text{N}/0{\cdot}0015\,\text{m} = 6{\cdot}67 \times 10^5\,\text{N/m}$$

The mass to be included is the sum of the mass of the lorry and the equivalent vibrating mass of the bridge. In chapter 2 it is shown that for simply supported beams of uniform section the equivalent mass is approximately equal to half the total one. Hence the equivalent lumped mass is

$$M = 10\,000\,\text{kg} + 0{\cdot}5 \times 200\,000\,\text{kg} = 110\,000\,\text{kg}$$

Finally, substitution of the values for $M$ and $K$ into eq. (1.7) yields

$$f = \frac{1}{2\pi}\sqrt{(6{\cdot}67 \times 10^5/110\,000)} = \underline{0{\cdot}392\,\text{Hz}}$$

**Example 1.2** It has been decided to undertake a preliminary study of the dynamic response characteristics of the cable-stayed cantilever roof shown in Fig. 1.1 by calculating the response of a one DOF system representing the vibration of the free end of the roof. Static testing of the roof had shown that it will deflect 1·0 mm when a force of 1·0 kN is applied at the free end, and dynamic testing that the frequency of vibration is 1·6 Hz. Calculate the stiffness and the mass of the equivalent one DOF mass–spring system needed for initial dynamic investigation.

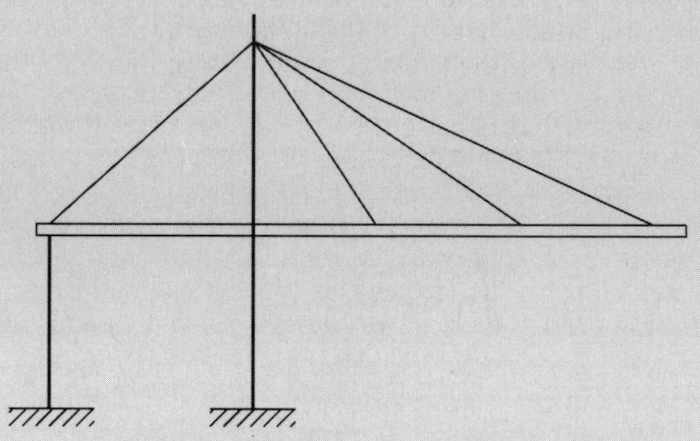

*Fig. 1.1. Cable-stayed cantilever roof*

The stiffness of the cantilever roof is

$$K = 1000\,\mathrm{N}/0{\cdot}002\,\mathrm{m} = 5{\cdot}0 \times 10^5\,\mathrm{N/m}$$

Hence the equivalent lumped mass of the structure is given by

$$M = K/(2\pi f)^2 = 5{\cdot}0 \times 10^5/(2\pi \times 1{\cdot}6)^2 = \underline{4947{\cdot}32\,\mathrm{kg}}$$

## Nature and dynamic effect of man-made and environmental forces

As mentioned above, the significance of the natural frequencies is that if a structure is excited by a pulsating force with the same frequency as one of the structural ones it will begin to vibrate with increasing amplitudes, which can be many times greater and therefore more destructive than the deflection that would have been caused by a static force of the same magnitude as the maximum pulsating force. When this is the case the structure is said to be vibrating in resonance. Thus a dynamic force or force component

$$P(t) = P_0 \sin(\omega t) \qquad (1.9)$$

will give rise to large amplitude vibration if $\omega = \omega_n$.

A commonly encountered form of dynamic excitation is that caused by unbalanced rotating machines and motors. This form of dynamic force can generally be expressed by

$$P(t) = me\omega_i^2 \sin(\omega_i t) \tag{1.10}$$

where $m$ is the total unbalanced mass, $e$ is the eccentricity of the mass $m$, and $\omega_i$ is the speed of the motor. Even small values of the product $me$ can lead to problems if $\omega = \omega_n$, unless designed against.

Figure 1.2 shows the recorded histories of wind velocities at different heights along a mast; Fig. 1.3 shows the recorded history of ground acceleration due to an earthquake. Such samples have been subjected to Fourier analysis, which has shown that both forms of motion can be considered to consist of the summation of a large number of sinusoidal

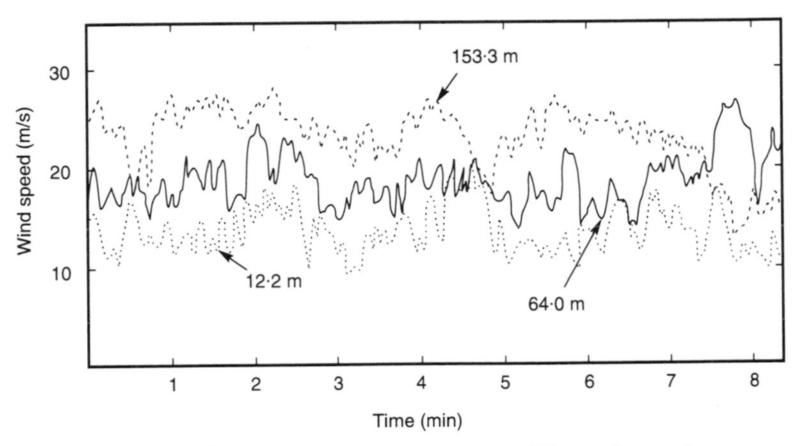


*Fig. 1.2. Records of wind speeds at three levels of a 153 m tall guyed mast*

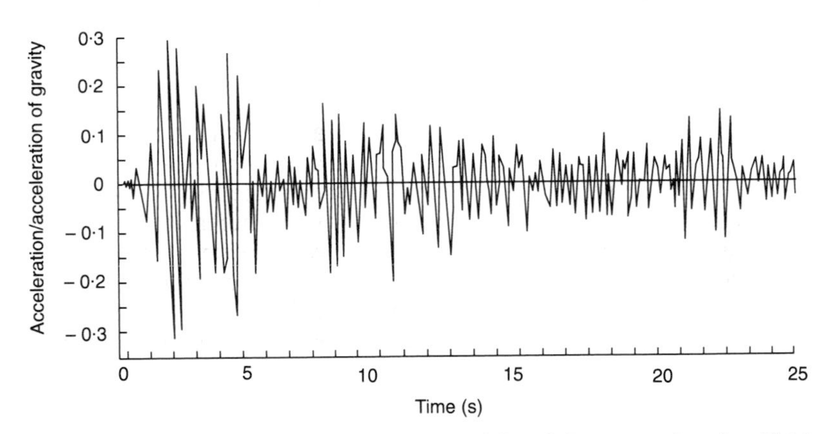


*Fig. 1.3. Accelerogram of the NS component of the El Centro earthquake, 18 May 1940*

waves with varying frequencies and amplitudes. Thus the velocity of wind at any time may be written as

$$V(t) = \tilde{V} + \sum_{i=1}^{N} \alpha_i \sin(\omega_i t + \phi_i) \tag{1.11}$$

where $\tilde{V}$ is the mean wind speed, $\alpha_i$ is the amplitude of fluctuations, $\omega_i$ is the angular frequency in rad/s, $\phi_i$ is the random phase angle in rad, and the subscript $i$ indicates the $i$th harmonic component.

The corresponding drag force acting on a structure may be written as

$$F_\mathrm{d}(t) = \tilde{F}_\mathrm{d} + \sum_{i=1}^{N} F_{\mathrm{d}i} \sin(\omega_i t + \phi_i) \tag{1.12}$$

The fluctuating drag force given by eq. (1.12) can give rise to quite significant amplitudes of vibration if the frequency of only one of its components is equal to a dominant structural frequency. The swaying motion of some very tall slender buildings with low first natural frequencies is a direct result of the fact that their dominant frequencies coincide with the frequency components of wind in the part of the wind frequency spectrum where wind possesses a considerable amount of energy. The same phenomenon may occasionally be observed in nature during periods of strong gusty winds, when for the same reason a single tree may suddenly vibrate violently, while the remainder merely bend in the along-wind direction.

Similarly, the acceleration of the strong motion of an earthquake may be expressed as

$$\ddot{x}_g(t) = \sum_{i=1}^{N} \ddot{x}_i \sin(\phi_i) \tag{1.13}$$

where $\ddot{x}_i$ is the amplitude of acceleration, $\omega_i$ is the angular frequency in rad/s, and $\phi_i$ is the random phase in rad. The corresponding exciting force acting on a one DOF structure of mass is

$$M\ddot{x}_g(t) = \sum_{i=1}^{N} M\ddot{x}_i \sin(\omega_i t + \phi_i) \tag{1.14}$$

During earthquakes it has been observed that buildings with a first natural frequency equal or close to the dominant frequency of the ground may vibrate quite violently while others with frequencies different from the dominant ground frequency vibrate less. This again underlines the fact that the larger amplitude vibrations will occur when the dominant frequencies of structures coincide with one of the frequency components in a random exciting force.

Wind and earthquakes as well as waves (whose effect is not considered in this book) can lead to vibrations with large amplitudes if one or more of the natural frequencies of a structure are equal to some of the angular frequencies $\omega_i$ in equations (1.10), (1.11) or (1.12). Large amplitude vibration can be very destructive, and even if the amplitudes are not large, continued vibration may lead to fatigue failures. Therefore, the possibility of vibration must be taken into account at the design stage.

## Methods of dynamic response analysis

There are basically two approaches for predicting the dynamic response of structures: time domain methods and frequency domain methods. The first method is used to construct time histories of such variables as forces, moments and displacements by calculating the response at the end of a succession of very small time steps. The second method is used to predict the maximum value of the same quantities by adding the response in each mode in which the structure vibrates. Time domain methods can be used to calculate the dynamic response of both linear and nonlinear structures and require that time histories for the dynamic forces be available or can be generated. Frequency domain analysis is limited to linear structures, as the natural frequencies of nonlinear structures vary with the amplitude of response. The method has won considerable popularity in spite of its limitations, as it permits the use of power and response spectra, which to date have been more easily available than time histories. Power spectra for wind are introduced in chapter 10, and response and power spectra for earthquakes in chapter 13. Methods for generating correlated wind and earthquake histories are presented in chapter 14.

## Single DOF and multi-DOF structures

In general even the simplest of structures such as simply supported beams and cantilevers are in reality multi-DOF systems with an infinite number of DOF. For practical purposes, however, many simple structures and structural elements may initially (as already mentioned above) be analysed as one DOF systems, by considering them as simple mass–spring systems with an equivalent lumped mass and an equivalent elastic spring. Some examples of this form of simplification are illustrated in Fig. 1.4.

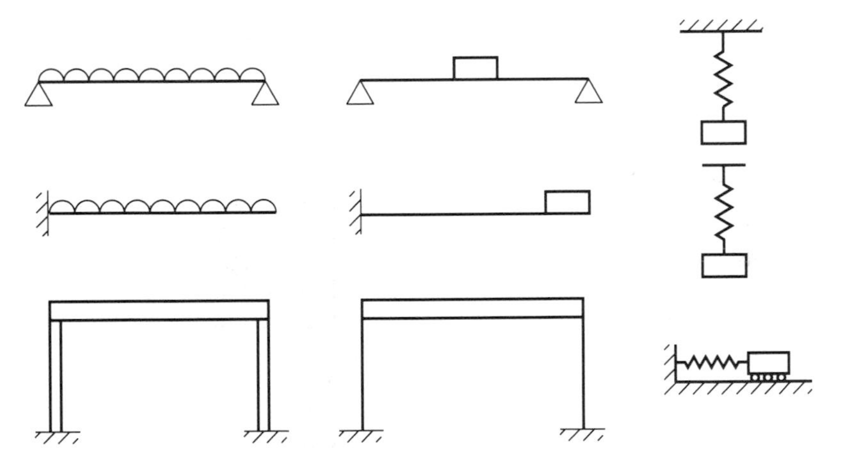

*Fig. 1.4. Equivalent one DOF mass–spring systems*

When a structure is reduced to a one DOF system, it is possible only to calculate the response in one mode, usually the dominant one. In order to study the vibration in several modes a structure has to be modelled as multi-DOF mass–spring systems. An example is shown in Fig. 1.5, where a three-storey portal frame structure, in which the floors are assumed to be rigid, is modelled as a three DOF mass–spring system. Figure 1.6 shows how a pinjointed frame may be modelled as a multi-DOF mass–spring system by lumping the mass of the members at the nodes and considering the stiffness of the members as weightless springs.

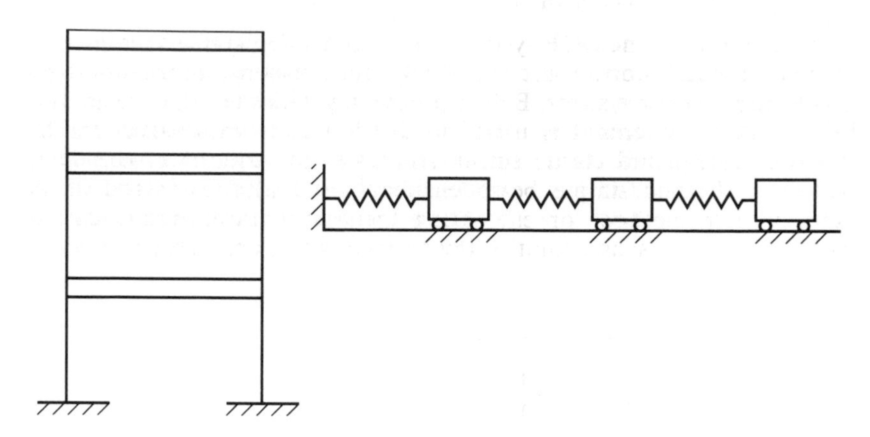

*Fig. 1.5. Three-storey portal frame modelled as a three DOF mass–spring system*

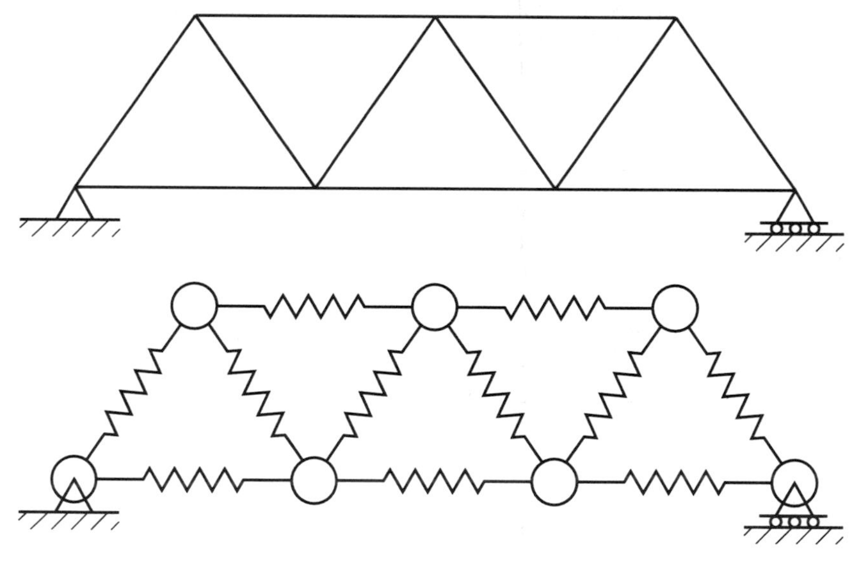

*Fig. 1.6. Modelling of a pinjointed frame as a multi-DOF mass–spring system*

The dynamic response of a large number of structures can, at least initially, be determined by modelling them as one DOF systems. In chapter 8 it is shown how the dynamic response of $N$ DOF structures can be determined by

- transforming them into $N$ one DOF mass–spring systems, each with a natural frequency equal to one of those of the original structure
- calculating the response of each of the one DOF systems
- transforming the responses of these one DOF systems to yield the global response of the original $N$ DOF structure.

Thus, not only one DOF systems, but also $N$ DOF systems require that engineers be fully conversant with the dynamic response analysis of single DOF mass–spring systems. Before proceeding with the response analysis of one DOF systems it is useful to develop some expressions for the lumped masses and elastic spring stiffnesses for equivalent one DOF systems of some simple beam elements, and also to introduce an approximate method for estimating the first natural frequencies of continuous beams and multistorey framed structures. This is done in chapter 2.

**Problem 1.1** A tapering tubular 20 m tall antenna-mast supports a disc weighing 10 kN at the top. Analysis of accelerometer reading shows that the dominant frequency of the mast is 2·3 Hz. A rope attached to the top of the mast deflects the point of attachment 5 mm horizontally when the horizontal component of the tension in the rope is 20 kN. Calculate the equivalent elastic spring stiffness and lumped mass of a mass–spring system which is to be used for studying the response at the top of the mast to wind.

**Problem 1.2** A continuous steel box girder bridge is designed with a central span of 50 m and two outer spans of 25 m each. The expressions for the mass and spring stiffness of a dynamically equivalent mass–spring system are $0{\cdot}89\,wL/g$ and $13{\cdot}7\,EI/L^3$ respectively. Calculate the dominant frequency of the bridge if $L = 25\,\text{m}$, $w = 120\,\text{kN/m}$, $E = 210\,\text{kN/mm}^2$ and $I = 0{\cdot}225\,\text{m}^4$.

## References

Bolt, B. A. *Earthquakes, a primer*. W. H. Freeman, San Francisco, 1978.

Brebbia, C. A. & Walker, C. *Dynamic analysis of offshore structures*. Newnes–Butterworth, 1979.

Buchholdt, H. A. *Introduction to cable roof structures*. Cambridge University Press, Cambridge, 1985.
Clough, R. W. & Penzien, J. *Dynamics of structures*. McGraw-Hill, London, 1975.
Harris, C. M. *Shock vibration*, 3rd edn. McGraw-Hill, London, 1988.
Key, D. E. *Earthquake design practice for buildings*. Thomas Telford, London, 1988.
Krishna, P. *Cable suspended roofs*. McGraw-Hill, New York, 1978.
Lawson, T. W. *Wind effects on buildings*, vols 1 and 2. Applied Science, 1990.
Simue, E. & Scalan, R. H. *Wind effects on structures*. Wiley, Chichester, 1978.
Warburton, G. B. *Reduction of vibrations*. Wiley, Chichester, 1992.
Wolf, J. H. *Dynamic soil–structure interaction*. Prentice-Hall, Englewood Cliffs, 1985.

# 2. Equivalent one degree of freedom systems

## Modelling of structures as one DOF systems

The natural frequency and approximate response of line-like structures such as tall slim buildings, masts, chimneys, bridges and towers may, as mentioned in chapter 1, be estimated by assuming that they mainly respond in the first mode, and by modelling them as single mass–spring systems. This in many cases is made relatively easy by the fact that the first mode of vibration of these types of structure has a mode shape very similar to the deflected form caused by the appropriate concentrated and/or distributed load. The modelling of such structures requires the evaluation of the equivalent or generalized mass $M$, spring stiffness $K$, damping coefficient $C$ and forcing function $P(t)$, such that the frequency of the model is the same as that for the structure itself, and the response of the mass is equal in magnitude to the movement of the point of the structure that one wishes to simulate.

Newton's law of motion states that force = mass × acceleration. Thus, if the mass and stiffness are denoted by $M$ and $K$ respectively, and the amplitude and acceleration at time $t$ are $x$ and $\ddot{x}$ respectively, then since force $F = kx$, it follows that

$$Kx = -M\ddot{x} \tag{2.1}$$

Since the vibration of structures may be assumed to resemble closely that of SHM, the displacement $x$ and acceleration $\ddot{x}$ may be written as

$$x = X\sin(\omega t)$$
$$\ddot{x} = -X\omega^2 \sin(\omega t)$$

Substitution for $x$ and $\ddot{x}$ into eq. (2.1) yields

$$KX = MX\omega^2 \tag{2.2}$$

which in turn yields

$$\omega = \sqrt{(K/M)} \tag{2.3}$$

$$f = \frac{1}{2\pi}\sqrt{(K/M)} \tag{2.4}$$

Multiplication of both sides of eq. (2.2) by $\frac{1}{2}X$ yields

$$\tfrac{1}{2}KX^2 = \tfrac{1}{2}MX^2\omega^2 \tag{2.5}$$

which means that the maximum strain energy $\frac{1}{2}KX^2$ is equal to the maximum kinetic energy $\frac{1}{2}MX^2\omega^2$. From this it follows that the spring stiffness and lumped mass of an equivalent mass–spring system may be found by determining the spring stiffness such that the energy stored in the spring will be the same as that stored in the structure when both are deflected an amount $X$, and that the lumped mass will have the same kinetic energy as the structure when both experience a maximum velocity of $X\omega$, given that the motion of the lumped mass–spring system represents the motion of the structure at the position where the maximum amplitude of vibration is $X$. A method based on this approach is presented in the following section.

## Theoretical modelling by equivalent one DOF mass–spring systems

In order to evaluate the expressions for the equivalent lumped mass, spring stiffness, damping coefficient and generalized dynamic force consider the cantilever column shown in Fig. 2.1, where the flexural rigidity, mass damping coefficient, dynamic force, motion at a distance

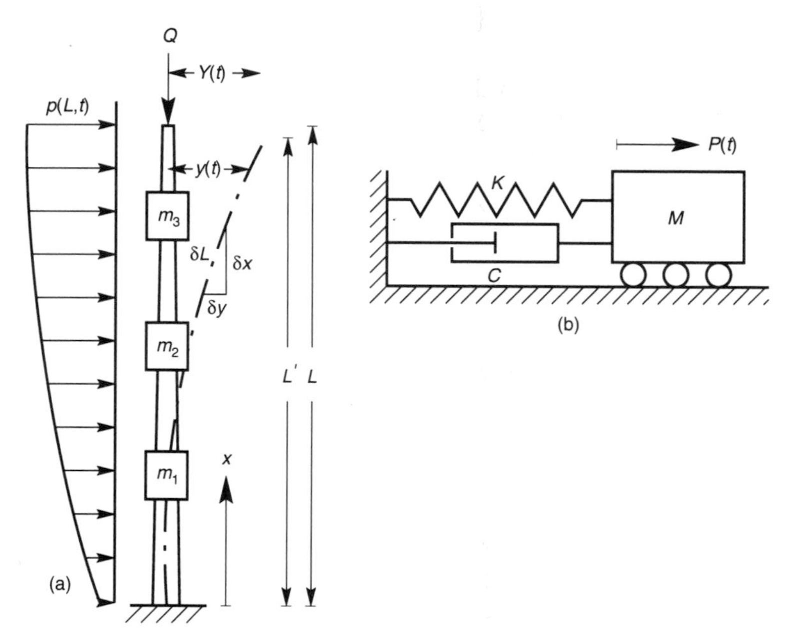


*Fig. 2.1. (a) Cantilever column with flexural rigidity $EI(x)$, mass $m(x)$ and damping coefficient $c(x)$ subjected to a dynamic load $p(x, t)$ at a distance $x$ above the base; (b) equivalent mass–spring system with stiffness $K$, mass $M$, damping coefficient $C$ and dynamic load $P(t)$*

$x$ from the base and motion at the top of the column are given by $EI(x)$, $m(x)$, $c(x)$, $p(x)$, $y(x, t)$ and $Y(t)$ respectively. The height of the undeformed column is $L$, and the height of the deformed column $L^*$. $Q$ is a constant axial force and $\phi(x)$ a *shape function* that defines the shape of the mode of vibration and is unity at the point of the structure at which motion is to be modelled by the mass–spring system. In case of the tower shown in Fig. 2.1 the shape function is assumed to be unity when $x = L$, in which case the model will simulate the movement at the top of the column.

Thus the relationship between $y(x, t)$ and $Y(t)$ may be expressed as

$$y(x, t) = \phi(x)Y(t) \tag{2.6a}$$

$$\dot{y}(x, t) = \phi(x)\dot{Y}(t) \tag{2.6b}$$

To develop an expression for the equivalent mass it is assumed that the spring is weightless, and the kinetic energy of the mass–spring system is equated to that of the column. Thus

$$\tfrac{1}{2}M\dot{Y}^2(t) = \tfrac{1}{2}\int_0^L m(x)\{\phi(x)\dot{Y}(t)\}^2\,\mathrm{d}x + \tfrac{1}{2}\sum_{i=1}^{N} m_i\{\phi(x_i)\dot{Y}(t)\}^2 \tag{2.7}$$

Hence

$$M = \int_0^L m(x)\{\phi(x)\}^2\,\mathrm{d}x + \sum_{i=1}^{N} m_i\{\phi(x_i)\}^2 \tag{2.8}$$

The expression for the equivalent elastic spring stiffness is similarly found by equating the strain energy stored in the spring to that stored in the column. Thus

$$\tfrac{1}{2}K_{\mathrm{E}}Y^2(t) = \tfrac{1}{2}\int_0^L M(x)\,\mathrm{d}\theta \tag{2.9}$$

Because

$$\mathrm{d}^2y/\mathrm{d}x^2 = M(x)/EI(x) = \mathrm{d}\theta/\mathrm{d}x \tag{2.10}$$

equation (2.9) may also be written as

$$\tfrac{1}{2}K_{\mathrm{E}}Y^2(t) = \int_0^L EI(x)\{\mathrm{d}^2y/\mathrm{d}x^2\}^2\,\mathrm{d}x \tag{2.11}$$

or

$$\tfrac{1}{2}K_{\mathrm{E}}Y^2(t) = \tfrac{1}{2}\int_0^L EI(x)\{(\mathrm{d}^2\phi/\mathrm{d}x^2)\,Y(t)\}^2\,\mathrm{d}x \tag{2.12}$$

Hence

$$K_{\mathrm{E}} = \int_0^L EI(x)\{\mathrm{d}^2\phi/\mathrm{d}x^2\}^2\,\mathrm{d}x \tag{2.13}$$

In order to take account of the constant axial force $Q$, it is necessary to define a new stiffness referred to as the equivalent geometric stiffness $K_G$ of the mass–spring system. The expression for this stiffness is obtained by equating the potential energy of the axial load $Q$ to the strain energy stored in the spring due to this load. Thus

$$\tfrac{1}{2}K_G Y^2(t) = Q(L - L^*) \tag{2.14}$$

To develop an expression for $K_G$ it is therefore first necessary to obtain an expression for $(L - L^*)$. Consider the element $\delta L$. The length of this element may be expressed as

$$\delta L = \sqrt{\left[1 + (\mathrm{d}y/\mathrm{d}x)^2\right]} \tag{2.15}$$

Expansion of the above square root by the binomial theorem and integration over the vertical projection of the height $L^*$ of the deformed column yield

$$L = \int_0^{L^*} \left\{1 + \tfrac{1}{2}(\mathrm{d}y/\mathrm{d}x)^2 - \tfrac{1}{8}(\mathrm{d}y/\mathrm{d}x)^4 + \ldots\right\} \mathrm{d}x \tag{2.16}$$

If one ignores the quartic and higher order terms, then

$$L - L^* = \int_0^{L^*} \tfrac{1}{2}(\mathrm{d}y/\mathrm{d}x)^2 \,\mathrm{d}x \tag{2.17}$$

The upper limit in eq. (2.17) may be changed from $L^*$ to $L$ if it can be assumed that $L^* \approx L$. Making this assumption and substituting the above expression for $L - L^*$ into eq. (2.14), one obtains

$$\tfrac{1}{2}K_G Y(t)^2 = \tfrac{1}{2}Q \int_0^L \{(\mathrm{d}\phi/\mathrm{d}x)\,Y(t)\}^2 \,\mathrm{d}x \tag{2.18}$$

Hence

$$K_G = Q \int_0^L \{\mathrm{d}\phi/\mathrm{d}x\}^2 \,\mathrm{d}x \tag{2.19}$$

The total spring stiffness is therefore given by

$$K + K_E + K_G \tag{2.20}$$

or

$$K = \int_0^L EI(x)\{\mathrm{d}^2\phi/\mathrm{d}x^2\}^2 \,\mathrm{d}x + Q \int_0^L \{\mathrm{d}\phi/\mathrm{d}x\}^2 \,\mathrm{d}x \tag{2.21}$$

The total stiffness $K$ increases with increasing axial force and decreases with increasing compressive force. Thus $Q$ is taken as positive if it causes tension and negative if it causes compression. The critical load $Q_{crit}$ has been reached when

$$K_E + K_G = 0 \tag{2.22}$$

The expression for the equivalent damping, indicated as a dash pot in Fig. 2.1(b), is found by equating the virtual work of the damping force in the mass–spring system to the virtual work of the damping forces in or acting on the column. In chapter 3 it is explained that the damping forces at a given time $t$ may be expressed as the product of a viscous damping coefficient and the velocity of the motion of the structure. Thus an expression for $C$, the damping coefficient for the equivalent mass–spring system, may be found from

$$C\dot{Y}(t)\delta y(t) = \int_0^L c(x)\{\phi(x)\dot{Y}(t)\}\,\mathrm{d}x + \sum_{i=1}^{N} c_i\{\phi(x_i)\dot{Y}(t)\}\{\phi(x_i)\delta Y\} \tag{2.23}$$

Hence

$$C = \int_0^L c(x)\{\phi(x)\}^2\,\mathrm{d}x + \sum_{i=1}^{N} c_i\{\phi(x_i)\}^2 \tag{2.24}$$

Similarly, the expression for the equivalent dynamic force that should be applied to the mass–spring system may be found by equating the virtual work of this force to that of the real forces. Thus

$$P(t)\delta Y(t) = \int_0^L p(x,t)\{\phi(x)\delta Y(t)\}\,\mathrm{d}x + \sum_{i=1}^{N} P_i\{\phi(x_i)\delta Y\} \tag{2.25}$$

Hence

$$P(t) = \int_0^L p(t)\phi(x)\,\mathrm{d}x + \sum_{i=1}^{N} P_i\phi(x_i) \tag{2.26}$$

The use of eqs (2.8), (2.13), (2.19), (2.21) and (2.26) will yield the equivalent mass, stiffness, damping and dynamic force for the modelling of a structure as a one DOF system, provided the mode shape of vibration is known. The latter can, as mentioned above, be found by assuming the mode of vibration to be geometrically similar to the deflected shape caused by a uniform or concentrated loads, or by determining the mode shape by an eigenvalue analysis as shown in chapter 6. In practice the use of eq. (2.24) is very limited, as the value for the damping coefficient $C$ is based on experimental data associated not only with the properties of the material used and the method of construction, but also with the modeshape of vibration. Thus a damping coefficient evaluated or assumed for a given mode can be used directly without using eq. (2.24).

In the following sections expressions for the equivalent mass, stiffness, critical load and natural frequencies are developed for some simple structures and structural elements which in the first mode vibrate with modeshapes that are geometrically similar to their statically deformed shapes.

## Equivalent one DOF mass–spring systems for linearly elastic line structures

### *Cantilevers and columns with uniformly distributed load*

Assume the mode shape of vibration of the cantilever shown in Fig. 2.2 to be geometrically similar to the deflected form $y(x)$ caused by the uniformly distributed load $wL$. The deflected form may be determined by integration of the expression for the bending moment $M(x)$ at a distance $x$ from the fixed end, where

$$M(x) = EI\,\mathrm{d}^2y/\mathrm{d}x^2 = \tfrac{1}{2}w(L-x)^2 \tag{2.27}$$

Integration of eq. (2.27) and imposition of the boundary conditions $y(x) = \mathrm{d}y/\mathrm{d}x = 0$ when $x = 0$ yields

$$y = (w/24EI)\{6L^2x^2 - 4Lx^3 + x^4\} \tag{2.28}$$

$$y_{x=L} = wL^4/8EI \tag{2.29}$$

For the equivalent mass–spring system to model the motion of the free end of the cantilever, the shape function must be unity at that point. This will be the case when

$$w = 8EI/L^4 \tag{2.30}$$

Substitution of this expression for $w$ into eq. (2.28) yields the following expressions for the shape function $\phi(x)$ and its first and second derivatives

$$\phi(x) = (1/3L^4)\{6L^2x^2 - 4Lx^3 + x^4\} \tag{2.31a}$$

$$\phi'(x) = (4/3L^4)\{3L^2x - 3Lx^2 + x^3\} \tag{2.31b}$$

$$\phi''(x) = (4/L^4\{l^2 - 2Lx + x^2\} \tag{2.31c}$$

Thus the weight of the equivalent lumped mass is given by

$$W = \int_0^L w\{\phi(x)\}^2\,\mathrm{d}x = \int_0^L w(1/3L^4)^2\{6L^2x^2 - 4Lx^3 + x^4\}^2\,\mathrm{d}x \tag{2.32}$$

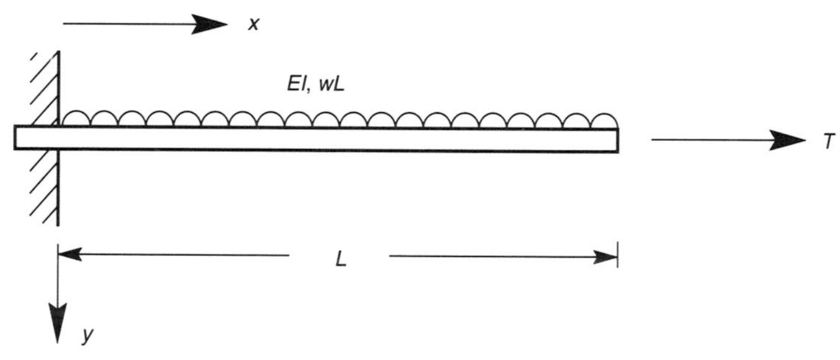


*Fig. 2.2. Cantilever with uniformly distributed load wL and axial tensile force T*

Hence

$$\underline{W = (728/2835)wL} \tag{2.33}$$

The equivalent elastic spring stiffness is given by

$$K_{\mathrm{E}} = \int_0^L EI\{\phi''(x)\}^2\,\mathrm{d}x = \int_0^L EI(4/L^4)^2\{L^2 - 2Lx + x^2\}^2\,\mathrm{d}x \tag{2.34}$$

Hence

$$\underline{K_{\mathrm{E}} = 16EI/L^3} \tag{2.35}$$

The equivalent geometrical spring stiffness is given by

$$K_{\mathrm{G}} = \int_0^L T\{\phi'(x)\}^2\,\mathrm{d}x = \int_0^L T(4/3L^4)^2\{3L^2x - 3Lx^2 + x^3\}^2\,\mathrm{d}x \tag{2.36}$$

Hence

$$\underline{K_{\mathrm{G}} = 8T/7L} \tag{2.37}$$

The critical value for the axial force occurs when

$$K = K_{\mathrm{E}} + K_{\mathrm{G}} = 16EI/5L^3 + 8T/7L = 0 \tag{2.38}$$

or

$$\underline{T = -14EI/5L^2} \tag{2.39}$$

If one ignores the geometrical stiffness, the natural frequency of the cantilever is given by

$$f = \frac{1}{2\pi}\sqrt{\left(K_{\mathrm{E}}g/W\right)} = \frac{1}{2\pi}\sqrt{\left(\frac{16EIg/5L^3}{728wL/2835}\right)} \tag{2.40}$$

or

$$\underline{f = 0{\cdot}5618313\sqrt{(EIg/wL^4)}} \tag{2.41}$$

When the correct modeshape is used the natural frequency is given by

$$\underline{f = 0{\cdot}5602254\sqrt{(EIg/wL^4)}} \tag{2.42}$$

Thus the error caused by assuming the modeshape to be similar to the deflected form is 0·287%.

### *Cantilevers and columns with a concentrated load at the free end*

Assume that the modeshape of the uniformly loaded cantilever subjected to a vertical concentrated load $P$ and an axial tensile force $T$ at the free end, as shown in Fig. 2.3, is geometrically similar to the deflected form due to $P$. The deflected shape $y(x)$ may be determined from the expression for the bending moment $M(x)$ at a distance $x$ from the fixed end, which is given by

$$M(x) = EI\mathrm{d}^2y/\mathrm{d}x^2 = P(L - x) \tag{2.43}$$

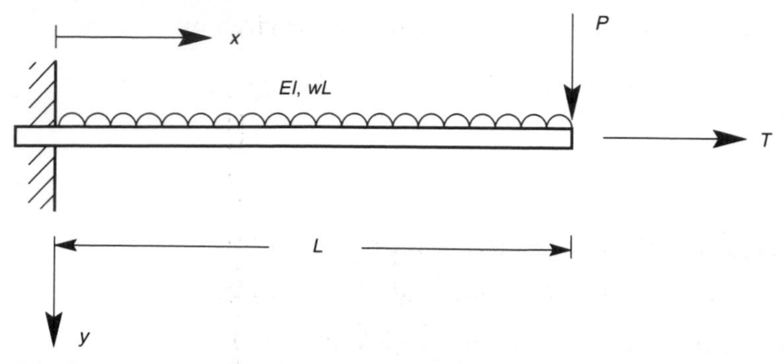


*Fig. 2.3. Cantilever beam column with uniformly distributed load, concentrated vertical load P and axial tensile load T*

Integration of eq. (2.43) twice and imposition of the boundary conditions $y(x) = dy/dx = 0$ when $x = 0$ yields

$$y(x) = (P/6EI)\{3Lx^2 - x^3\} \tag{2.44}$$

$$y_{x=L} = PL^3/3EI \tag{2.45}$$

For a mass–spring system to model the motion of the free end of the cantilever, the shape function must be unity at this point. When this is the case

$$P = 3EI/L^3 \tag{2.46}$$

Substitution of this value for $P$ into eq. (2.44) yields the following expressions for the shape function $\phi(x)$ and its first and second derivatives

$$\phi(x) = (1/2L^3)\{3Lx^2 - x^3\} \tag{2.47a}$$

$$\phi'(x) = (3/2L^3)\{2Lx - x^2\} \tag{2.47b}$$

$$\phi''(x) = (3/L^3)\{L - x\} \tag{2.47c}$$

Thus the weight of the equivalent lumped mass is given by

$$W = P + \int_0^L w\left\{\phi(x)^2\right\}^2 dx = \int_0^L w(1/2L^3)^2\{3Lx^2 - x^3\}^2 dx \tag{2.48}$$

Hence

$$\underline{W = P + (33/140)wL} \tag{2.49}$$

The equivalent elastic spring stiffness is given by

$$K_E = \int_0^L EI\{\phi''(x)\}^2 dx = \int_0^L EI(3/L^3)^2\{L - x\}^2 dx \tag{2.50}$$

Hence

$$\underline{K_E = 3EI/L^3} \tag{2.51}$$

The equivalent geometrical spring stiffness is given by

$$K_G = \int_0^L T\{\phi'(x)\}^2\,dx = \int_0^L T(3/2L^3)^2\{2Lx - x^2\}^2\,dx \qquad (2.52)$$

Hence

$$\underline{K_G = 6T/5L} \qquad (2.53)$$

The critical value for the axial force occurs when

$$K = K_E + K_G = 3EI/L^3 + 6T/5L = 0 \qquad (2.54)$$

or

$$\underline{T = -5EI/2L^3} \qquad (2.55)$$

Comparison of the expressions for the critical load given by eqs (2.55) and (2.39) reveals that the two assumed modeshapes lead to a difference in the value for $T$ of 12·0%

If the geometrical stiffness and the concentrated vertical load are ignored, the frequency of the cantilever with the assumed modeshape is

$$f = \frac{1}{2\pi}\sqrt{\left(\frac{K_E g}{W}\right)} = \frac{1}{2\pi}\sqrt{\left(\frac{3EIg/L^3}{33wL/140}\right)} \qquad (2.56)$$

or

$$\underline{f = 0{\cdot}56779\sqrt{(EIg/wL^4)}} \qquad (2.57)$$

Thus the error in the natural frequency caused by the assumed modeshape when the load is uniformly distributed is approximately 1·35%.

### *Simply supported beam with uniformly distributed load*

Assume the mode shape of vibration of the simply supported beam shown in Fig. 2.4 and subjected to an axial tensile force $T$ to be similar to the deflected form $y(x)$ caused by the distributed load $wL$. The deflected shape $y(x)$ is obtained from the expression for the bending moment at a distance $x$ from the left-hand support

$$M(x) = EI\,d^2y/dx^2 = -\tfrac{1}{2}wLx + \tfrac{1}{2}wx^2 \qquad (2.58)$$

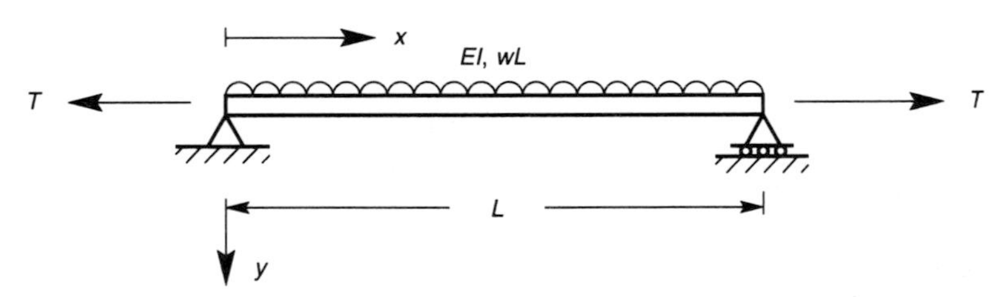


*Fig. 2.4. Simply supported beam with uniformly distributed load wL and axial tensile force T*

Integration of eq. (2.58) twice and imposition of the boundary conditions that $y(0) = y(L) = 0$ yields

$$y(x) = (w/24EI)\{x^4 - 2Lx^3 + L^3x\} \tag{2.59}$$

$$y_{x=L/2} = 5wL^4/384EI \tag{2.60}$$

If the mass–spring system is to model the motion at the centre of the beam, then the modeshape at this point must be equal to unity. When this is the case

$$w = 384EI/5L^4 \tag{2.61}$$

Substitution of this value of $w$ into eq. (2.59) yields the following expressions for the shape function and its first and second derivatives

$$\phi(x) = (16/5L^4)\{x^4 - 2Lx^3 + Lx^3\} \tag{2.62a}$$

$$\phi'(x) = (16/5L^4)\{4x^3 - 6Lx^2 + L^3\} \tag{2.62b}$$

$$\phi''(x) = (16/5L^4)\{12x^2 - 12Lx\} \tag{2.62c}$$

Thus the weight of the equivalent lumped mass is given by

$$W = \int_0^L w\{\phi(x)\}^2 \,\mathrm{d}x = \int_0^L w(16/5L^4)^2\{x^4 - 2Lx^3 + L^3x\}^2 \,\mathrm{d}x \tag{2.63}$$

Hence

$$\underline{W = (3968/7875)wL} \tag{2.64}$$

The equivalent elastic spring stiffness is given by

$$K_\mathrm{E} = \int_0^L EI\{\phi''(x)\}^2 \mathrm{d}x = \int_0^L EI(16/5L^4)^2\{12x^2 - 12Lx\}^2 \,\mathrm{d}x \tag{2.65}$$

Hence

$$\underline{K_\mathrm{E} = 6144EI/125L^3} \tag{2.66}$$

The equivalent geometrical spring stiffness is given by

$$K_\mathrm{G} = \int_0^L T\{\phi'(x)\}^2 \,\mathrm{d}x = \int_0^L T(16/5L^4)^2\{4x^3 - 6Lx^2 + L^3\}^2 \mathrm{d}x \tag{2.67}$$

Hence

$$\underline{K_\mathrm{G} = 4353T/875L} \tag{2.68}$$

The critical value for the axial force occurs when

$$K = K_\mathrm{E} + K_\mathrm{G} = 6144EI/125L^3 + 4352T/875L = 0 \tag{2.69}$$

or

$$\underline{T = -9{\cdot}8824EI/L^2} \tag{2.70}$$

The natural frequency for the beam, ignoring the axial load, is given by

$$f = \frac{1}{2\pi}\sqrt{\left(\frac{K_E g}{W}\right)} = \frac{1}{2\pi}\sqrt{\left(\frac{6144EIg/125L^3}{3968wL/7875}\right)} \quad (2.71)$$

or

$$\underline{f = 1{\cdot}571919\sqrt{(EIg/wL^4)}} \quad (2.72)$$

When the correct modeshape is used, the expression for the natural frequency is

$$\underline{1{\cdot}5707963\sqrt{(EIg/wL^4)}} \quad (2.73)$$

Thus the error in the natural frequency resulting from assuming the modeshape to be geometrically similar to the deflected form caused by the self-weight of the beam is 0·07%.

### *Simply supported beam with a concentrated load at midspan*

If a beam supports a concentrated load in addition to its own weight, as shown in Fig. 2.5, it may be assumed that the modeshape of vibration is similar to the deflected form caused by $P$. The deflected shape $y(x)$ is obtained from the expression for the bending moment $M(x)$ at a distance $x$ from the left-hand support

$$M(x) = EI\,\mathrm{d}^2y/\mathrm{d}x^2 = -\tfrac{1}{2}Px \quad (2.74)$$

Integration of eq. (2.74) twice and imposition of the boundary conditions $y = 0$ when $x = 0$ and $\mathrm{d}y/\mathrm{d}x = 0$ when $x = L/2$ yields

$$y(x) = (P/48EI)\{3L^2x - 4x^3\} \quad (2.75)$$

$$y_{x=L/2} = PL^3/48EI \quad (2.76)$$

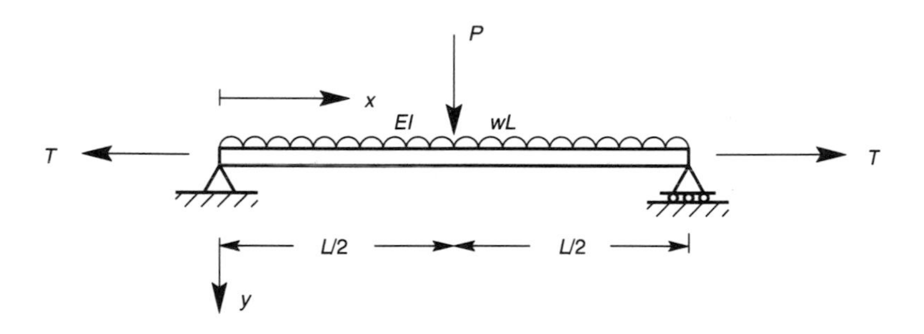


*Fig. 2.5. Simply supported beam with concentrated load P at midspan and axial tensile force T*

If the mass–spring system is to model the motion at the centre of the beam then the shape function at this point must, as previously, be unity. When this is the case

$$P = 48EI/L^3 \tag{2.77}$$

Substitution of the above expression for $P$ into eq. (2.75) yields the following expressions for the shape function and its derivatives

$$\phi(x) = (1/L^3)\{3L^2x - 4x^3\} \tag{2.78a}$$

$$\phi'(x) = (1/L^3)\{3L^2 - 12x^2\} \tag{2.78b}$$

$$\phi''(x) = (1/L^3)\{-24x\} \tag{2.78c}$$

Thus the weight of the equivalent lumped mass is given by

$$W = P + 2\int_0^{L/2} w\{\phi(x)\}^2\,\mathrm{d}x = P + 2\int_0^{L/2} w(1/L^3)^2\{3L^2x - 4x^3\}^2\,\mathrm{d}x \tag{2.79}$$

Hence

$$\underline{W = P + (17/35)wL} \tag{2.80}$$

The equivalent elastic spring stiffness is given by

$$K_\mathrm{E} = 2\int_0^{L/2} EI\{\phi''(x)\}^2\,\mathrm{d}x = 2\int_0^{L/2} EI\{-24x\}^2\,\mathrm{d}x \tag{2.81}$$

Hence

$$\underline{K_\mathrm{E} = 48EI/L^3} \tag{2.82}$$

The equivalent geometrical spring stiffness is given by

$$K_\mathrm{G} = 2\int_0^{L/2} T\{\phi'(x)\}^2\,\mathrm{d}x = 2\int_0^{L/2} T(1/L^3)^2\{3L^2 - 12x^2\}^2\,\mathrm{d}x \tag{2.83}$$

Hence

$$\underline{K_\mathrm{G} = 24T/5L} \tag{2.84}$$

The critical value for the axial force occurs when

$$K = K_\mathrm{E} + K_\mathrm{G} = 48EI/L^3 + 24T/5L = 0 \tag{2.85}$$

or

$$\underline{T = -10EI/L^2} \tag{2.86}$$

Comparison of the expressions for the critical axial load given by eqs (2.70) and (2.86) shows that the two differently assumed modeshapes lead to a difference of 1·17% in the values for $T$.

The natural frequency for the assumed modeshape, ignoring the concentrated load $P$ and the axial force $T$ is given by

$$f = \frac{1}{2\pi}\sqrt{\left(\frac{K_G g}{W}\right)} = \frac{1}{2\pi}\sqrt{\left(\frac{46EIg/L^3}{17wL/35}\right)} \tag{2.87}$$

or

$$\underline{f = 1{\cdot}5821597\sqrt{(EIg/wL^4)}} \tag{2.88}$$

The error caused by the assumed modeshape is therefore in this case equal to 0·723%.

*Built-in beam with uniformly distributed load*

Assume the modeshape of vibration of the uniformly loaded built-in beam, subjected to an axial load $T$ and having a constant flexural rigidity $EI$, as shown in Fig. 2.6, to be geometrically similar to the deflected form caused by the load $wL$. An expression for the deflected form can be found from the expression for the bending moment at a distance $x$ from the left-hand support, which is given by

$$M(x) = EI\,\mathrm{d}^2y/\mathrm{d}x^2 = M_A - \tfrac{1}{2}wLx + \tfrac{1}{2}wx^2 \tag{2.89}$$

Integration of eq. (2.89) twice and imposition of the boundary conditions $y - \mathrm{d}y/\mathrm{d}x = 0$ when $x = 0$, and $y = 0$ when $x = L$, yields

$$y(x) = (w/24EI)\{x^4 - 2Lx^3 + L^2x^2\} \tag{2.90}$$

$$y_{x=1/2} = wL^4/384EI \tag{2.91}$$

To model the motion at midspan, the shape function must be unity at this point. When this is the case

$$w = 384EI/L^4 \tag{2.92}$$

Substitution of this value for $w$ into eq. (2.90) yields the following expressions for the shape function and its first and second derivatives

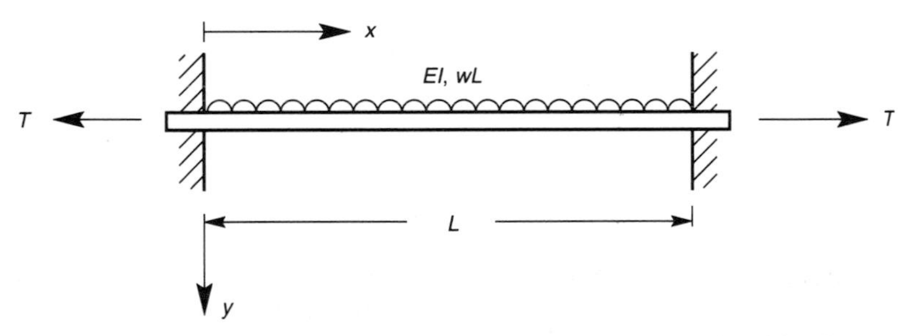


*Fig. 2.6. Built-in beam with uniformly distributed load wL and axial load T*

$$\phi(x) = (16/L^4)\{x^4 - 2Lx^3 + L^2x^2\} \tag{2.93a}$$

$$\phi'(x) = (32/L^4)\{2x^3 - 3Lx^2 + L^2x\} \tag{2.93b}$$

$$\phi''(x) = (32/L^4)\{6x^2 - 6Lx + L^2\} \tag{2.93c}$$

Thus the weight of the equivalent lumped mass is given by

$$W = \int_0^L w\{\phi(x)\}^2\,\mathrm{d}x = \int_0^L w(16L^4)^2\{x^4 - 2Lx^3 + L^2x^2\}^2\,\mathrm{d}x \tag{2.94}$$

Hence

$$\underline{W = (128/315)wL} \tag{2.95}$$

The equivalent elastic spring stiffness is given by

$$K_{\mathrm{E}} = \int_0^L EI\{\phi''(x)\}^2\,\mathrm{d}x = \int_0^L EI(32/L^4)^2\{6x^2 - 6Lx + L^2\}^2\,\mathrm{d}x \tag{2.96}$$

Hence

$$\underline{K_{\mathrm{G}} = 1024EI/5L^3} \tag{2.97}$$

The equivalent geometrical spring stiffness is given by

$$K_{\mathrm{G}} = \int_0^L T\{\phi'(x)\}^2\,\mathrm{d}x = \int_0^L T(32/L^4)^2\{2x^3 - 3Lx^2 + L^2x\}^2\,\mathrm{d}x \tag{2.98}$$

Hence

$$\underline{K_{\mathrm{G}} = 512T/105L} \tag{2.99}$$

The critical value for the axial force occurs when

$$K_{\mathrm{E}} + K_{\mathrm{G}} = 1024EI/5L^3 + 512T/105L = 0 \tag{2.100}$$

or

$$\underline{T = -42EI/L^2} \tag{2.101}$$

Ignoring the axial force, the natural frequency associated with the assumed mode shape is given by

$$f = \frac{1}{2\pi}\sqrt{\left(\frac{K_{\mathrm{E}}g}{W}\right)} = \frac{1}{2\pi}\sqrt{\left(\frac{1024EIg/5L^3}{128wL/315}\right)} \tag{2.102}$$

or

$$\underline{f = 3{\cdot}5730196\sqrt{(EIg/wL^4)}} \tag{2.103}$$

Using the correct mode shape

$$\underline{f = 3{\cdot}5608213\sqrt{(EIg/wL^4)}} \tag{2.104}$$

Thus the error in this case is 0·34%.

*Built-in beam with concentrated load at midspan*

Consider the beam shown in Fig. 2.7. Because of the concentrated load at midspan it is assumed that the modeshape of vibration is geometrically similar to the deflected form due to $P$. The deflected form itself is found from the expression for the bending moment at $x$, between the left-hand support and $P$, which is given by

$$M(x) = EI\,\mathrm{d}^2y/\mathrm{d}x^2 = M_\mathrm{A} - \tfrac{1}{2}Px \qquad (2.105)$$

Integration of eq. (2.105) twice and imposition of the boundary conditions $y = \mathrm{d}y/\mathrm{d}x = 0$ when $x = 0$ and $\mathrm{d}y/\mathrm{d}x = 0$ when $x = L/2$ yields

$$y(x) = (P/48EI)\{3Lx^2 - 4x^3\} \qquad (2.106)$$

Hence

$$y_{x=L/2} = PL^3/192EI \qquad (2.107)$$

For the equivalent mass–spring system to represent the motion at midspan the shape function at this point must be unity. This requires that

$$P = 192EI/L^3 \qquad (2.108)$$

Substitution of this expression for $P$ into eq. (2.106) yields the required shape function and hence its first and second differentials

$$\phi(x) = (4/L^3)\{3LX^2 - 4x^3\} \qquad (2.109a)$$

$$\phi'(x) = (24/L^3)\{Lx - 2x^2\} \qquad (2.109b)$$

$$\phi''(x) = (24/L^3)\{L - 4x\} \qquad (2.109c)$$

Thus the weight of the equivalent lumped mass is given by

$$W = P + 2\int_0^{L/2} w\{\phi(x)\}^2\,\mathrm{d}x = P + 2\int_0^{L/2} w(4/L^3)^2\{3Lx^2 - 4x^3\}^2\mathrm{d}x \qquad (2.110)$$

Hence

$$\underline{W = (13/35)wL} \qquad (2.111)$$

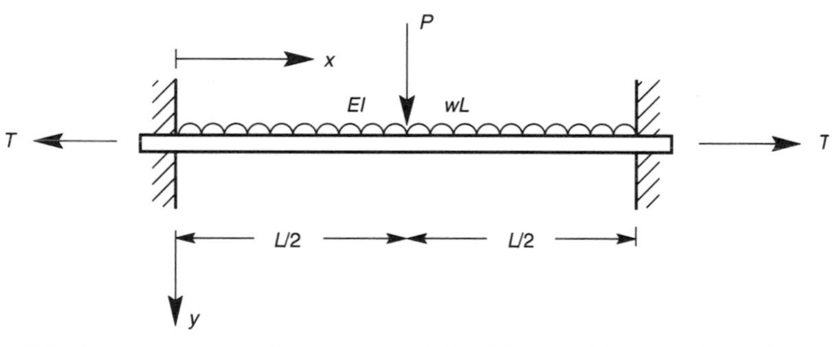


*Fig. 2.7. Built-in beam with concentrated load P at midspan and axial tension load T*

The equivalent elastic spring stiffness is given by

$$K_{\mathrm{E}} = 2\int_0^{L/2} EI\{\phi''(x)\}^2\,\mathrm{d}x = 2\int_0^{L/2} EI(24/L^3)^2\{L-4x\}^2\,\mathrm{d}x \quad (2.112)$$

Hence

$$\underline{K_{\mathrm{E}} = 192EI/L^3} \quad (2.113)$$

The equivalent geometrical spring stiffness is given by

$$K_{\mathrm{G}} = 2\int_0^{L/2} T\{\phi'(x)\}^2\,\mathrm{d}x = 2\int_0^{L/2} T(24/L^3)^2\{Lx-2x^2\}^2\,\mathrm{d}x \quad (2.114)$$

Hence

$$\underline{K_{\mathrm{G}} = 24T/5L} \quad (2.115)$$

The critical value for the axial force occurs when

$$K = K_{\mathrm{E}} + K_{\mathrm{G}} = 192EI/L^3 + 24T/5L = 0 \quad (2.116)$$

or

$$\underline{T = -40EI/L^2} \quad (2.117)$$

Comparison of the expressions for the critical axial load given by eqs (2.101) and (2.117) shows that the two differently assumed modeshapes lead to a difference of 5%.

Ignoring the concentrated load $P$ and the axial load $T$ yields the following frequency for a uniformly loaded built-in beam

$$f = \frac{1}{2\pi}\sqrt{\left(\frac{K_{\mathrm{E}}g}{W}\right)} = \frac{1}{2\pi}\sqrt{\left(\frac{192EIg/L^2}{13wL/35}\right)} \quad (2.118)$$

or

$$\underline{f = 3{\cdot}6185376\sqrt{(EIg/wL^4)}} \quad (2.119)$$

Thus the error resulting from this form of modeshape is 1·62%.

### *Uniformly loaded beam with one end simply supported and one end built in*

Consider the axially loaded beam shown in Fig. 2.8 supporting a uniformly distributed load $wL$. In order to model the point of the beam that will vibrate with the greatest amplitude as a mass–spring system, it is first necessary to determine the modeshape of vibration. This is most easily done by assuming it to be geometrically similar to the deflected form caused by the distributed load. The deflected shape $y(x)$ can be found from the bending moment at a distance $x$ from the hinged end

$$M(x) = EI\,\mathrm{d}^2y/\mathrm{d}x^2 = -Rx + \tfrac{1}{2}wx^2 \quad (2.120)$$

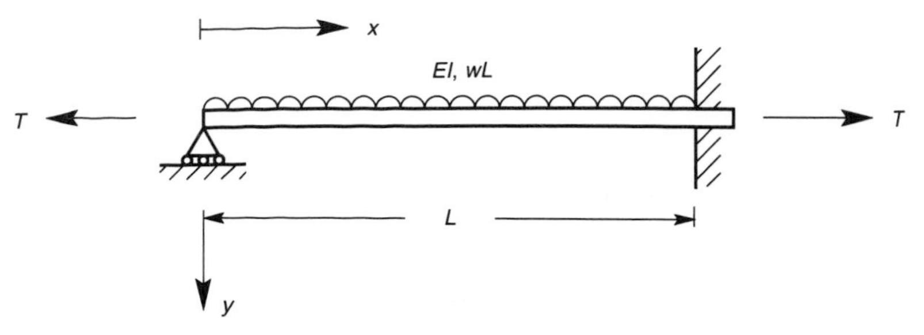


*Fig. 2.8. Beam with one end simply supported and one built-in, supporting a distributed load wL and an axial load T*

Integration of eq. (2.121) twice and imposition of the boundary conditions $y(x) = 0$ when $x = 0$ and $x = L$, and $\mathrm{d}y/\mathrm{d}x = 0$ when $x = L$ yields

$$y(x) = (w/48EI)\{L^3x - 3Lx^3 + 2x^4\} \qquad (2.121)$$

The maximum deflection occurs when $\mathrm{d}y/\mathrm{d}x = 0$. This is, since $\mathrm{d}y/\mathrm{d}x = 0$ when $x = L$, when

$$8x^3 - 9Lx^2 + L^3 = (x - x_1)(x - x_2)(x - L) = 0 \qquad (2.122)$$

Division of the left-hand side of eq. (2.122) by $(x - L)$ and solution of the resulting quadratic equation with respect to $x$ yields

$$x_1 = 0{\cdot}4215351L$$
$$x_2 = -0{\cdot}2965351L$$

The negative value for $x$ obviously has no practical meaning, thus the maximum displacement is found by substitution of the value for $x_1$ into eq. (2.121). This yields

$$y(x)_{\max} = 0{\cdot}2599738(wL^4/48EI) \qquad (2.123)$$

For the equivalent mass–spring system to model the motion at position $x = x_1$, the displacement at this point must be unity. When this is the case

$$w/48EI = 3{\cdot}8465403/L^4 \qquad (2.124)$$

Substitution of the above value for $w/48EI$ into eq. (2.121) yields the shape function and hence its first and second derivatives

$$\phi(x) = (3{\cdot}8465403/L^4)\{L^3x - 3Lx^3 + 2x^4\} \qquad (2.125a)$$
$$\phi'(x) = (3{\cdot}8465403/L^4)\{L^3 - 9Lx^2 + 8x^3\} \qquad (2.125b)$$
$$\phi''(x) = (23{\cdot}079242/L^4)\{-3Lx + 4x^2\} \qquad (2.125c)$$

Thus the weight of the equivalent lumped mass is given by

$$\begin{aligned} W &= \int_0^L w\{\phi(x)\}^2\,\mathrm{d}x \\ &= \int_0^L w(3{\cdot}8465403/L^4)^2\{L^3x - 3Lx^3 + 2x^4\}^2\,\mathrm{d}x \end{aligned} \tag{2.126}$$

Hence

$$\underline{W = 0{\cdot}4462246wL} \tag{2.127}$$

The equivalent elastic stiffness is given by

$$\begin{aligned} K_\mathrm{E} &= \int_0^L EI\{\phi''(x)\}^2\,\mathrm{d}x \\ &= \int_0^L EI(23{\cdot}079242/L^4)^2\{-3Lx + 4x^4\}^2\,\mathrm{d}x \end{aligned} \tag{2.128}$$

Hence

$$\underline{K_\mathrm{E} = 106{\cdot}53028EI/L^3} \tag{2.129}$$

The equivalent geometrical stiffness is given by

$$\begin{aligned} K_\mathrm{G} &= \int_0^L T\{\phi'(x)\}^2\,\mathrm{d}x \\ &= \int_0^L T(3{\cdot}8465403/L^4)^2\{L^3 - 9Lx^2 + 8x^3\}^2\,\mathrm{d}x \end{aligned} \tag{2.130}$$

Hence

$$\underline{K_\mathrm{G} = 5{\cdot}0728704T/L^2} \tag{2.131}$$

The critical value for the axial load occurs when

$$K = K_\mathrm{E} + K_\mathrm{G} = 106{\cdot}53028EI/L^3 + 5{\cdot}0728704T/L^2 = 0 \tag{2.132}$$

or

$$\underline{T = -21EI/L^3} \tag{2.133}$$

If one ignores the effect of the axial force, the natural frequency of the beam is given by

$$f = \frac{1}{2\pi}\sqrt{\left(\frac{K_\mathrm{E}g}{W}\right)} = \frac{1}{2\pi}\sqrt{\left(\frac{106{\cdot}53028EIg/L^3}{0{\cdot}4462246wL}\right)} \tag{3.134}$$

or

$$\underline{f = 2{\cdot}4591211\sqrt{(EIg/wL^4)}} \tag{2.135}$$

Use of the correct modeshape yields

$$\underline{f = 2{\cdot}4509861\sqrt{(EIg/wL^4)}} \tag{2.136}$$

Thus the error in the natural frequency caused by a uniformly distributed load is 0·332%.

In the preceding development the mass–spring system was modelled to represent the motion of the beam at the position of maximum static displacement. If it is desired to study the motion of the beam at, say, midspan, then the shape function at this point must be unity. This is achieved by determining the value for $w$ that will cause the deflection at midspan to be equal to 1. From eq. (2.121) the deflection at midspan is obtained as

$$y_{x=L/2} = wL^4/192EI \tag{2.137}$$

When $y_{x=L/2} = 1$

$$w = 192EI/L^4 \tag{2.138}$$

Substitution of this expression for $w$ into eq. (2.121) yields

$$\phi(x) = (4/L^4)\{L^3x - 3Lx^3 + 2x^4\} \tag{2.139a}$$

$$\phi'(x) = (4/L^4)\{L^3 - 9Lx^2 + 8x^3\} \tag{2.139b}$$

$$\phi''(x) = (24/L^4)\{-3Lx + 4x^3\} \tag{2.139c}$$

thus the weight of the equivalent lumped mass is given by

$$W = \int_0^L w\{\phi(x)\}^2\,\mathrm{d}x = \int_0^L w(4/L^4)^2\{L^3x - 3Lx^3 + 2x^4\}^2\,\mathrm{d}x \tag{2.140}$$

Hence

$$\underline{W = (152/315)wL} \tag{2.141}$$

The equivalent elastic spring stiffness is given by

$$K_\mathrm{E} = \int_0^L EI\{\phi''(x)\}^2\,\mathrm{d}x = \int_0^L EI(24/L^4)\{-3Lx^3 + 4x^3\}^2\,\mathrm{d}x \tag{2.142}$$

Hence

$$\underline{K_\mathrm{E} = 576EI/5L^3} \tag{2.143}$$

The equivalent geometrical spring stiffness is given by

$$K_\mathrm{G} = \int_0^L T\{\phi'(x)\}^2\,\mathrm{d}x = \int_0^L T(4/L^4)^2\{L^3 - 9Lx^2 + 8x^3\}^2\,\mathrm{d}x \tag{2.144}$$

Hence

$$\underline{K_\mathrm{G} = 192T/35L} \tag{2.145}$$

These expressions for $W$, $K_\mathrm{E}$ and $K_\mathrm{G}$ result in the same values for the critical axial load and natural frequency as given by eqs (2.133) and (2.136) respectively.

### *Uniformly loaded beam with one end simply supported, one end built in and a concentrated load at midspan*

Assume the modeshape of the propped cantilever shown in Fig. 2.9 to be geometrically similar to the deflected form caused by the concentrated load at midspan. The deflected shape is found by integration of the expression for the bending moment at section $x$, which is given by

$$EI\,\mathrm{d}^2y/\mathrm{d}x^2 = -Rx + P\left(x - \tfrac{1}{2}L\right) \tag{2.146}$$

and imposition of the boundary conditions $y = 0$ when $x = 0$ and $x = L$, and $\mathrm{d}y/\mathrm{d}x = 0$ when $x = L$. This yields

$$y(x) = (P/96EI)\left\{3L^2x + 16\left(x - \tfrac{1}{2}L\right)^3 - 5x^3\right\} \tag{2.147}$$

$$y_{x=L/2} = 7PL^3/768EI \tag{2.148}$$

When $y_{x=L/2} = 1$

$$P = 768EI/7L^3 \tag{2.149}$$

Substitution of this expression for $P$ into eq. (2.148) yields the following expressions for the shape function $\phi(x)$ and its first and second derivatives

$$\phi(x) = \left(8/7L^3\right)\left\{3L^2x + 16\left(x - \tfrac{1}{2}L\right)^3 - 5x^3\right\} \tag{2.150a}$$

$$\phi'(x) = \left(8/7L^3\right)\left\{3L^2 + 48\left(x - \tfrac{1}{2}L\right)^2 - 15x^2\right\} \tag{2.150b}$$

$$\phi''(x) = \left(8/7L^3\right)\left\{96\left(x - \tfrac{1}{2}L\right) - 30x\right\} \tag{2.150c}$$

The weight of the equivalent lumped mass at the centre of the beam is now given by

$$W = P + \int_0^{L/2} w\{\phi(x)\}^2\,\mathrm{d}x + \int_{L/2}^{L} w\{\phi(x)\}^2\,\mathrm{d}x \tag{2.151}$$

or

$$\begin{aligned} W = {} & P + \int_0^{L/2} w\left(8/7L^3\right)^2\{3L^2x - 5x^3\}^2\,\mathrm{d}x + \\ & \int_{L/2}^{L} w\left(8/7L^3\right)^2\{3L^2x + 16\left(x - \tfrac{1}{2}L\right)^3 - 5x^3\}^2\,\mathrm{d}x \end{aligned} \tag{2.152}$$

This yields

$$W = P + 0{\cdot}2813411wL + 0{\cdot}1641398wL \tag{2.153}$$

Hence

$$\underline{W = P + 0{\cdot}4454809wL} \tag{2.154}$$

The corresponding equivalent elastic spring stiffness is given by

$$K_{\mathrm{E}} = \int_0^{L/2} EI\{\phi''(x)\}^2\,\mathrm{d}x + \int_{L/2}^{L} EI\{\phi''(x)\}^2\,\mathrm{d}x \tag{2.155}$$

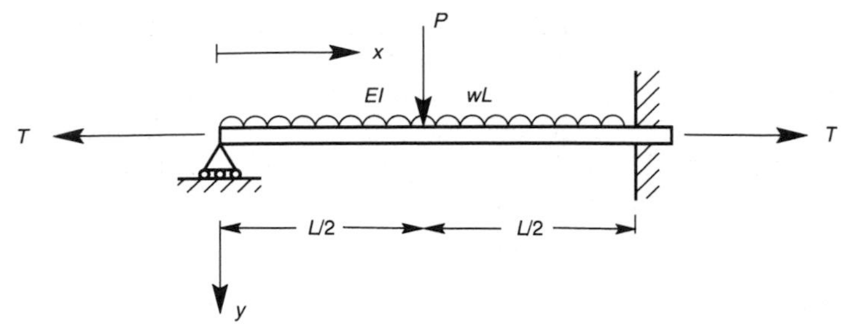


*Fig. 2.9. Uniformly loaded propped cantilever subjected to a load P at midspan and an axial tensile force T*

or

$$K_E = \int_0^{L/2} EI(8/7L^3)^2\{-30x\}^2\,dx + \int_{L/2}^{L} EI(8/7L^3)^2\{96(x-\tfrac{1}{2}L)-30x\}^2\,dx \tag{2.156}$$

This yields

$$K_E = 48{\cdot}979592EI/L^3 + 60{\cdot}734694EI/L^3 \tag{2.157}$$

Hence

$$\underline{K_E = 109{\cdot}714286EI/L^3} \tag{2.158}$$

The equivalent geometrical spring stiffness is given by

$$K_G = \int_0^{L/2} T\{\phi'(x)\}^2\,dx + \int_{L/2}^{L} T\{\phi'(x)\}^2\,dx \tag{2.159}$$

or

$$K_G = \int_0^{L/2} T(8/7L^3)^2\{3L^2-15x^2\}^2\,dx + \int_{L/2}^{L} T(8/7L^3)^2\{3L^2-48(x-L/2)^2-15x^2\}^2\,dx \tag{2.160}$$

This yields

$$K_G = 2{\cdot}8163265T/L + 3{\cdot}2897959T/L \tag{2.161}$$

Hence

$$\underline{K_G = 5{\cdot}1061224T/L} \tag{2.162}$$

Thus the critical value for the axial load occurs when

$$K_E + K_G = 109{\cdot}714286EI/L^3 + 5{\cdot}1061224T/L = 0 \tag{2.163}$$

This yields

$$\underline{T = -21{\cdot}486812EI/L^2} \tag{2.164}$$

If one ignores the axial force $T$ and the concentrated load $P$, the assumed modeshape yields

$$\underline{f = 2{\cdot}4976822\sqrt{(EIg/wL^4)}} \qquad (2.165)$$

Thus the error in the natural frequency caused by assuming the mode shape to be geometrically similar to the deflected form caused by a concentrated load at midspan is 1·905%.

### *Built-in beam with uniformly distributed load and one end vibrating vertically*

Assume the modeshape of vibration to be similar to the deflected form $y(x)$ caused by a vertical displacement $Y$ of the right-hand support. The deflected form due to this displacement is found from the expression of the bending moment at a section $x$ from the left-hand support

$$EI\,\mathrm{d}^2y/\mathrm{d}x^2 = -M_\mathrm{A} + V_\mathrm{A} - \tfrac{1}{2}x^2 \qquad (2.166)$$

Integration of eq. (2.166) twice and imposition of the boundary conditions $y(x) = \mathrm{d}y/\mathrm{d}x = 0$ when $x = 0$, and $x = L$ yield

$$y(x) = (Y/L^3)\{3Lx^2 - 2x^3\} \qquad (2.167)$$

For a mass–spring system to model the motion of the right-hand support, the shape function and its first and second derivatives are given by

$$\phi(x) = (1/L^3)\{3Lx^2 - 2x^3\} \qquad (2.168a)$$

$$\phi'(x) = (6/L^3)\{Lx - x^2\} \qquad (2.168b)$$

$$\phi''(x) = (6/L^3)\{L - x\} \qquad (2.168c)$$

Thus the weight of the equivalent lumped mass system is given by

$$W = \int_0^L w\{\phi(x)\}^2\,\mathrm{d}x = \int_0^L w(1/L^3)^2\{3Lx^2 - 2x^3\}^2\,\mathrm{d}x \qquad (2.169)$$

or

$$\underline{W = (13/35)wL} \qquad (2.170)$$

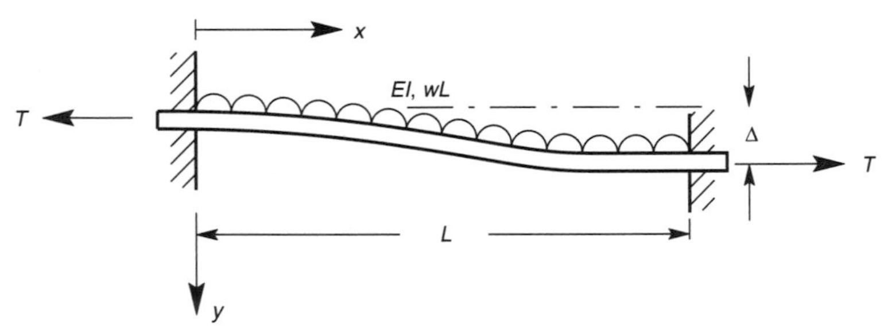


*Fig. 2.10. Built-in beam with uniformly distributed load wL, axial force T and one support vibrating vertically*

The equivalent elastic spring stiffness is given by

$$K_E = \int_0^L EI\{\phi''(x)\}^2\, dx = \int_0^L EI(6/L^3)^2\{L-2x\}^2\, dx \qquad (2.171)$$

or

$$\underline{K_E = 12EI/L^3} \qquad (2.172)$$

and the equivalent geometrical stiffness by

$$K_G = \int_0^L T\{\phi'(x)\}^2\, dx = \int_0^L T(6/L^3)^2\{Lx-x^2\}^2\, dx \qquad (2.173)$$

or

$$\underline{K_G = 6T/5L} \qquad (2.174)$$

The critical value for the axial force occurs when

$$K = K_E + K_G = 12EI/L^3 + 6T/5L \qquad (2.175)$$

or

$$\underline{T = -10EI/L^2} \qquad (2.176)$$

*Beam with uniformly distributed load, one end hinged and the built-in end vibrating vertically*

Assume the modeshape of vibration of the propped cantilever shown in Fig. 2.11 to be geometrically similar to the deflected form $y(x)$ caused by a vertical displacement $Y$ of the built-in support. The deflected shape may be found from the expression for the bending moment at a distance $x$ from the left-hand support

$$M = EI\, d^2y/dx^2 = Vx - \tfrac{1}{2}wx^2 \qquad (2.177)$$

Integration of eq. (2.177) twice and imposition of the boundary conditions $y(x) = 0$ when $x = 0$, $y(x) = Y$ when $x = L$, and $dy/dx = 0$ when $x = L$ yield

$$y(x)(Y/2L^3)\{3L^2x - x^3\} \qquad (2.178)$$

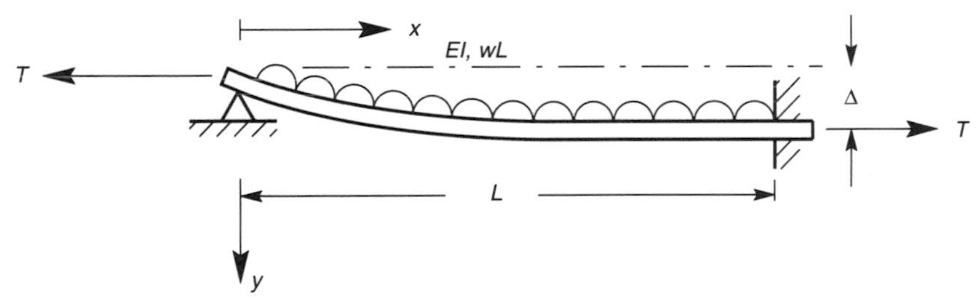


*Fig. 2.11. Propped cantilever with uniformly distributed load wL, axial force T and the built-in end vibrating vertically*

For an equivalent mass–spring system to model the vertical motion of the built-in end, the shape function at that point must be unity. This requires that $Y = 1$. When this is the case the shape function and its first and second derivatives are given by

$$\phi(x) = (1/2L^3)\{3L^2x - x^3\} \tag{2.179a}$$

$$\phi'(x) = (3/2L^3)\{L^2 - x^2\} \tag{2.179b}$$

$$\phi''(x) = (3/2L^3)\{-2x\} \tag{2.179c}$$

Thus the weight of the equivalent lumped mass is given by

$$W = \int_0^L w\{\phi(x)\}^2\,\mathrm{d}x = \int_0^L w(1/2L^3)^2\{3L^2x - x^3\}^2\,\mathrm{d}x \tag{2.180}$$

or

$$\underline{W = (17/35)wL} \tag{2.181}$$

The equivalent elastic stiffness is given by

$$K_{\mathrm{E}} = \int_0^L EI\{\phi''(x)\}^2\,\mathrm{d}x = \int_0^L EI(3/2L^3)^2\{-2x\}^2\,\mathrm{d}x \tag{2.182}$$

or

$$K_{\mathrm{E}} = 3EI/L^3 \tag{2.183}$$

and the equivalent geometrical stiffness by

$$K_{\mathrm{G}} = \int_0^L T\{\phi'(x)\}^2\,\mathrm{d}x = \int_0^L T(3/2L^3)^2\{L^2 - x^2\}^2\,\mathrm{d}x \tag{2.184}$$

or

$$\underline{K_{\mathrm{G}} = 6T/5L} \tag{2.185}$$

The critical value for the axial force occurs when

$$K = K_{\mathrm{E}} + K_{\mathrm{G}} = 3EI/L^3 + 6T/5L = 0 \tag{2.186}$$

or

$$\underline{T = -5EI/2L^2} \tag{2.187}$$

## Equivalent one DOF mass–spring systems for linearly elastic continuous beams

The method used in the preceding sections to develop expressions for equivalent lumped masses and spring stiffnesses for single span beams by use of eqs (2.8), (2.13) and (2.19), and assuming the modeshapes to be geometrically similar to a deflected form, can be extended to continuous beams by, for example, assuming the modeshape to be similar to the deflected shape caused by a uniformly distributed load acting alternately downwards and upwards on succeeding spans, or by assuming the mode shapes to be geometrically similar to the deflected form caused by a concentrated load at the point where the response is to be studied.

Generally, however, this method of approach is not practical because of the time involved in developing the required shape functions and the subsequent integrations. Thus, when it is necessary to determine the natural frequencies of multi-span beams it is better to use one of the many structural analysis programs available that include the solution of the eigenvalue problem. At the design stage, however, and in order to check the output from a computer analysis, it is useful to have available a simple method that enables a quick estimate of the first and perhaps even the second natural frequencies of continuous beams. Such estimates can be made by assuming the modeshapes of the individual spans to be similar to the modeshapes of corresponding simply supported beams, or if one end or both ends of a continuous beam are rigidly encased, by beams being simply supported at one end and built-in at the other. Thus each span can be modelled by the expressions for the equivalent lumped masses and spring stiffnesses given by eqs (2.64) and (2.66), or (2.127) and (2.129). The equivalent lumped mass for a continuous beam is found by equating the maximum kinetic energy of the equivalent lumped masses of the individual spans. Similarly, the equivalent spring stiffness is determined by equating the maximum strain energy of the spring to the sum of the maximum strain energies in the equivalent springs for each span.

Consider a continuous beam with $N$ spans. The maximum kinetic energy of the beam in terms of the equivalent lumped masses of the individual spans is given by

$$\tfrac{1}{2}M_{\mathrm{E}}\dot{Y}^2 = \tfrac{1}{2}\sum_{i=1}^{N} M_i\phi(x_i)^2\dot{Y}^2 \tag{2.188}$$

where $M_{\mathrm{E}}$ is the equivalent lumped mass for the continuous beam, $M_i$ is the equivalent lumped mass for the $i$th span as given by eq. (2.64) or (2.127), $\dot{Y}$ is the maximum velocity of mass $M_{\mathrm{E}}$, and $\phi(x_i)$ is the value of the shape function at position $x_i$ of $M_i$. Thus

$$M_{\mathrm{E}} = \sum_{i=1}^{N} M_i\phi(x_i)^2 \tag{2.189}$$

Similarly, the maximum strain energy of the continuous beam as a function of the equivalent elastic springs representing the stiffness of each span is given by

$$\tfrac{1}{2}K_{\mathrm{E}}Y^2 = \tfrac{1}{2}\sum^{N} K_{\mathrm{E}i}\phi(x_i)^2Y^2 \tag{2.190}$$

where $K_{\mathrm{E}}$ is the equivalent elastic spring stiffness of the continuous beam, $K_{\mathrm{E}i}$ is the equivalent spring stiffness of the $i$th span as given by eq. (2.66) or (2.129), and $Y$ is the maximum amplitude of the equivalent mass $M$. Thus

$$K_{\mathrm{E}} = \sum_{i=1}^{N} K_{\mathrm{E}i}\phi(x_i)^2 \tag{2.191}$$

The natural frequency of a continuous or multi-span beam is therefore given by

$$f = \left\{ \sum_{i=1}^{N} K_{Ei}\phi(x_i)^2 \bigg/ \sum_{i=1}^{N} M_i\phi(x_i)^2 \right\}^{1/2} \qquad (2.192)$$

The degree of accuracy obtained by use of eq. (2.192) depends very much on the estimates of the relative values of $\phi(x_i)$. In practice such estimates can be difficult, and the most accurate values for the frequencies are most easily obtained when the spans are of approximately the same length, which is often the case in real structures. This statement is demonstrated by the following two examples.

**Example 2.1** The continuous beam ABCDE shown in Fig. 2.12 has four equal spans of length $L$ and is built-in at E. The section of the beam is constant throughout, having a flexural rigidity $EI$, and weighs $w$ per unit length. Develop first expressions for the weight of the equivalent lumped mass and spring stiffness of the beam corresponding to vibration in the first mode, and hence an expression for its first natural frequency, and then expressions for the equivalent lumped mass, spring stiffness and natural frequency corresponding to vibration in the second mode shown.

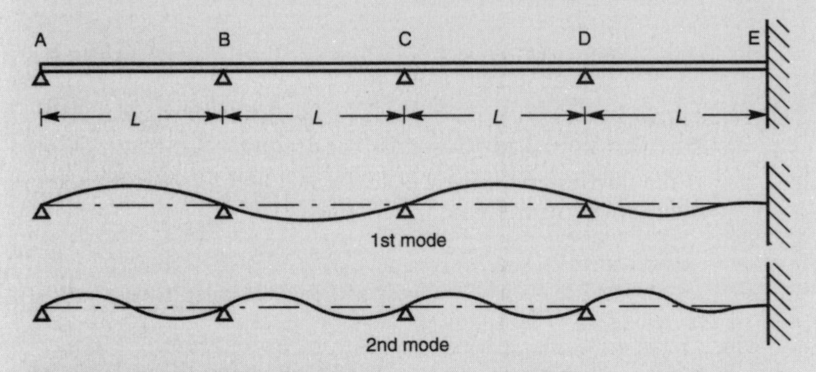


*Fig. 2.12. Continuous beam with four equal spans and one end built-in at E*

In order to develop expressions for the equivalent weight and spring stiffness of the first mode shown in Fig. 2.12, assume that the maximum amplitude of vibration for each of the spans AB, BC and CD is equal to unity, and that the maximum amplitude of span DE is proportional to the maximum amplitude of the other three spans. Thus the amplitude of span DE is given by

$$\phi(x)^{DE}_{max} = \frac{5{\cdot}41403 \times 10^{-3} wL^4/EI}{13{\cdot}0208 \times 10^{-3} wL^4/EI} = 0{\cdot}4157985$$

Thus the assumed modeshape of vibration for the first mode is

$$\phi(x) = \{\, 1{\cdot}0 \quad -1{\cdot}0 \quad 1{\cdot}0 \quad -0{\cdot}415958 \,\}$$

Substitution of these values and the expression for the weights of the equivalent lumped masses for each of the four spans given by eqs (2.64) and (2.129) into eq. (2.189) yields

$$M = \left\{3 \times (3968/7875)wL \times 1{\cdot}0^2 + (0{\cdot}4462246)wL \times 0{\cdot}415958^2\right\}/g$$
$$= 1{\cdot}588252wL/g$$

Substitution of the same values for $\phi(x)$ and the expressions for the equivalent elastic spring stiffnesses for each of the four spans given by eqs (2.66) and (2.129) into eq. (2.191) yields

$$K_E = \left\{3 \times \left(6144EI/125L^3\right) \times 1{\cdot}0^2 + \left(106{\cdot}53628EI/L^3\right) \times 0{\cdot}415958^2\right\}$$
$$= 165{\cdot}88902EI/L^3$$

Thus the first natural frequency is

$$f_1 = \frac{1}{2\pi}\sqrt{\left(\frac{165{\cdot}88902EI/L^3}{1{\cdot}588252wL/g}\right)}$$

or

$$\underline{f_1 = 1{\cdot}6265568\sqrt{(EIg/wL^4)}}$$

The correct value is

$$\underline{f_1 = 1{\cdot}6392959\sqrt{(EIg/wL^4)}}$$

Thus the error in this case is 0·78%. This of course is very good, but it is rather fortuitous as in a real modeshape the ratio of the amplitudes of spans CD and DE would tend to be less than one, if it could be assumed that the slopes of the tangents of the two spans at D were equal. Also, the amplitude of span CD would be less than the amplitudes of spans AB and BC because of the built-in end at E. In example 2.2, therefore, it is shown that modeshapes based on assumptions of the relative displacements of noncontinuous individual spans can lead to considerable errors.

The frequency of the beam corresponding to the second modeshape shown in Fig. 18 can be determined by assuming the beam to have eight spans, each of length $L/2$. Thus the assumed modeshape vector is given by

$$\phi(x) = \{\, 1{\cdot}0 \quad -1{\cdot}0 \quad 1{\cdot}0 \quad -1{\cdot}0 \quad 1{\cdot}0 \quad -1{\cdot}0 \quad 1{\cdot}0 \quad -0{\cdot}415958 \,\}$$

Substitution of these values and the equivalent lumped mass for each half span into eq. (2.189) yields

$$M_E = \{7 \times (3968/7875)wL/2 \times 1{\cdot}0^2 + (0{\cdot}4462246)wL/2 \times 0{\cdot}415958^2\}/g$$
$$= 1{\cdot}8021587wL$$

Substitution of the same values for $\phi(x)$ and the equivalent spring stiffnesses for each half span into eq. (2.191) yields

$$K_E = \{7 \times (6144 \times 8EI/125L^3) \times 1{\cdot}0^2 + (106{\cdot}53628 \times 8EI/L^3) \times 0{\cdot}415958^2\} = 2899{\cdot}9762EI/L^3$$

Thus the frequency for this mode is given by

$$f_2 = \frac{1}{2\pi}\sqrt{\left(\frac{2899{\cdot}9762EIg/L^3}{1{\cdot}8021587wL}\right)}$$

Hence

$$\underline{f_2 = 6{\cdot}38441089\sqrt{(EIg/wL^4)}}$$

The correct value is

$$\underline{f_2 = 6{\cdot}4378174\sqrt{(EIg/wL^4)}}$$

Thus, also in this case the error is small—0·83%.

**Example 2.2** The continuous beam ABCD shown in Fig. 2.13 is of uniform section with constant flexural rigidity $EI$ and self-weight $w$ per unit length. The lengths of spans AB and BC are equal to $L$; and that of span BC is equal to $2L$. Determine the first natural frequency of the beam. Assume that the first modeshape is geometrically similar to that caused by a uniformly distributed load acting upwards on spans AB and CD and downwards on span BC.

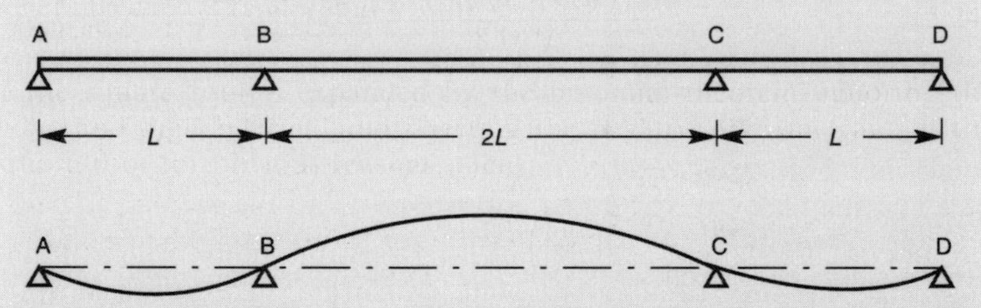


*Fig. 2.13. Continuous beam with the length of the central span twice the length of the two outer ones*

If it is assumed that the beam is noncontinuous at supports B and C, then the following modeshape vector is obtained for the amplitudes of the central points of each span

$$\phi(x) = \{0{\cdot}125 \quad -1{\cdot}0 \quad 0{\cdot}125\}$$

Substitution of the above values for $\phi(x)$ and the appropriate weight of the equivalent lumped masses of the individual spans into eq. (2.189) yields

$$M_E = \{2 \times (3968/7875)wL \times 0{\cdot}125^2 + (3968/7875)w2L \times 1{\cdot}0^2\}/g$$
$$= 1{\cdot}0234921wL/g$$

The equivalent elastic spring stiffness is determined by substitution of the values for $\phi(x)$ and the equivalent spring stiffnesses for the three spans into eq. (2.191). This yields

$$K_E = \{2 \times (6144EI/125L^3) \times 0{\cdot}125^2 + (6144EI/125 \times 8L^3) \times 1{\cdot}0^2\}$$
$$= 7{\cdot}68EI/L^3$$

Hence

$$f = \frac{1}{2\pi}\sqrt{\left(\frac{7{\cdot}68EI/L^3}{1{\cdot}0234921wL/g}\right)}$$

or

$$\underline{f = 0{\cdot}4359719\sqrt{(EIg/wL^4)}}$$

The correct value is

$$\underline{f = 0{\cdot}613798\sqrt{(EIg/wL^4)}}$$

Thus the error in the estimated value is $-28{\cdot}97\%$. This is obviously not very good, but is expected as the overall stiffness of the beam is reduced by assuming noncontinuity at the supports.

An alternative estimate may be achieved by assuming that the points of contraflexure occur at positions B and C. This yields the modeshape vector

$$\phi(x) = \{0{\cdot}5 \quad -1{\cdot}0 \quad 0{\cdot}5\}$$

Use of the above values for $\phi(x)$ yield

$$M_E = \{2 \times (3968/7875)wL \times 0{\cdot}5^2 + (3968/7875)w2L \times 1{\cdot}0^2\}/g$$
$$= 1{\cdot}2596825wL/g$$
$$K_E = \{2 \times (6144EI/125L^3) \times 0{\cdot}5^2 + (6144EI/125 \times 8L^3) \times 1{\cdot}0^2\}$$
$$= 30{\cdot}72EI/L^3$$

Thus

$$f = \frac{1}{2\pi}\sqrt{\left(\frac{30{\cdot}72EI/L^3}{1{\cdot}2596825wL/g}\right)}$$

or

$$\underline{f = 0{\cdot}7859595\sqrt{(EIg/wL^4)}}$$

The error is this case is therefore +28·05%. This is not very good either, but is again expected as the assumption made implies that the $EI$ value for the central span is much greater than for the outer ones. By use of the theorem of three moments it can in fact be shown that the points of contraflexure lie to the right of B and to the left of C. This will result in a maximum deflection of the outer spans relative to the central one somewhere between 0·125 and 0·5. If it is assumed that these amplitudes are equal to $\frac{1}{2}(1{\cdot}25 + 0{\cdot}5)$, one obtains the modeshape vector

$$\phi(x) = \{0{\cdot}3125 \quad -1{\cdot}0 \quad 0{\cdot}3125\}$$

Thus

$$M_{\mathrm{E}} = \{2 \times (3968/7875)wL \times 0{\cdot}3125^2 + (3968/7875)w^2L \times 1{\cdot}0^2\}/g$$
$$= 1{\cdot}1061587wL/g$$
$$K_{\mathrm{E}} = \{2 \times (6144EI/125L^3) \times 0{\cdot}3125^2 + (6144EI/125 \times 8L^3) \times 1{\cdot}0^2\}$$
$$= 15{\cdot}744EI/L^3$$

Hence

$$\underline{f = 0{\cdot}600439\sqrt{(EIg/wL^4)}}$$

The error in this case is therefore only −2·145%, which is acceptable. Obviously, the reasons for the discrepancies are the geometries of the assumed modeshapes. The above modeshapes can be compared with the modeshape vector

$$\phi(x) = \{0{\cdot}326 \quad -1{\cdot}0 \quad 0{\cdot}326\}$$

which was obtained from a computer analysis, and which when substituted into eqs (2.189) and (2.191) together with the appropriate values for equivalent masses and spring stiffness yields

$$\underline{f = 0{\cdot}6139793\sqrt{(EIg/wL^4)}}$$

Thus when the correct modeshape is used the error resulting from considering each beam as a mass–spring system is only 0·03%. A relatively simple way to decide on a modeshape is to determine the points of contraflexure by first constructing the bending moment diagram and then sketching the corresponding deflected form knowing that the amplitude of the central span is 1·0 and that of the outer spans lies between 0·125 and 0·5 m. It is left to the reader to try this out. When the author tried it he obtained the modeshape vector

$$\phi(x) = \{0{\cdot}32 \quad -1{\cdot}0 \quad 0{\cdot}32\}$$

**Example 2.3** Determine the first natural frequency of the beam in example 2.2 by assuming that the modeshape is geometrically similar to that caused by a point load applied at the centre of span BC.

With the origin at A, the deflected form of section AB can be shown to be

$$y = (3P/96P)\{x^3 - L^2x\} \tag{2.193a}$$

and with the origin at B the deflected shape of section BC can be shown to be

$$y = (P/96EI)\{-8x^3 + 9Lx^2 + 6L^2x\} \tag{2.193b}$$

Thus the shape functions and their derivatives for spans AB, BC and CD are given by

$$\phi_{AB}(x) = \phi_{CD}(x) = (3/7L^3)\{x^3 - L^2x\} \tag{2.194a}$$

$$\phi'_{AB}(x) = \phi'_{CD}(x) = (3/7L^3)\{3x^2 - L^2\} \tag{2.194b}$$

$$\phi''_{AB}(x) = \phi''_{CD}(x) = (18/7L^3\{x\} \tag{2.194c}$$

$$\phi_{BC}(x) = (1/7L^3)\{-8x^3 + 9Lx^2 + 6L^2x\} \tag{2.194d}$$

$$\phi'_{BC}(x) = (2/7L^3)\{-12x^2 + 9Lx + 3L^2\} \tag{2.194e}$$

$$\phi''_{BC}(x) = (6/7L^3)\{-8x + 3L\} \tag{2.194f}$$

The expression for the weight of the equivalent lumped mass is

$$W_E = P + 2\int_0^L w\{\phi_{AB}(x)\}^2\,dx + 2\int_0^L w\{\phi_{BC}(x)\}^2\,dx \tag{2.195}$$

Substitution of the expression for $\phi_{AB}(x)$ and $\phi_{BC}(x)$ and integration between limits yield

$$\begin{aligned} W_E &= P + (48/1715)wL + (1480/1715)wL \\ &= \underline{P + (1528/1715)wL} \end{aligned} \tag{2.196}$$

The expression for the equivalent spring stiffness is

$$K_E = 2\int_0^L EI\{\phi''_{AB}(x)\}^2\,dx + 2\int_0^L EI\{\phi''_{CD}(x)\}^2\,dx \tag{2.197}$$

Substitution of the expressions for $\phi''_{AB}(x)$ and $\phi''_{CD}(x)$ and integration between limits yield

$$K_E = 216EI/49L^3 + 456EI/49L^3 = \underline{96EI/7L^3} \tag{2.198}$$

Thus, if one ignores the concentrated load $P$, the natural frequency is given by

$$f = \frac{1}{2\pi}\sqrt{\left(\frac{96EI/7L^3}{1528wL/1715}\right)}$$

or

$$\underline{f = 0{\cdot}6244204\sqrt{(EIg/wL^4)}} \qquad (2.199)$$

This approach therefore yields an error of only 1·73%. The amount of work involved, however, is considerable.

The equivalent geometrical stiffness due to an axial tensile force $T$ that is applied to either end of the beam is given by

$$K_G = 2\int_0^L T\{\phi'_{AB}(x)\}^2\,dx + 2\int_0^L T\{\phi'_{BC}(x)\}^2\,dx \qquad (2.200)$$

Substitution for $\phi'_{AB}(x)$ and $\phi'_{BC}(x)$ and integration between limits yields

$$K_G = 72T/245L + 552/245L = \underline{624T/245L} \qquad (2.201)$$

**Example 2.4** Dynamic testing of a continuous beam of the same proportions as shown in Fig. 2.13, but with supports that at either end were partially restrained from moving horizontally, yielded a first resonance frequency of $F = 0{\cdot}753\sqrt{(EIg/wL^4)}$. Assuming the modeshape of vibration to be geometrically similar to the deflected form caused by a concentrated load at the midpoint of the central span, estimate the additional equivalent spring stiffness or geometrical stiffness and the axial force caused by these restraints.

The geometrical stiffness may be found from the relationship

$$f = \frac{1}{2\pi}\sqrt{\left(\frac{K_E + K_G}{M_E}\right)} = 0{\cdot}753\sqrt{\left(\frac{EIg}{wL^4}\right)}$$

Thus

$$K_G = (0{\cdot}753 \times 2\pi)^2(EIg/wL^4) \times (1528wL/1715g) - (96EI/7L^3)$$
$$= \underline{6{\cdot}2295606EI/L^3}$$

From eq. (2.201)

$$T = (245L/624) \times K_G = (245L/624) \times (6{\cdot}2295606EI/L^3)$$
$$= \underline{2{\cdot}4459012EI/L^2}$$

## First natural frequency of sway structures

The most common types of sway structure are towers, chimneys and tall multistorey buildings. The dominant frequency of the first two can usually be assessed by considering them as cantilever columns with constant or tapering sections, while the dominant mode and hence the frequency of the last can be calculated by considering the sway of the columns between each floor level as a lumped mass. A good approximation to the modeshapes of multistorey buildings is to assume them to be geometrically similar to the deflected forms caused by concentrated loads, each applied horizontally at floor level and in magnitude equal to the weight of the floor. The modeshape having been determined in this manner, the natural frequency corresponding to this modeshape can be determined by equating the maximum kinetic energy of the lumped mass system to the maximum strain energy stored in the columns. The details of the method are most easily explained through examples.

### *Multistorey shear structures*

Shear structures are structures in which the floors are so stiff compared to the columns that they be assumed to be rigid.

**Example 2.5** Use an approximate method to determine the shape of the first mode of vibration and the first natural frequency for the three-storey shear structure shown in Fig. 2.14, if the shear stiffness $k = 12EI/L^3$ is the same for all the columns, and the weight of each of the three floors is $w$ per unit length. If $w = 20{\cdot}0$ kN/m calculate the value of the flexural rigidity $EI$ to yield a dominant frequency of 3·0 Hz. The weight of the columns may be ignored.

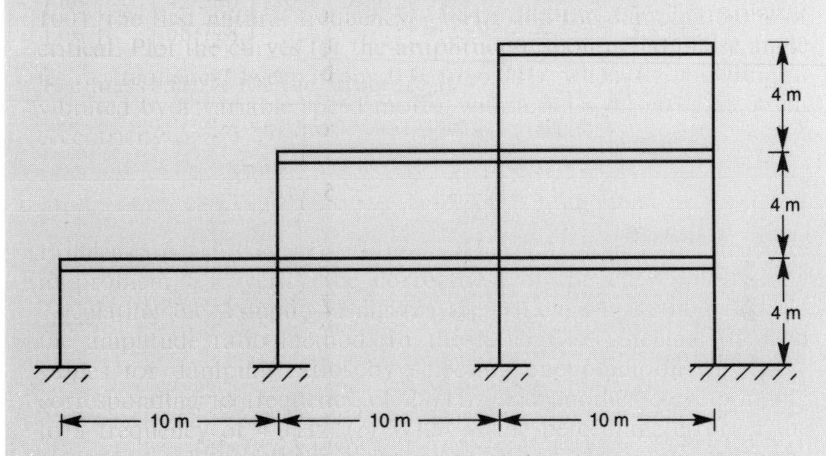


*Fig. 2.14. Three-storey shear structure with EI constant for all columns, and the weight of all floors equal to w/m*

The modeshape is determined by applying a force at each floor level proportional to the weight of each floor. Thus the following horizontal force vector may be used

$$P = \{3{\cdot}0 \quad 2{\cdot}0 \quad 1{\cdot}0\}$$

The displacement due to $P$ at level 1 is

$$x_1 = (3{\cdot}0 + 2{\cdot}0 + 1{\cdot}0)/4k = 1{\cdot}5/k$$

at level 2 is

$$x_2 = 1{\cdot}5/k + (2{\cdot}0 + 1{\cdot}0)/3k = 2{\cdot}5/k$$

and at level 3 is

$$x_3 = 2{\cdot}5/k + 1{\cdot}0/2k = 3{\cdot}0/k$$

Therefore the assumed modeshape vector is

$$x = \{1{\cdot}5 \quad 2{\cdot}5 \quad 3{\cdot}0\}$$

The maximum kinetic energy is given by

$$\begin{aligned} \mathrm{KE_{max}} &= \tfrac{1}{2}\sum_{i=1}^{3} m_i \omega^2 x_i^2 \\ &= \tfrac{1}{2}\omega^2 (w/g)\{30 \times 1{\cdot}5^2 + 20 \times 2{\cdot}5^2 + 10 \times 3{\cdot}0^2\} \\ &= 141{\cdot}25\omega^2 w/g \end{aligned}$$

The maximum strain energy is given by

$$\begin{aligned} U_{\max} &= \tfrac{1}{2}\sum_{i=1}^{3} k_i (x_i - x_{i-1})^2 \\ &= \tfrac{1}{2}k\left\{4 \times 1{\cdot}5^2 + 3 \times (2{\cdot}5 - 1{\cdot}5)^2 + 2 \times (3{\cdot}0 - 2{\cdot}5)^2\right\} \\ &= 6{\cdot}25k \end{aligned}$$

Equating the maximum kinetic energy and the maximum strain energy yields

$$6{\cdot}25 \times 12EI/4{\cdot}0^3 = 141{\cdot}25\omega^2 w/g$$

Hence

$$\underline{f = 0{\cdot}0144966\surd(EIg/w)}$$

The correct value is

$$\underline{f = 0{\cdot}0143625\surd(EIg/w)}$$

Thus the error in this case is only 0·94%. The flexural rigidity of the columns is now found by substitution of the value of 3·0 Hz for $f$ in the above expression for the natural frequency of the structure. Thus

$$3{\cdot}0 = 0{\cdot}0144966\sqrt{(EIg/20{\cdot}0)}$$

and hence

$$\underline{EI = 87\,311{\cdot}447\ \text{kN}\,\text{m}^2}$$

For low-rise shear structures an even simpler, but less accurate, method for estimating the dominant frequency is to lump the masses of all the floors together as an equivalent mass at the first floor level, thus reducing the structure to a one DOF system. For the structure in Fig. 2.14 the equivalent mass is

$$M_{\text{E}} = (w/g)\{10 \times 12 + 20 \times 8 + 30 \times 4\}/4 = 100w/g$$

The total shear stiffness of the columns below the first floor level is

$$K_{\text{E}} = 4 \times 12EI/4{\cdot}0^3 = 3EI/4{\cdot}0$$

Hence

$$\underline{f = 0{\cdot}0137832\sqrt{(EIg/w)}}$$

Thus the error is −4·28%.

An alternative method of estimating the first natural frequency is to consider the structure as an equivalent one DOF system where the lumped mass is at the roof level. The equivalent mass at this level is given by

$$M_{\text{E}} = (w/g)\{30 \times 4{\cdot}0 + 20 \times 8{\cdot}0 + 10 \times 12{\cdot}0\}/12 = 400w/12g$$

The equivalent spring stiffness $K_{\text{E}}$ is determined from

$$\frac{1}{K_{\text{E}}} = \frac{1}{k_1} + \frac{1}{k_2} + \frac{1}{k_3} = \frac{1}{4k} + \frac{1}{3k} + \frac{1}{2k} = \frac{13}{12k}$$

which yields

$$K_{\text{E}} = \frac{12k}{13} = \frac{12}{13} \times \frac{12EI}{4{\cdot}0^3} = \frac{9EI}{52}$$

Hence

$$f = \frac{1}{2\pi}\sqrt{\left(\frac{9EI \times 12g}{52 \times 400w}\right)} = \underline{0{\cdot}0114683\sqrt{(EIg/w)}}$$

Thus the error in this case is −20·15%. The reason for this magnitude of error is that the implied modeshape assumed in this simplification differs significantly from the real one.

**Example 2.6** Use an approximate method to estimate the equivalent mass, stiffness and dominant frequency of a ten-storey shear structure, the data for which are given in Table 1. Assume the distance between each floor to be 3·0 m.

*Table 1. Example 2.6 data*

| Level | Mass × $10^{-3}$ (kg) | Stiffness × $10^{-6}$ (N/m) |
|---|---|---|
| 1 | 1225·0 | 2500·0 |
| 2 | 1225·0 | 2500·0 |
| 3 | 1225·0 | 2500·0 |
| 4 | 1225·0 | 1024·0 |
| 5 | 1225·0 | 1024·0 |
| 6 | 1225·0 | 1024·0 |
| 7 | 1225·0 | 1024·0 |
| 8 | 1225·0 | 324·0 |
| 9 | 1225·0 | 324·0 |
| 10 | 1225·0 | 324·0 |

The first step is to calculate the modeshape vector for the first mode by applying a horizontal load at each floor level proportional to the weight of the floor. As in this case the weight of all the floors is the same, 1·0 kN may be applied at each level. Hence the displacements at the various levels are

$$\text{level 1:}\quad x_1 = 10 \times 1{\cdot}0/2{\cdot}5 \times 10^{-6} = 4{\cdot}000 \times 10^{-6}\,\text{m}$$

$$\text{level 2:}\quad x_2 = 4{\cdot}000 \times 10^{-6} + 9{\cdot}0 \times 1{\cdot}0/2{\cdot}500 \times 10^{-6} = 7{\cdot}600 \times 10^{-6}\,\text{m}$$

$$\text{level 3:}\quad x_3 = 7{\cdot}600 \times 10^{-6} + 8{\cdot}0 \times 1{\cdot}0/2{\cdot}500 \times 10^{-6} = 10{\cdot}800 \times 10^{-6}\,\text{m}$$

$$\text{level 4:}\quad x_4 = 10{\cdot}800 \times 10^{-6} + 7{\cdot}0 \times 1{\cdot}0/1{\cdot}024 \times 10^{-6} = 17{\cdot}636 \times 10^{-6}\,\text{m}$$

$$\text{level 5:}\quad x_5 = 17{\cdot}636 \times 10^{-6} + 6{\cdot}0 \times 1{\cdot}0/1{\cdot}024 \times 10^{-6} = 23{\cdot}495 \times 10^{-6}\,\text{m}$$

$$\text{level 6:}\quad x_6 = 23{\cdot}495 \times 10^{-6} + 5{\cdot}0 \times 1{\cdot}0/1{\cdot}024 \times 10^{-6} = 28{\cdot}378 \times 10^{-6}\,\text{m}$$

$$\text{level 7:}\quad x_7 = 28{\cdot}378 \times 10^{-6} + 4{\cdot}0 \times 1{\cdot}0/1{\cdot}024 \times 10^{-6} = 32{\cdot}284 \times 10^{-6}\,\text{m}$$

$$\text{level 8:}\quad x_8 = 32{\cdot}284 \times 10^{-6} + 3{\cdot}0 \times 1{\cdot}0/0{\cdot}324 \times 10^{-6} = 41{\cdot}543 \times 10^{-6}\,\text{m}$$

$$\text{level 9:}\quad x_9 = 41{\cdot}543 \times 10^{-6} + 2{\cdot}0 \times 1{\cdot}0/0{\cdot}324 \times 10^{-6} = 47{\cdot}716 \times 10^{-6}\,\text{m}$$

$$\text{level 10:}\quad x_{10} = 47{\cdot}716 \times 10^{-6} + 1{\cdot}0 \times 1{\cdot}0/0{\cdot}324 \times 10^{-6} = 50{\cdot}802 \times 10^{-6}\,\text{m}$$

Dividing all the above displacements by $x_{10}$ yields the modeshape vector

$$x = \{0{\cdot}0787 \quad 0{\cdot}1496 \quad 0{\cdot}2126 \quad 0{\cdot}3471 \quad 0{\cdot}4625 \quad 0{\cdot}5586 \quad 0{\cdot}6355 \quad 0{\cdot}8177 \quad 0{\cdot}9392 \quad 1{\cdot}0000\}$$

Substitution of these values together with the given values for masses and shear stiffnesses yields the following values for the maximum kinetic energy and maximum strain energy

$$KE_{max} = \tfrac{1}{2}\sum_{i=1}^{10} m_i \omega_i^2 x_i^2 = 2{\cdot}25081\omega^2 \times 10^6\,\mathrm{N\,m}$$

$$U_{max} = \tfrac{1}{2}\sum_{i=1}^{10} k_i (x_i - x_{i-1})^2 = 50{\cdot}63286 \times 10^6\,\mathrm{N\,m}$$

Because $x_{10}$ is assumed to be unity the equivalent mass, spring stiffness and natural frequency for studying the motion of the top of the building are respectively

$$\underline{M = 4{\cdot}50162 \times 10^6\,\mathrm{kg}}$$

$$\underline{K_E = 101{\cdot}26572 \times 10^6\,\mathrm{N/m}}$$

$$\underline{f_1 = 0{\cdot}7549\,\mathrm{Hz}}$$

The value for the first natural frequency obtained from an eigenvalue analysis is $f = 0{\cdot}7419$ Hz. Thus the error is 1·75%. Although the method yields a surprisingly accurate value for the natural frequency, it is not advisable to use the assumed modeshape when calculating the bending stresses in the columns. The reason for this will become clear when the elements in the modeshape vector $x$ are compared with those in the vector $\tilde{x}$ below, which are calculated by an eigenvalue analysis.

$$x = \{0{\cdot}0787 \quad 0{\cdot}1496 \quad 0{\cdot}2126 \quad 0{\cdot}3471 \quad 0{\cdot}4625 \quad 0{\cdot}5586 \quad 0{\cdot}6355 \quad 0{\cdot}8177 \quad 0{\cdot}9392 \quad 1{\cdot}0000\}$$

$$\tilde{x} = \{0{\cdot}0488 \quad 0{\cdot}0971 \quad 0{\cdot}1443 \quad 0{\cdot}2559 \quad 0{\cdot}3608 \quad 0{\cdot}4564 \quad 0{\cdot}5401 \quad 0{\cdot}7602 \quad 0{\cdot}9117 \quad 1{\cdot}0000\}$$

Simplification of the structure to a one DOF system with the mass concentrated at the first floor level yields

$$M_E = 1{\cdot}225 \times 10^6\{30{\cdot}0 + 27{\cdot}0 + 24{\cdot}0 + 21{\cdot}0 + 18{\cdot}0 + 15{\cdot}0 + 12{\cdot}0 + 9{\cdot}0 + 6{\cdot}0 + 3{\cdot}0\}/3{\cdot}0 = 67{\cdot}375 \times 10^6\,\mathrm{kg}$$

$$K_E = 2500 \times 10^6\,\mathrm{N/m}$$

Hence

$$f = \frac{1}{2\pi}\sqrt{(2500 \times 10^6/67{\cdot}375 \times 10^6)} = \underline{0{\cdot}9695\,\mathrm{Hz}}$$

The error is therefore +30·68%, which is obviously not good enough.

Simplification of the structure to a one DOF system with the mass concentrated at roof level yields

$$M_{\mathrm{E}} = 1{\cdot}225 \times 10^6\{30{\cdot}0 + 27{\cdot}0 + 24{\cdot}0 + 21{\cdot}0 + 18{\cdot}0 + 15{\cdot}0 + 12{\cdot}0 + 9{\cdot}0 + 6{\cdot}0 + 3{\cdot}0\}/30{\cdot}0$$

Hence

$$M_{\mathrm{E}} = 6{\cdot}7375 \times 10^6\,\mathrm{kg}$$

The equivalent spring stiffness $K_{\mathrm{E}}$ is found from

$$\frac{1}{K_{\mathrm{E}}} = \frac{1}{k_1} + \frac{1}{k_2} + \frac{1}{k_3} + \frac{1}{k_4} + \frac{1}{k_5} + \frac{1}{k_6} + \frac{1}{k_7} + \frac{1}{k_8} + \frac{1}{k_9} + \frac{1}{k_{10}}$$

$$\frac{1}{K_{\mathrm{E}}} = \frac{3}{2500 \times 10^6} + \frac{4}{1024 \times 10^6} + \frac{3}{324 \times 10^6} = 0{\cdot}0143655 \times 10^{-6}$$

Hence

$$K_{\mathrm{E}} = 69{\cdot}611176 \times 10^6\,\mathrm{N/m}$$

$$f = \frac{1}{2\pi}\sqrt{(69{\cdot}611176 \times 10^6/6{\cdot}7375 \times 10^6)} = \underline{0{\cdot}5118\,\mathrm{Hz}}$$

Thus the error in this case is −31·01%.

### *Multistorey structures with flexible floors*

Estimates of the dominant frequencies of all buildings that cannot be regarded as shear structures require that the joint rotations and deformation of the floors, which reduce the overall shear stiffness of a structure, be taken into account. In such cases it is first necessary to calculate equivalent shear stiffnesses by modifying the shear stiffness of the columns at each level by multiplying them by a reduction factor, which is a function of both the column and the beam rigidities, and then assuming the modeshapes to be similar to the deflected forms caused by concentrated loads applied horizontally at each floor level, whose magnitudes are equal to the weight of the floor at which they are applied. With some assumptions the expression for the reduction factor can be shown to be

$$\mathrm{RF} = \frac{\sum(EI/L)_{\mathrm{beams}}}{\sum(EI/L)_{\mathrm{beams}} + \frac{1}{2}\sum(EI/L)_{\mathrm{columns}}} \tag{2.202}$$

The error in the above factor depends on the degree of fixity at the foundations, the distribution of the $EI/L$ values of beams and columns at different floor levels, and the size of the frame. For normally proportioned structures the experience is that eq. (2.202) gives reasonable values.

In the following example the above method is used to determine the first natural frequencies of two structural models whose frequencies had previously been determined by eigenvalue analysis using computers, and resonance testing.

**Example 2.7** Fig. 2.15 shows a steel frame model of a three-storey structure. The values of the $EI/L$ in N m are marked against each member. The mass at the first floor level is 0·4228 kg, at the second level 0·3979 kg and at roof level 0·2985 kg. Calculate the equivalent mass and spring stiffness at roof level and hence the first natural frequency of the frame.

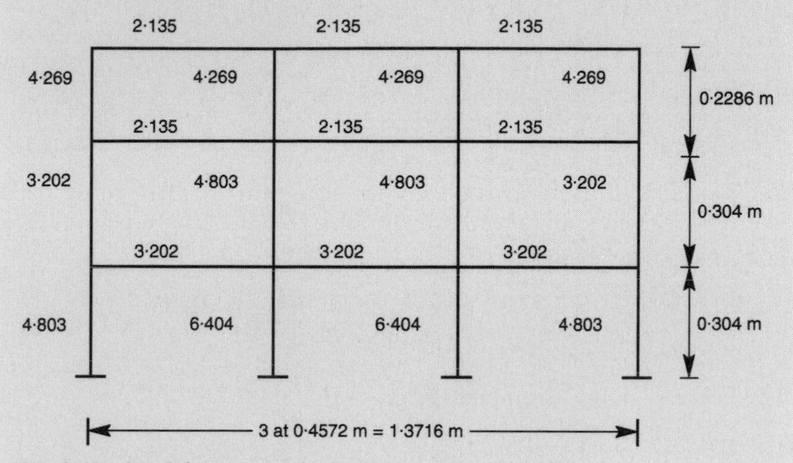


*Fig. 2.15. Steel frame model of a three-storey structure*

The sway stiffness $k$ at each level is found by first calculating the sum of the shear stiffnesses of the columns at each level, assuming zero joint rotations, and then multiplying this stiffness by a reduction factor calculated in accordance with eq. (2.202). Hence

$$k_r = \left\{ \sum_{i=1}^{N} \frac{12EI_i}{L_i^3} \right\} \text{RF}$$

where $N$ is the number of columns between floor levels $r$ and $(r-1)$. Thus at level 1

$$k_1 = \{12(2 \times 4{\cdot}803 + 2 \times 6{\cdot}404)/0{\cdot}3048^2\} \times \{3 \times 3{\cdot}202/(3 \times 3{\cdot}202 + 4{\cdot}803 + 6{\cdot}404)\} = 1336{\cdot}2219\ \text{N/m}$$

at level 2

$$k_2 = \{12(2 \times 3{\cdot}202 + 2 \times 4{\cdot}803)/0{\cdot}3048^2\} \times \{3 \times 2{\cdot}135/(3 \times 2{\cdot}135 + 3{\cdot}202 + 4{\cdot}803)\} = 919{\cdot}1742\ \text{N/m}$$

and at roof level

$$k_3 = \{12(4 \times 4{\cdot}269/0{\cdot}2286^2\} \times \{3 \times 2{\cdot}135/(3 \times 2{\cdot}135 + 2 \times 4{\cdot}269)\}$$
$$= 1680{\cdot}7235\ \mathrm{N/m}$$

Assuming the first modeshape to be geometrically similar to the deflected form caused by the load vector

$$P = \{0{\cdot}4228 \quad 0{\cdot}3979 \quad 0{\cdot}2985\} \times 9{\cdot}81\ \mathrm{N}$$

yields

$$x_1 = (0{\cdot}4228 + 0{\cdot}3979 + 0{\cdot}2985) \times 9{\cdot}81/1336{\cdot}2219 = 8{\cdot}2167 \times 10^{-3}\ \mathrm{m}$$
$$x_2 = 8{\cdot}21671 \times 10^{-3} + (0{\cdot}3979 + 0{\cdot}2985) \times 9{\cdot}81/919{\cdot}1742$$
$$= 15{\cdot}6491 \times 10^{-3}\ \mathrm{m}$$
$$x_3 = 15{\cdot}6491 \times 10^{-3} + 0{\cdot}2985 \times 9{\cdot}81/1680{\cdot}7235 = 17{\cdot}3914 \times 10^{-3}\ \mathrm{m}$$

Thus the modeshape vector if the amplitude at level three is taken as unity is

$$x = \{0{\cdot}4725 \quad 0{\cdot}8998 \quad 1{\cdot}0000\}$$

Hence the kinetic energy is

$$\mathrm{KE} = \tfrac{1}{2}(0{\cdot}4228 \times 0{\cdot}4725^2 + 0{\cdot}3979 \times 0{\cdot}8998^2 + 0{\cdot}2985 \times 1{\cdot}0^2)\omega^2$$
$$= 0{\cdot}3575242\omega^2\ \mathrm{N/m}$$

The corresponding strain energy is

$$U = \tfrac{1}{2}\Big\{1336{\cdot}2219 \times 0{\cdot}4725^2 + 919{\cdot}1742(0{\cdot}8998 - 0{\cdot}4725)^2 +$$
$$1680{\cdot}7235(1{\cdot}0 - 0{\cdot}8998)^2\Big\} = 241{\cdot}5111\ \mathrm{N\,m}$$

$$f_n = \frac{1}{2\pi}\sqrt{\left(\frac{241{\cdot}5111}{0{\cdot}3575242}\right)} = \underline{4{\cdot}1365\ \mathrm{Hz}}$$

The correct value obtained from a computer analysis is 3·630 Hz. Thus the error is +13·953%, which is not particularly good.

An alternative method of estimating the first natural frequency is to determine the stiffness and mass of an equivalent one DOF system. The equivalent mass at roof level is given by

$$M_E = 0{\cdot}2985 + 0{\cdot}3979 \times 0{\cdot}6096/0{\cdot}8382 + 0{\cdot}4228 \times 0{\cdot}3048/0{\cdot}8382$$
$$= 0{\cdot}7435\ \mathrm{kg}$$

The equivalent spring stiffness $K_E$ is found from

$$\frac{1}{K_E} = \frac{1}{k_1} + \frac{1}{k_2} + \frac{1}{k_3} = \frac{1}{1336{\cdot}2219} + \frac{1}{919{\cdot}1742} + \frac{1}{1680{\cdot}7235}$$
$$= 411{\cdot}30368\ \mathrm{N/m}$$

and the frequency by

$$f = \frac{1}{2\pi}\sqrt{\left(\frac{411{\cdot}30368}{0{\cdot}7435}\right)} = \underline{3{\cdot}743\,\text{Hz}}$$

Thus the error in this case is only +2·83%.

**Example 2.8** The overall dimensions of the five-storey model shown in Fig. 2.16 are 1000 × 200 × 200 mm. Floor levels are 200 mm apart. The mass at each level is represented by four steel cubes of dimensions 40 × 40 × 40 mm, each cube weighing 0·483 kg. The columns consist of 5 mm dia. steel rods and the beams of 3 mm dia. rods. The total weight of the columns is 0·628 kg and that of the beams is 0·200 kg. The second moment of inertia of the columns is 30·7 mm$^4$ and that of the beams is 3·98 mm$^4$.

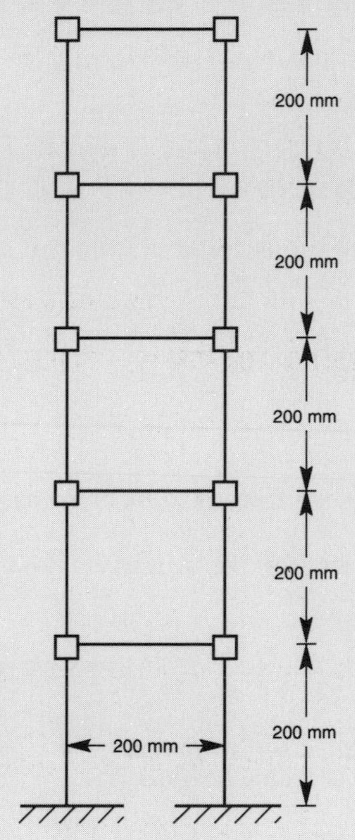


*Fig. 2.16. Five-storey flexible steel model with four lumped masses attached at each floor level*

Determine the dominant frequency by first modelling the structure as a one DOF system with the lumped mass at the top of the model, and then by assuming the modeshape to be geometrically similar to the deflected form caused by unit point loads applied horizontally at each floor level. Assume the modulus of elasticity for steel to be 205 kN/mm$^2$. The measured natural frequency was 2·105 Hz and the calculated one using a computer was 1·95 Hz.

The calculation of the equivalent lumped mass at the top of the model is

$$M_E = \frac{10{\cdot}488}{5}\left\{\frac{1000}{1000}+\frac{800}{1000}+\frac{600}{1000}+\frac{400}{1000}+\frac{200}{1000}\right\} = \underline{6{\cdot}2928\,\text{kg}}$$

for which an expression is given by eq. (2.202). Thus

$$\text{RF} = \frac{2\times(205\times 3{\cdot}98)/200}{2\times(205\times 3{\cdot}98)/200+\frac{1}{2}\times 4\times(205\times 30{\cdot}7)/200} = 0{\cdot}1147635$$

The stiffness at each floor level is given by

$$K = 4\times\frac{12EI}{L^3}\times\text{RF} = 4\times\frac{12\times 205\times 30{\cdot}7}{200^3}\times 0{\cdot}1147635$$
$$= 4{\cdot}33358\times 10^{-3}\,\text{kN/mm}$$

Hence the equivalent spring stiffness at the top of the model can be determined from the equality

$$\frac{1}{K_E} = \frac{1}{K}+\frac{1}{K}+\frac{1}{K}+\frac{1}{K}+\frac{1}{K} = \frac{5}{K}$$

Thus

$$K_E\frac{4{\cdot}33358\times 10^{-3}}{5} = 8{\cdot}66714\times 10^{-4}\text{kN/mm} = \underline{866{\cdot}714\,\text{N/m}}$$

The natural frequency of the equivalent mass–spring system is therefore

$$f = \frac{1}{2\pi}\sqrt{\left(\frac{866{\cdot}714}{6{\cdot}2928}\right)} = \underline{1{\cdot}8678\,\text{Hz}}$$

Thus the percentage error compared with the theoretical value obtained from an eigenvalue analysis is

$$\text{percentage error} = \frac{1{\cdot}8678 - 1{\cdot}9490}{1{\cdot}9490} = \underline{-4{\cdot}166\%}$$

The modeshape assumed to be geometrically similar to the deflected form caused by a unit horizontal point load applied at each floor level is given by

$$\begin{aligned} x_1 &= 5/K &&= 5/K \\ x_2 &= 5/K + 4/K &&= 9/K \\ x_3 &= 5/K + 4/K + 3/K &&= 12/K \\ x_4 &= 5/K + 4/K + 3/K + 2/K &&= 14/K \\ x_5 &= 5/K + 4/K + 3/K + 2/K + 1/K &&= 15/K \end{aligned}$$

Division by $5/K$ yields the modeshape vector

$$x = \{1{\cdot}0 \quad 1{\cdot}8 \quad 2{\cdot}4 \quad 2{\cdot}8 \quad 3{\cdot}0\}$$

The maximum kinetic energy corresponding to this modeshape is given by

$$\begin{aligned} \text{KE} &= \frac{1}{2} \times \frac{10{\cdot}488}{5} \omega^2 [1{\cdot}0^2 + 1{\cdot}8^2 + 2{\cdot}4^2 + 2{\cdot}8^2 + 3{\cdot}0^2] \\ &= 28{\cdot}149792\omega^2 \, \text{N m} \end{aligned}$$

The corresponding maximum strain energy is

$$U = \frac{1}{2} \times 4333{\cdot}58 \times \left[1{\cdot}0^2 + (1{\cdot}8 - 1{\cdot}0)^2 + (2{\cdot}4 - 1{\cdot}8)^2 + (2{\cdot}8 - 2{\cdot}4)^2 + (3{\cdot}0 - 2{\cdot}8)^2\right] = 4766{\cdot}938 \, \text{N m}$$

Hence

$$f = \frac{1}{2\pi} \sqrt{\left(\frac{4766{\cdot}938}{28{\cdot}149792}\right)} = \underline{2{\cdot}0711 \, \text{Hz}}$$

$$\text{Percentage error} = \frac{2{\cdot}0711 - 1{\cdot}9490}{1{\cdot}9490} = \underline{+6{\cdot}265\%}$$

### Plates

Theoretically it is possible to determine the first frequency of a plate in the same manner as for beams by assuming the modeshape to be geometrically similar to that caused by a uniformly and/or concentrated applied load. However, because both the geometry and the support conditions can vary considerably from plate to plate, the above approach is not practical except for circular plates, and rectangular plates simply supported at each corner. For these two cases it is possible to calculate a first frequency by assuming the modeshape to be sinusoidal. For plates, therefore, the reader is referred to handbooks on vibration such as Harris (1988) which give extensive lists of expressions for the frequencies of plates with different geometries and support conditions.

## Summary and conclusions

The first part of this chapter presents a method for determining the dominant frequencies of beams with uniform and concentrated loading by assuming that the modeshape is geometrically similar to the deflected form caused by a concentrated and distributed load. The assumed modeshape is then used to determine the lumped mass and spring stiffness of an equivalent one DOF system using eqs (2.8) and (2.13). It is then shown the equivalent masses and spring stiffnesses for beams can be used to estimate the dominant frequencies of continuous beams and how the accuracies of such calculations are critically dependent on the assumed modeshapes. The method used is extended to sway structures with both rigid and flexural floors. In all the examples the degree of accuracy obtained is given. These examples indicate that the accuracy in the calculated frequencies depends on the assumed modeshapes. Thus, for the structure in example 2.8 it is found that the first modeshape obtained using a finite element program closely approximates the average of the two modeshapes assumed. When this modeshape is used the percentage difference between the theoretically correct frequency and that obtained by equating the maximum kinetic energy to the maximum strain energy is −1·35%. The accurate calculation of structural frequencies requires numerical modelling and the use of a computer. Data preparation takes time, and experience has shown that it is useful to be able to make an initial estimate of the fundamental frequency at the design stage, as well as when checking the values from any computer analysis. One question remains: why should a designer wish to change a structure that is designed to withstand the static forces stipulated in various design codes? The answer is given in chapter 1, where it is pointed out that dynamic forces such as wind, waves and earthquakes may be considered to consist of a large number of harmonic components with different frequencies and varying levels of energy. Therefore structures with dominant frequencies that lie within a high energy frequency band may respond in resonance. This may have, and has had, catastrophic consequences.

It having shown how many structures and structural elements can be reduced to one DOF mass–spring system, the next four chapters are devoted to the free and forced vibration of such systems.

**Problem 2.1** A simply supported beam of length $L$, flexural rigidity $EI$ and self-weight $wL$ supports three concentrated loads, each of weight $wL$, at positions $L/4$, $L/2$ and $3L/4$ along the span. Assume the mode of vibration to be geometrically similar to the deflected shape caused by a uniformly distributed dead load, and develop an expression for the natural frequency of the equivalent mass–spring system in terms of $w$, $L$, $EI$ and $g$.

**Problem 2.2** In the portal frame structure shown in Fig. 2.17 the floor BDF is assumed to be rigid. The columns AB, CD and EF are uniform with the same flexural rigidity $EI$. The weight of the floor is $w$ per m length, and the weight of the columns is $w/10$ per m length. Develop an expression for the natural frequency of the structure in terms of $EI$, $w$ and $g$.

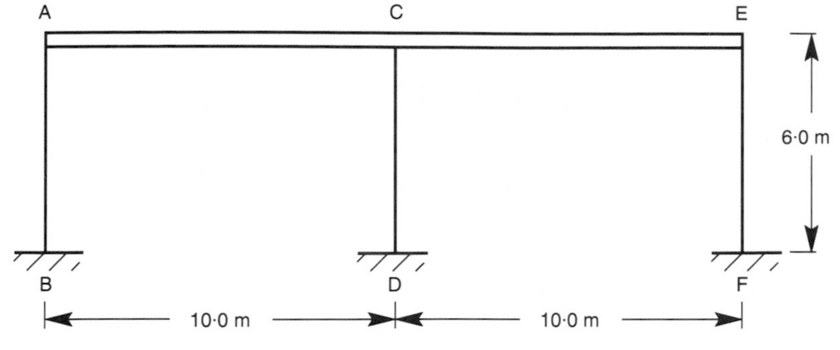


*Fig. 2.17. Portal frame structure with rigid floor slab*

**Problem 2.3** A continuous beam ABCD of uniform cross-section, flexural rigidity $EI$ and self-weight $w$ per unit length has three equal spans, each of length $L$. The beam is simply supported at B and C, but is built-in at A and D. Use an approximate method to estimate the first and second natural frequencies of the beam.

**Problem 2.4** Develop an expression for the dominant frequency of the five-storey shear structure shown in Fig. 2.17 in terms of the weight of the floor slabs and the flexural rigidity of the columns. Each floor slab weighs $w$ per unit length of span, and all columns have the same flexural rigidity $EI$.

**Problem 2.5** Determine the first modeshape of the structural model in example 2.8 by assuming it to be geometrically similar to the average of the deflected forms caused by only one horizontally applied load at the top of the model, and the form caused by equal horizontal concentrated loads applied at all five levels. Hence calculate the natural frequency of the model by equating the maximum kinetic energy to the maximum strain energy occurring for the given modeshape

**Problem 2.6** What is the expression for the dominant frequency for the structure shown in Fig. 2.18, if it is assumed that the flexural rigidities of all the floors are the same and in turn equal to five, ten and 20 times those of the columns. By what percentage will the frequency be reduced in each case relative to the frequency obtained for the case when all the floors are assumed to be rigid?

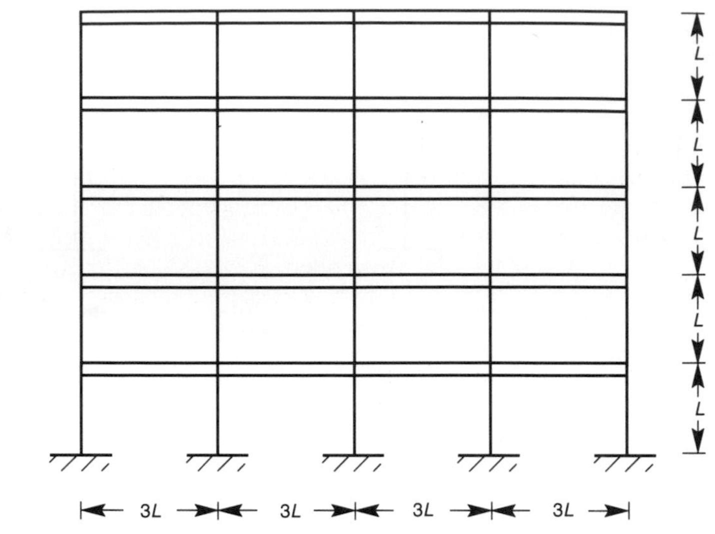


*Fig. 2.18. Five-storey shear structure*

## References

Clough, R.W. & Penzien J. *Dynamics of structures.* McGraw-Hill, London, 1975.

Craig, R. R., Jr. *Structural dynamics.* Wiley, Chichester, 1981.

Harris, C. M. *Shock vibration*, 3rd edn. McGraw-Hill, London, 1988.

Irvine, H. M. *Structural dynamics for the practising engineer.* Allen & Unwin, London, 1986.

Stroud, K. A. *Engineering mathematics.* Macmillan, London, 1970.

Timoshenko, S. P. & Gere, J. M. *Mechanics of materials.* Van Nostrand Reinhold, New York, 1972.

# 3. Free vibration of one degree of freedom systems

## Introduction

In the previous chapter it is shown how a large number of different types of structures and structural elements could be modelled as lumped mass–spring systems. In this chapter the free vibrations of such systems, both with and without damping, are investigated. Expressions for equivalent structural damping are developed and how it can be measured and how damping affects the natural frequencies are examined.

## Free undamped rectilinear vibration

Consider the system shown in Fig. 3.1. If the static deflection due to the weight of the lumped mass is

$$\Delta = W/K = Mg/K \tag{3.1}$$

then, from using Newton's law of motion

$$M\ddot{x} = W - K(\Delta + x) \tag{3.2}$$

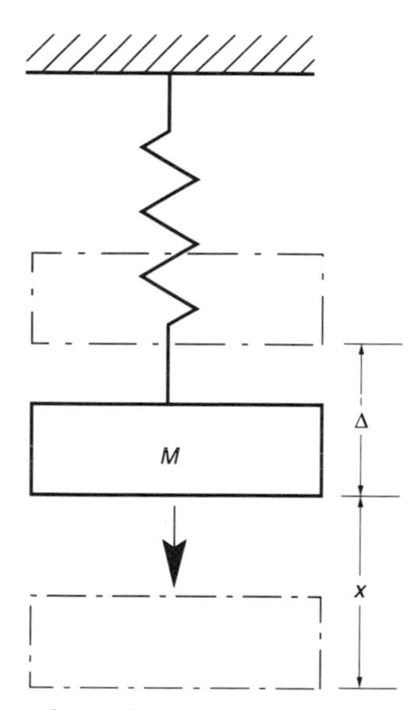


*Fig. 3.1. System of free undamped rectilinear vibration*

Hence

$$M\ddot{x} + Kx = 0 \tag{3.3}$$

or

$$\ddot{x} + \omega_n{}^2 x = 0 \tag{3.4}$$

where

$$\omega_n{}^2 = K/M \tag{3.5}$$

The solution to eq. (3.4) is given by

$$x = A\cos(\omega_n t) + B\sin(\omega_n t) \tag{3.6}$$

From eq. (3.5) it can be seen that the vertical motion of the mass $M$ has a vibratory character, since both $\sin(\omega_n t)$ and $\cos(\omega_n t)$ are periodic functions which repeat themselves after an interval of time $T$ such that

$$\omega_n(T + t) - \omega_n t = 2\pi \tag{3.7}$$

The time $T$ is called the period or periodic time of vibration. From eqs (3.7), (3.5) and (3.1) it follows that

$$T = \frac{2\pi}{\omega_n} = 2\pi\sqrt{\left(\frac{M}{K}\right)} = 2\pi\sqrt{\left(\frac{\Delta}{g}\right)} \tag{3.8}$$

The number of cycles per second is, as seen in chapter 2, referred to as the frequency of vibration, and 1 cycle/s is called 1 Hz. Then

$$\underline{f = \frac{1}{T} = \frac{\omega_n}{2\pi} = \frac{1}{2\pi}\sqrt{\left(\frac{K}{M}\right)} = \frac{1}{2\pi}\sqrt{\left(\frac{g}{\Delta}\right)}} \tag{3.9}$$

where $f$ is the frequency and $\omega_n$ is the natural *angular* frequency in rad/s.

### *Examination of eq. (3.6)*

For structures that vibrate with SHM the velocity $\dot{x}$ is a maximum when $x = 0$, and the acceleration $\ddot{x}$ is a maximum when $x = x_{max} = x_0$. With this information the values for the constants $A$ and $B$ in eq. (3.6) can be determined by choosing convenient starting points for the motion.

*$t = 0$ when $x = 0$*
Substitution of $x = 0$ and $t = 0$ into eq. (3.6) yields $A = 0$. Hence

$$x = B\sin(\omega_n t) \tag{3.10}$$

The amplitude $x$ is a maximum when $\sin(\omega t) = 1{\cdot}0$. Therefore $B = x_0$. Hence

$$x = x_0 \sin(\omega_n t) \tag{3.11a}$$

$$\dot{x} = x_0\omega_n \cos(\omega_n t) \tag{3.11b}$$

$$\ddot{x} = -x_0\omega_n{}^2 \sin(\omega_n t) \tag{3.11c}$$

*$t = 0$ when $x = x_0$*
When $x = x_0$ and $t = 0$, $A = x_0$. Substitution of this value for $A$ into eq. (3.6) and differentiation with respect to $t$ yields

$$\dot{x} = -x_0\omega_n \sin(\omega_n t) + B\omega_n \cos(\omega_n t) \tag{3.12}$$

When $t = 0$ and $\dot{x} = 0$, $B = 0$. Hence

$$x = x_0 \cos(\omega_n t) \tag{3.13a}$$

$$\dot{x} = -x_0\omega \sin(\omega_n t) \tag{3.13b}$$

$$\ddot{x} = -x_0\omega_n{}^2 \cos(\omega_n t) \tag{3.13c}$$

*$t = 0$ when $x = \alpha$, $0 < \alpha < |x_0|$*
When $t = 0$ and $x = \alpha$, $A = \alpha$. Substitution of this value for $A$ into eq. (3.6) and differentiation with respect to $t$ yields

$$\dot{x} = -\alpha\omega_n \sin(\omega_n t) + B\omega_n \cos(\omega_n t) \tag{3.14}$$

When $t = 0$ and $\dot{x} = \dot{\alpha}$, $B = \dot{\alpha}/\omega_n$. Hence

$$x = \alpha \cos(\omega_n t) + \frac{\dot{\alpha}}{\omega_n} \sin(\omega_n t) \tag{3.15}$$

i.e the amplitude $x$ at time $t$ is a function of the amplitude $\alpha$ and velocity $\dot{\alpha}$ at time $t = 0$.

### *Equivalent viscous damping*

During vibration energy is dissipated in the form of heat, and a steady amplitude of vibration cannot be maintained without its continuous replacement. The heat is generated in a number of different ways: by dry and fluid friction, by hysteresis effect in individual components (internal friction), in concrete by the opening and closing of hair cracks, and by magnetic, hydrodynamic and aerodynamic forces. The different forces that contribute to the damping of a structure may vary with amplitude, velocity, acceleration and stress intensity, and are difficult if not impossible to model mathematically. However, ideal models of damping have been conceived which represent more or less satisfactory approximations. Of these, the viscous damping model—in which the damping force is proportional to the velocity—leads to the simplest mathematical treatment and in general the most satisfactory results, provided damping forces caused by hydrodynamic and aerodynamic forces, when significant, are taken into account separately.

As a viscous damping force is proportional to the velocity of vibration at any time $t$, it can be expressed as

$$F_d = C\dot{x} \tag{3.16}$$

where $C$ is the constant of proportionality or coefficient of viscous damping, and $\dot{x}$ is the velocity of vibration at time $t$. The coefficient of viscous damping is numerically equal to the damping force when the velocity is unity. Thus the unit of $C$ is M s/m. Symbolically viscous damping is designated by a dash pot as shown in Fig. 3.2.

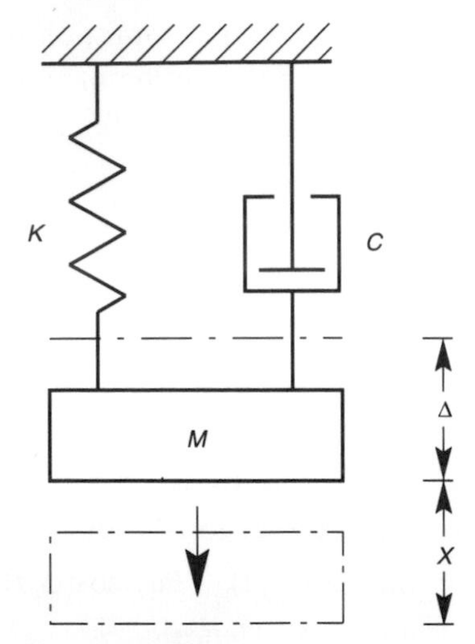


*Fig. 3.2. Lumped mass–spring system with viscous damping*

**Free rectilinear vibration with viscous damping**

Consider a vibrating damped lumped mass–spring system. If the motion of the mass is resisted by forces that are proportional to the velocity of the mass, then the resisting forces may be assumed to have viscous characteristics, in which case the damping mechanism is denoted by a dash pot as shown in Fig. 3.2.

From Newton's law of motion

$$M\ddot{x} = W - K(\Delta + x) - C\dot{x} \tag{3.17}$$

or

$$M\ddot{x} + C\dot{x} + Kx = 0 \tag{3.18}$$

Assume that eq. (3.18) is satisfied by a function of the form

$$x = Ae^{st} \tag{3.19}$$

Substitution of this function into eq. (3.18) yields

$$Ms^2Ae^{st} + CsAe^{st} + KAe^{st} = 0 \tag{3.20}$$

Division of each term in eq. (3.20) by $MAe^{st}$ yields

$$s^2 + 2ns + \omega_n{}^2 = 0 \tag{3.21}$$

where

$$n = C/2M$$

$$\omega\sqrt{(K/M)}$$

From eq. (3.21) it follows that

$$s = -n \pm \sqrt{(n^2 - \omega_n{}^2)} \tag{3.22}$$

Thus, when one is studying free damped vibration further it is necessary to consider the cases

$$n = \omega_n,\ n > \omega_n \text{ and } n < \omega_n$$

Or alternatively, since $n = C/2M$ and $\omega_n = \sqrt{(K/M)}$,

when
$$C = 2\sqrt{(KM)}$$
$$C > 2\sqrt{(KM)}$$
$$C < 2\sqrt{(KM)}$$

The term $2\sqrt{(KM)}$ is referred to as *critical damping* and is denoted by $C_c$. The ratio $C/C_c$ is called the *damping ratio* and is written as $\xi$. Therefore $n = \xi\omega_n$, and consequently

$$\underline{C = 2\xi\omega_n M} \tag{3.23}$$

Thus $n = \xi\omega_n$. Substitution of this value for $n$ into eq. (3.22) yields

$$s = -\xi\omega_n \pm \omega_n\sqrt{(\xi^2 - 1)} \tag{3.24}$$

or

$$s = -\xi\omega_n \pm \omega_n^i\sqrt{(1 - \xi^2)} \tag{3.25}$$

From eqs (3.24) and (3.25) it follows that the motion of free damped vibration needs to be studied for the cases when

$$\xi > 1{\cdot}0$$
$$\xi = 1{\cdot}0$$
$$\xi < 1{\cdot}0$$

*When $\xi > 1{\cdot}0$*

$$s = -\xi\omega_n \pm \omega_n p$$

Hence

$$\underline{x = e^{-\xi\omega_n t}\{Ce^{\omega_n pt} + De^{-\omega_n pt}\}} \tag{3.26}$$

Equation (3.26) is not a periodic function and therefore does not represent a periodic motion. If displaced from its position of equilibrium the mass will gradually return to its original position.

*When $\xi = 1{\cdot}0$*

$$s = -\xi\omega_n$$

Hence

$$\underline{x = Ce^{-\xi\omega_n t}} \tag{3.27}$$

Equation (3.27) does not contain a periodic function either, and therefore again does not represent a periodic motion. The mass will return to its position of equilibrium, if displaced, but more quickly than in the previous case ($\xi > 1{\cdot}0$).

*When $\xi < 1{\cdot}0$*

The concept of *damped angular natural frequency* is now introduced, defined by

$$\omega_d = \omega_n\sqrt{(1-\xi^2)} \tag{3.28}$$

Hence

$$s = -\xi\omega_n \pm i\omega_d$$

$$x = e^{-\xi\omega_n t}\{Ce^{i\omega_d t} + De^{-i\omega_d t}\} \tag{3.29}$$

Since

$$e^{+i\omega_d t} = \cos(\omega_d t + \sin(\omega_d t)$$

$$e^{-i\omega_d t} = \cos(\omega_d t) - \sin(\omega_d t)$$

it follows that

$$x = e^{-\xi\omega_n t}\{A\cos(\omega_d t) + B\sin(\omega_d t)\} \tag{3.30}$$

Equation (3.30) represents a periodic function. Comparison of the period of this function with that for undamped free vibration given by eq. (3.6) shows that the period $T$ increases from $2\pi/\omega_n$ to $(2\pi/\omega_n)\sqrt{(1-\xi^2)}$. When the value of $\xi$ is small compared with $1{\cdot}0$ the increase is negligible, and in most practical problems it can be assumed with sufficient accuracy that the damping does not affect the period of vibration, which can be assumed to be equal to $2\pi/\omega_n$.

In order to determine the constants $A$ and $B$ in eq. (3.30), let $x = x_{max} = x_0$ and $\dot{x} = 0$ when $t = 0$. This yields

$$x = x_0 e^{-\xi\omega_n t}\left\{\cos(\omega_d t) + \frac{\xi}{\sqrt{(1-\xi^2)}}\sin(\omega_d t)\right\} \tag{3.31}$$

From eq. (3.31) it can be seen that for every period $T$, or for every cycle of vibration, the amplitude is diminished by the ratio

$$x_0 e^{-\xi\omega_n T} : x_0 \tag{3.32}$$

Thus, if $x_r$ = the amplitude at the end of the $r$th oscillation, and $x_s$ = the amplitude at the end of the $s$th oscillation, then

$$\frac{x_r}{x_s} = \{x_0 e^{-\xi\omega_n rT}\}/\{x_0 e^{-\xi\omega_n sT}\} \tag{3.33}$$

or

$$\frac{x_r}{x_s} = e^{\xi\omega_n(s-r)T} \tag{3.34}$$

Hence

$$\xi\omega_n T = \frac{1}{s-r}\ln\frac{x_r}{x_s} \tag{3.35}$$

The product $\xi\omega_n T$ is referred to as the *logarithmic decrement of damping* and is usually denoted by $\delta$. Thus

$$\delta = \frac{1}{s-r}\ln\frac{x_r}{x_s} \tag{3.36}$$

When $r = 0$ and $s = n$

$$\underline{\delta = \frac{1}{n}\ln\frac{x_0}{x_n}} \tag{3.37}$$

As the period for damped vibration is

$$T = \frac{2\pi}{\omega_n\sqrt{(1-\xi^2)}} \tag{3.38}$$

it follows that

$$\delta = \frac{2\pi\xi}{\sqrt{(1-\xi^2)}} \tag{3.39}$$

from which is obtained

$$\underline{\xi = \frac{\delta}{\sqrt{(4\pi^2+\delta^2)}}} \tag{3.40}$$

Usually $\delta <<< 2\pi$. Equation (3.40) may therefore be simplified and written as

$$\underline{\xi = \frac{\delta}{2\pi}} \tag{3.41}$$

The motions for systems with $\xi > 1{\cdot}0$ and $\xi < 1{\cdot}0$ are shown in Figs 3.3 and 3.4 respectively.

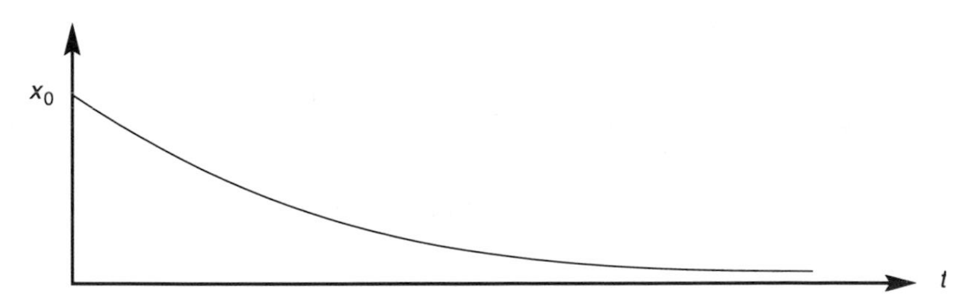


*Fig. 3.3. Motion of a lumped mass–spring system with viscous damping ratio* $\xi > 1{\cdot}0$

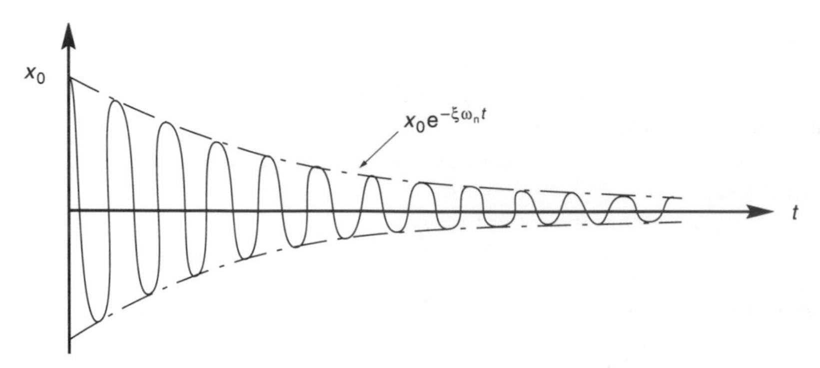


*Fig. 3.4. Diagram showing the motion of a lumped mass–spring system with viscous damping ratio* $\xi < 1{\cdot}0$

## Evaluation of logarithmic decrement of damping from the decay function

Decay functions of the type shown in Fig. 3.4 can be obtained by the sudden release of a load from a structure, or by vibrating it at resonance and then stopping the vibrator and recording the ensuing motion of a pen-recorder or by means of a computer, using either an accelerometer or a displacement transducer.

The expression for the logarithmic decrement of damping given by eq. (3.41) assumes that the resulting structural damping mechanism has the characteristics of viscous damping, i.e. that the damping force resisting the motion at any time is proportional to the velocity of vibration. This assumption can be checked by plotting the values of $\ln(x_0/x_n)$ against $n$, the number of oscillations, as shown in Fig. 3.5. The plotted values will lie along a straight line if the damping is proportional to the velocity, and along a curve if it is not. When the former is the case the damping is independent of, and if not it is dependent on, the amplitude of response.

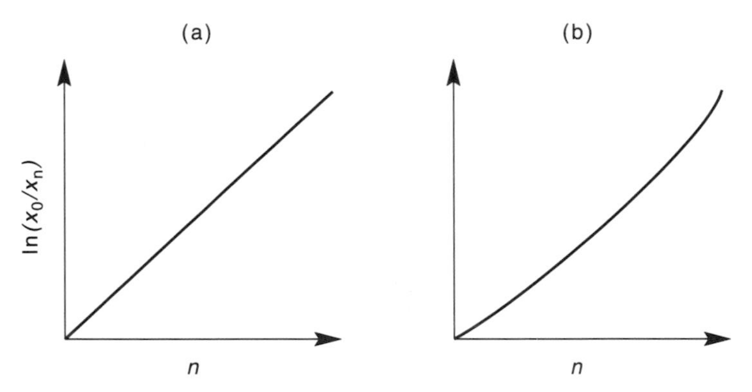


*Fig. 3.5.* $\mathrm{Ln}(x_0/x_n)$ *plotted against n for structures with (a) viscous and (b) nonviscous damping*

Since $\delta = \ln(x_0/x_n)/n$, it follows that the slope of the straight line is equal to $\delta$, and that the slope of any tangent to a curved line is the value of $\delta$ for the amplitude of vibration corresponding to the contact point between the tangent and the curve.

**Example 3.1** A tubular steel antenna-mast supporting a 3·0 m dia. disc is deflected by tensioning a rope attached to its top, and then set in motion by cutting the rope. The first part of the subsequent motion, recorded by the use of an accelerometer, is shown in Fig. 3.6. The estimated spring stiffness of at the top of the mast is 30·81 kN/m. Calculate first the logarithmic decrement of damping, the damping ratio, and the damped and undamped natural frequencies of the mast, and then the equivalent mass and damping coefficient for the generalized mass–spring system.

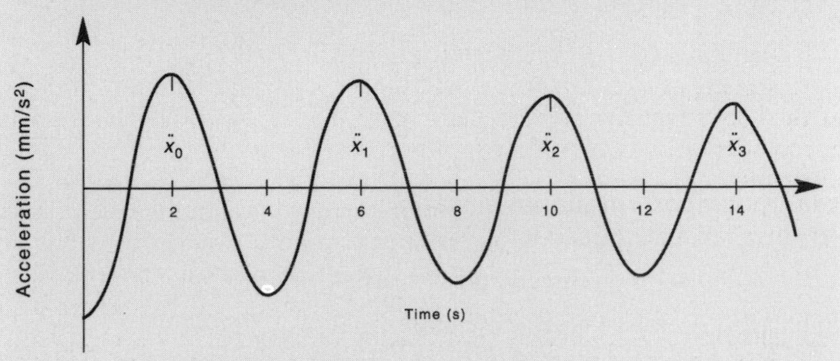


*Fig. 3.6. Pen-recorder trace of the motion at the top of a 20 m tall antenna-mast:* $\ddot{x}_0 = 258\ mm/s^2$; $\ddot{x}_1 = 226\ mm/s^2$; $\ddot{x}_2 = 199\ mm/s^2$; $\ddot{x}_3 = 176\ mm/s^2$

The logarithmic decrement of damping $\delta$ is given by

$$\delta = \frac{1}{n}\ln\frac{x_0}{x_n} = \frac{1}{3}\ln\frac{258}{176} = 0{\cdot}1274918$$

hence

$$\underline{\delta = 12{\cdot}75\%}$$

The damping ratio $\xi$ is given by

$$\xi = \frac{\delta}{2\pi} = \frac{0{\cdot}1274918}{2\pi} = 0{\cdot}0202909$$

hence

$$\underline{\xi = 2{\cdot}03\%}$$

The damped and undamped natural frequencies are given by

$$f_\mathrm{d} = \frac{3}{14 - 2} = \underline{0{\cdot}25\,\mathrm{Hz}}$$

$$f_\mathrm{n} = \frac{f_\mathrm{d}}{\sqrt{(1 - \xi^2)}} = \frac{0{\cdot}25}{\sqrt{(1 - 0{\cdot}0202909^2)}} = \underline{0{\cdot}2501029\,\mathrm{Hz}}$$

Thus the damping has only a negligible effect on the natural frequency, and the difference between the two values is certainly less than any expected accuracy in the measured value.

The equivalent lumped mass at the top of the mast is

$$M = \frac{K}{4\pi^2 f_\mathrm{n}^2} = \frac{30\,810}{4\pi^2 \times 0{\cdot}2501029^2} = \underline{12\,476{\cdot}543\,\mathrm{kg}}$$

The damping coefficient $C$ for the equivalent mass–spring system is therefore given by

$$C = 2\xi\omega_\mathrm{n} M = 2 \times 0{\cdot}0202909 \times (0{\cdot}2501029 \times 2\pi) \times 12\,476{\cdot}543$$
$$= \underline{795{\cdot}654\,\mathrm{N\ s/m}}$$

## Free undamped rotational vibration

Newton's law of motion states that

$$\text{force} = \text{mass} \times \text{acceleration}$$

Similarly, d'Alembert's principle states that

$$\text{moment of force} = \text{polar moment of inertia} \times \text{angular acceleration}$$

Thus

$$T = -I_\mathrm{p}\ddot{\theta} \tag{3.42}$$

If it is assumed that the forcing moment $T$ strains a bar element in pure St Venant torsion and rotates one end of it through an angle $\theta$, then

$$T = K_\mathrm{t}\theta \tag{3.43}$$

where $K_\mathrm{t}$ is the torsional stiffness. Substitution of this expression for the moment of force into eq. (3.42) yields

$$I_\mathrm{p}\ddot{\theta} + K_\mathrm{t}\theta = 0 \tag{3.44}$$

If it is further assumed that the motion represented by eq. (3.44) is simple harmonic, i.e. that

$$\theta = \phi \sin(\omega t)$$
$$\ddot{\theta} = -\phi\omega^2 \sin(\omega t)$$

it follows that

$$-I_\mathrm{p}\phi\omega^2 + K_\mathrm{t}\phi = 0 \tag{3.45}$$

and hence that

$$\omega_n = \sqrt{\left(\frac{K_t}{I_p}\right)} \tag{3.46}$$

Multiplication of each term in eq. (3.45) by $\frac{1}{2}\phi$ yields

$$\tfrac{1}{2}K_t\phi^2 = \tfrac{1}{2}I_p\phi^2\omega^2 \tag{3.47}$$

Thus, as in the case of rectilinear vibration, the maximum strain energy is equal to the maximum kinetic energy when energy losses due to damping are ignored.

For a cylindrical element of length $L$, radius $R$, mass $M$, and specific density $\rho$ the polar moment of inertia is

$$I_p = \tfrac{1}{2}\rho\pi LR^4 = \tfrac{1}{2}MR^2 \tag{3.48}$$

$$K_t = \frac{\pi GR^4}{2L} \tag{3.49}$$

For a bar element with length $L$, mass $M$ and rectangular cross-sectional area of dimensions $a \times b$ the polar moment of inertia about the central axis is

$$I_p = \tfrac{1}{12}M\left(a^2 + b^2\right) \tag{3.50}$$

$$K_t = \frac{Gab\left(a^2 + b^2\right)}{12L} \tag{3.51}$$

The strain energy stored in the same element when subjected to a forcing moment or torque $T$ causing a differential end rotation $\phi$ is

$$U = \frac{T\phi}{2} = \frac{T^2L}{2GJ} = \frac{GJ\phi^2}{2L} \tag{3.52}$$

where $J$ is the second polar moment of area about the central axis. Equation (3.52) implies that $I_p = JL\rho$.

### Polar moment of inertia of equivalent lumped mass–spring system of bar element with one free end

The equivalent polar moment of inertia of the mass–spring system shown in Fig. 3.7(b)—whose rotational vibration is to represent the vibration of the free end of the bar shown in Fig. 3.7(a)—can be determined by equating the rotational kinetic energy of the lumped mass–spring system to that of the bar. If the angular velocity at the free end of the bar at time $t$ is $\dot{\theta}(t)$ and $I_{pe}$ is the equivalent polar moment of inertia of the lumped mass, then

$$\tfrac{1}{2}I_{pe}\dot{\theta}^2(t) = \int_0^L \tfrac{1}{2}\frac{I_p}{L}\left\{\frac{\dot{\theta}(t)x}{L}\right\}^2 \mathrm{d}x \tag{3.53}$$

Hence

$$I_{pe} = \tfrac{1}{3}I_p \tag{3.54}$$

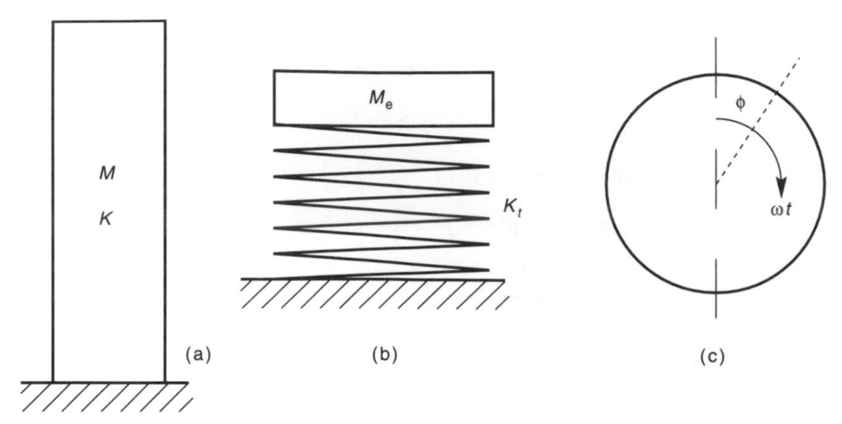


*Fig. 3.7. Equivalent rotational mass–spring system of linear elastic element vibrating about its own central axis*

**Example 3.2** An 8 m dia. circular post-tensioned concrete platform 0·25 m thick, is supported centrally on a 5·0 m tall circular hollow concrete column with an external diameter of 2·0 m and an internal diameter of 1·5 m. Calculate the natural rotational frequency. The specific density of concrete is 2400 kg/m³ and the modulus of rigidity is 12·0 kN/mm².

The rotational stiffness of the hollow column is given by

$$K_t = \frac{\pi G(R^4 - r^4)}{2L} = \frac{\pi \times 12{\cdot}0 \times 10^6(1{\cdot}0^4 - 0{\cdot}75^4)}{2 \times 5{\cdot}0}$$
$$= 2{\cdot}5770877 \times 10^6 \text{ kN m/rad}$$

The mass of the 8·0 m dia. platform is

$$M = \rho\pi R^2 t = 2400{\cdot}0 \times \pi \times 4{\cdot}0^2 \times 0{\cdot}25 = 30\,159{\cdot}289 \text{ kg}$$

Thus the polar moment of inertia of the platform about its central support is

$$I_p = \tfrac{1}{2}MR^2 = \tfrac{1}{2} \times 30\,159{\cdot}289 \times 4{\cdot}0^2 = 241\,274{\cdot}31 \text{ kg m}^2$$

The additional equivalent polar moment of inertia of the column is

$$I_{pe} = \tfrac{1}{3}I_p = \tfrac{1}{3} \times \tfrac{1}{2}\rho\pi(R^4 - r^4)L$$
$$= \tfrac{1}{3} \times \tfrac{1}{2} \times 2400{\cdot}0 \times \pi \times (1{\cdot}0^4 - 0{\cdot}75)^4 \times 5{\cdot}0 = 4295{\cdot}1462 \text{ kg m}^2$$

The rotational stiffness of the column is

$$K_t = \frac{\pi G(R^4 - r^4)}{2L} = \frac{\pi \times 12{\cdot}0 \times 10^6 (1{\cdot}0^4 - 0{\cdot}75^4)}{2 \times 5{\cdot}0}$$

$$= 2{\cdot}5770877 \times 10^6 \text{ N m/rad}$$

Thus the natural rotational angular frequency is

$$\omega_n \sqrt{\left(\frac{K_t}{I_p}\right)} = \sqrt{\left(\frac{2{\cdot}5770877 \times 10^6}{241\,274{\cdot}31 + 4295{\cdot}1462}\right)} = 3{\cdot}2394958 \text{ rad/s}$$

Hence

$$f_n = \frac{\omega_n}{2\pi} = \underline{0{\cdot}5155817 \text{ Hz}}$$

**Example 3.3** The square 40 m × 40 m platform shown in Fig. 3.8 is supported on four 45 m tall hollow circular concrete columns. The external diameters of the columns are 4·0 m and the internal diameters 3·0 m. The columns are spaced 20 m centre to centre as shown in the figure, and may be considered to be rigidly fixed at the base and to the platform. The mass of the platform is $3{\cdot}84 \times 10^6$ kg, and it is so stiff that is may be assumed to be rigid. Calculate the natural lateral and rotational frequencies of the structure. The specific density of concrete is 2400 kg/m$^3$, the modulus of elasticity is 30·0 kN/mm$^2$ and the shear modulus of elasticity 12·0 kN/mm$^2$.

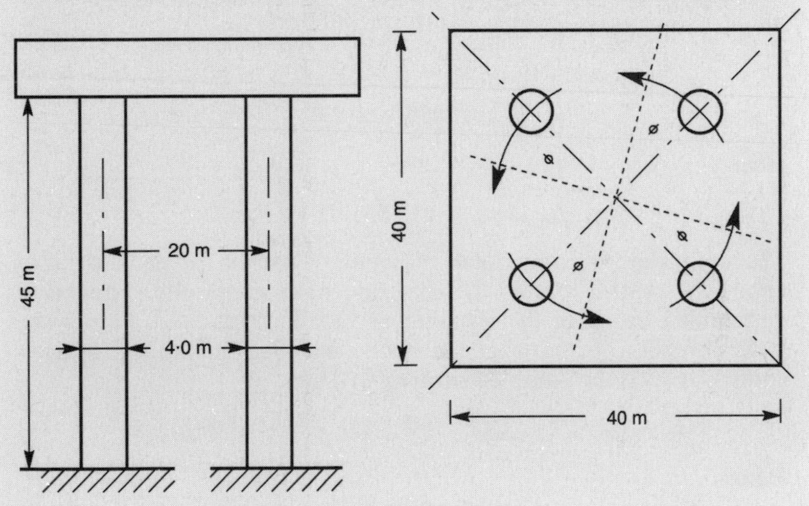


*Fig. 3.8. Elevation and plan view of 40 m × 40 m platform in example 3.3*

*Determination of the natural frequency of the lateral motion*
The second moments of the cross-sectional areas of the columns are

$$I = \tfrac{1}{4}\pi(R^4 - r^4) = \tfrac{1}{4}\pi(2{\cdot}0^4 - 1{\cdot}5^4) = 8{\cdot}5902924\,\text{m}^4$$

The shear stiffness of each column is

$$K = \frac{12EI}{L^3} = \frac{12 \times 30{\cdot}0 \times 10^6 \times 8{\cdot}5902924}{45{\cdot}0^3} = 33{\cdot}936958 \times 10^3\,\text{kN/m}$$

The equivalent mass at the top of each column is

$$M_e = (13/35)wL = (13/35)\pi(2{\cdot}0^2 - 1{\cdot}5^2) \times 2400{\cdot}0 \times 45{\cdot}0$$
$$= 220\,539{\cdot}8\,\text{kg}$$

Thus the total equivalent mass at a height of 45·0 m is

$$M_e = 3{\cdot}84 \times 10^6 + 4 \times 0{\cdot}2205398 \times 10^6 = 4{\cdot}7221592 \times 10^6\,\text{kg}$$

Hence the lateral natural frequency of the platform is

$$F_{n,\,\text{lateral}} = \frac{1}{2\pi}\sqrt{\left(\frac{4 \times 33{\cdot}936958 \times 10^6}{4{\cdot}7221592 \times 10^6}\right)} = \underline{0{\cdot}8533\,\text{Hz}}$$

*Determination of the natural frequency of the rotational motion*
The polar moment of inertia of the platform deck is given by

$$I_{p,\,\text{deck}} = \tfrac{1}{12}M(a^2 + b^2) = \tfrac{1}{12} \times 3{\cdot}84 \times 10^6(40{\cdot}0^2 + 40{\cdot}0^2)$$
$$= 1024{\cdot}0 \times 10^6\,\text{kg m}^2$$

The equivalent second polar moment of inertia of each column about its own axis at 45 m above the supports is

$$I_{pe,\,\text{column}} = \tfrac{1}{3}I_{p,\,\text{column}} = \tfrac{1}{3} \times \tfrac{1}{2}\rho\pi(R^4 - r^4)L$$
$$= \tfrac{1}{3} \times \tfrac{1}{2} \times 2400 \times \pi(2{\cdot}0^4 - 1{\cdot}5^4) \times 45{\cdot}0$$

Hence

$$I_{pe,\,\text{column}} = 618\,501{\cdot}05\,\text{kg m}^2$$

The equivalent second polar moment of inertia of each column about the central axis of the structure 45 m above the supports is determined by using the theorem of parallel axes, and adding the polar moment of inertia of the equivalent mass at the top of each column due to the shear deformation. Thus

$$I_{p,\,\text{column}} = I_{pe,\,\text{column}} + \tfrac{1}{3}M_{\text{column}} \times h^2 + \tfrac{13}{35}M_{\text{column}}h^2$$

Hence

$$I_{p,\,\text{column}} = I_{pe} + \tfrac{74}{105}M_{\text{column}} \times h^2$$

$$I_{p,\,column} = 618\,501{\cdot}05 + \tfrac{74}{105} \times 2400 \times \pi \times \left(2{\cdot}0^2 - 1{\cdot}5^2\right) \times 45 \times \left(10{\cdot}0^2 + 10{\cdot}0^2\right)$$
$$= 84{\cdot}310529 \times 10^6 \text{ kg m}^2$$

Thus

$$I_{p,\,total} = 1024{\cdot}0 \times 10^6 + 4 \times 84{\cdot}310529 \times 10^6$$
$$= 1361{\cdot}2421 \times 10^6 \text{ kg m}^2$$

The equivalent rotational spring stiffness is most conveniently determined by equating the strain energy stored in the spring to that stored in the columns, due to bending and torsion, when both the structure and the spring are rotated through and angle $\phi$. Thus

$$\tfrac{1}{2}K_{t,\,spring}\phi^2 = 4\left\{\tfrac{1}{2} \times \frac{12EI}{L^3}(h\phi)^2 + \tfrac{1}{2} \times \frac{\pi G\left(R^4 - r^4\right)}{2L}\phi^2\right\}$$

$$K_{t,\,spring} = 4\left\{\frac{12EIh^2}{L^3} + \frac{\pi G\left(R^4 - r^4\right)}{2L}\right\}$$

$$K_{t,\,spring} = 4\left\{\frac{12 \times 30{\cdot}0 \times 10^6 \times 8{\cdot}5902924 \times 200}{45{\cdot}0^3} + \frac{\pi \times 12{\cdot}0 \times 10^6 \times \left(2{\cdot}0^4 - 1{\cdot}5^4\right)}{2 \times 45{\cdot}0}\right\}$$
$$= 45{\cdot}475523 \times 10^6 \text{ kN m/rad}$$

The rotational frequency can now be found by substitution of the values for the equivalent rotational spring stiffness and total polar moment of inertia into eq. (3.46). Thus

$$f_{n,\,rotational} = \frac{1}{2\pi}\sqrt{\left(\frac{45{\cdot}475523 \times 10^9}{1{\cdot}3612421 \times 10^9}\right)} = \underline{0{\cdot}9199 \text{ Hz}}$$

**Free rotational vibration with viscous damping**

From d'Alembert's principle it follows that

$$I_p\ddot{\theta} = -K_t\theta - C_t\dot{\theta} \tag{3.55}$$

or

$$I_p\ddot{\theta} + C_t\dot{\theta} + K_t\theta = 0 \tag{3.56}$$

where $C_t\dot{\theta}$ is a viscous damping force which is proportional to the angular velocity and $C_t$ is a viscous damping coefficient with units of N/s/rad. The critical damping coefficient $C_{tc}$ and damping ratio $\xi_t$ can be shown, in the same way as for rectilinear motion, to be

$$C_{tc} = 2\sqrt{(K_tI_p)} \tag{3.57}$$
$$\xi_t = C_t/Ctc \tag{3.58}$$

Thus

$$C_t = 2\xi_t I_p \omega_n \tag{3.59}$$

Substitution of eq. (3.59) into eq. (3.56) yields

$$I_p \ddot{\theta} + 2\xi_t I_p \omega_n \dot{\theta} = K_t \theta \tag{3.60}$$

It can further be shown that

$$\xi_t = \frac{\delta_t}{\sqrt{(4\pi^2 + \delta_t^2)}} \tag{3.61}$$

where $\delta_t$ is the logarithmic decrement of damping, which is determined from decay functions in exactly the same way as for transverse and lateral motions. Usually $\delta_t <<< 2\pi$, hence eq. (3.61) may be written as

$$\xi_t = \frac{\delta_t}{2\pi} \tag{3.62}$$

**Problem 3.1** The vibration of an elastic system consisting of a weight $W = 100$ N and a spring with stiffness $K = 8{\cdot}0$ N/mm is to be damped with a viscous dash pot so that the ratio of two successive amplitudes is 1·00 to 0·85. Determine: (*a*) the natural frequency of the undamped system; (*b*) the required value of the logarithmic decrement of damping; (*c*) the required damping ratio; (*d*) the corresponding damping coefficient; (*e*) the resulting damped natural frequency; (*f*) the amplitude after the tenth oscillation, if the first amplitude of free vibration is 5·00 mm.

**Problem 3.2** The amplitude of vibration of an elastic mass–spring system is observed to decrease by 5·0% with each successive cycle of the motion. Determine the damping coefficient $C$ of the system if the spring stiffness of the system $K = 35{\cdot}0$ Kn/m and the mass $M = 4{\cdot}5$ kg.

**Problem 3.3** A structure is modelled as a viscously damped oscillator with a spring constant $K = 5900{\cdot}0$ kN/m and undamped natural frequency $\omega_n = 25{\cdot}0$ rad/s. Experimentally it was found that a force of 0·5 kN produced a relative velocity of 50 mm/s in the damping element. Determine: (*a*) the damping ratio $\xi$; (*b*) the damped period $T_d$; (*c*) the logarithmic decrement of damping $\delta$; (*d*) the ration between two consecutive amplitudes.

**Problem 3.4** Repeat the calculation of the natural lateral and rotational frequencies of the platform in example 3.3 with the external column diameters assumed to be 5·0 m and the internal one 4·0 m.

## References

Clough, R. W. & Penzien, J. *Dynamics of structures.* McGraw-Hill, London, 1975.

Craig, R. R. Jr. *Structural dynamics.* Wiley, Chichester, 1981.

Harris, C. M. *Shock vibration*, 3rd edn. McGraw-Hill, London, 1988.

Irvine, H. M. *Structural dynamics for the practising engineer.* Allen & Unwin, London, 1986.

Paz, M. *Structural dynamics.* Van Nostrand Reinhold, 1980.

Stroud, K. A. *Engineering mathematics.* Macmillan, London, 1970.

Timoshenko, S. P. & Gere, J. M. *Mechanics of materials.* Van Nostrand, New York, 1972.

# 4. Forced harmonic vibration of one degree of freedom systems

## Introduction

In chapters 1–3 it is pointed out that rotating machines tend to generate harmonic pulsating forces when not properly balanced, and that Fourier analyses of random forces such as wind, waves and earthquakes show that they can be considered as sums of harmonic components, as indeed can explosions and impulse forces. It is also pointed out that a large number of civil engineering structures respond mainly in the first mode, and that it is possible in many cases to reduce such structures to equivalent one DOF mass–spring systems; also, the different forms of structural damping can be modelled as viscous damping in the form of a dash pot. This chapter takes a step forward and considers the response of viscously damped one DOF mass–spring systems when subjected to harmonic forcing functions. A thorough knowledge of how damped equivalent one DOF systems respond to harmonic excitation is fundamental to an understanding of how structures exposed to random dynamic forces are likely to behave.

## Rectilinear response of one DOF system with viscous damping to harmonic excitation

Consider the motion of the damped mass–spring system shown in Fig. 4.1 when subjected to the harmonic exciting force

$$P(t) = P_0 \sin(\omega t) \tag{4.1}$$

from Newton's law of motion

$$M\ddot{x} = W - K(\Delta + x) - C\dot{x} + P_0 \sin(\omega t) \tag{4.2}$$

Hence

$$M\ddot{x} + C\dot{x} + Kx = P_0 \sin(\omega t) \tag{4.3}$$

which, on division of each of the elements by $M$, yields

$$\ddot{x} + 2\xi\omega_n\dot{x} + \omega_n{}^2 x = q_0 \sin(\omega t) \tag{4.4}$$

where

$$2\xi\omega_n = C/M$$

$$\omega_n{}^2 = K/M$$

$$q_0 = P_0/M$$

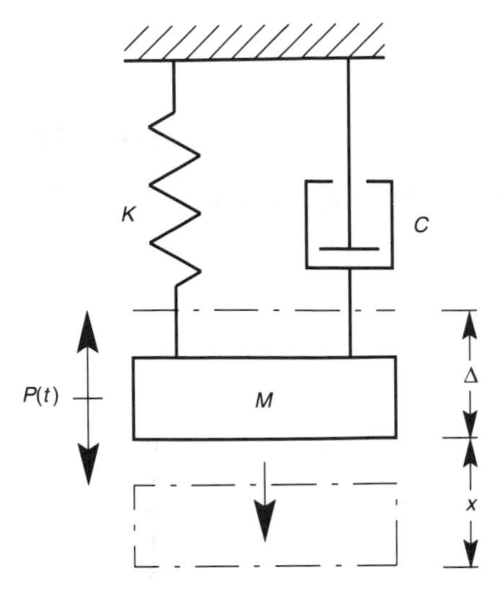


*Fig. 4.1. Forced vibration of damped mass–spring system*

If it is assumed that $\omega_n > \xi\omega_n$, then from eq. (3.29) the complementary function for eq. (4.4) is of the form

$$x = e^{-\xi\omega_n t}\{A\cos(\omega_d t) + B\sin(\omega_d t)\} \tag{4.5}$$

$$\omega_d = \omega_n\sqrt{(1-\xi^2)} \tag{4.6}$$

The particular integral is found by assuming that

$$x = C\sin(\omega t) + D\cos(\omega t)$$
$$\dot{x} = D\omega\cos(\omega t) - D\omega\sin(\omega t)$$
$$\ddot{x} = -\,C\omega^2\sin(\omega t) - D\omega^2\cos(\omega t)$$

Substitution of the expressions for $x$, $\dot{x}$ and $\ddot{x}$ into eq. (4.4) yields

$$-\,C\omega^2\sin(\omega t) - D\omega^2\cos(\omega t) + 2C\xi\omega_n\omega\cos(\omega t)$$
$$-\,2D\xi\omega_n\omega\sin(\omega t) + C\omega_n{}^2\sin(\omega t) + D\omega_n{}^2\cos(\omega t) = q_0\sin(\omega t) \tag{4.7}$$

Equating the coefficients of the $\cos(\omega t)$ and $\sin(\omega t)$ terms in eq. (4.7) yields

$$-\,C\omega^2 - 2D\xi\omega_n\omega + C\omega_n{}^2 = q_0 \tag{4.8a}$$
$$-\,D\omega^2 + 2C\xi\omega_n\omega + D\omega_n{}^2 = 0 \tag{4.8b}$$

Thus

$$C = \frac{q_0(1-r^2)}{\omega_n{}^2(1-r^2)^2 + 4\xi^2r^2} \tag{4.9a}$$

$$D = \frac{2q_0\xi r}{\omega_n{}^2(1-r^2)^2 + 4\xi^2r^2} \tag{4.9b}$$

Hence the complete solution to eq. (4.4) is

$$x = e^{-\xi\omega_n t}\{A\cos(\omega_d t) + B\sin(\omega_d t)\} + \{C\sin(\omega t) + D\cos(\omega t)\} \quad (4.10)$$

where the constants $C$ and $D$ are given by eqs (4.9a) and (4.9b). As the value of $t$ increases the first term on the right-hand side of eq. (4.10) will gradually decrease until it becomes negligible. The free vibration component represented by the first term is called the transient vibration. The second term containing the disturbing force represents the forced vibration.

The expression for forced damped vibration given by the second term in eq. (4.10) can be simplified by considering steady-state response and by using rotating vectors as shown in Fig. 4.2, from which it can be seen that an alternative expression for the dynamic displacement $x$ is

$$x = E\sin(\omega t - \alpha) \quad (4.11)$$

where

$$\tan(\alpha) = \frac{D}{C} + \frac{2\xi r}{1 - r^2} \quad (4.12)$$

From Fig. 4.2

$$E = \sqrt{(C^2 + D^2)} = \frac{q_0}{\omega_n{}^2\sqrt{\left[(1 - r^2)^2 + 4\xi^2 r^2\right]}} \quad (4.13)$$

and, hence, since $q_0 = P_0/M$, $\omega_n{}^2 = K/M$ and $x_{st} + P_0/K$

$$E = \frac{x_{st}}{\sqrt{\left[(1 - r^2)^2 + 4\xi^2 r^2\right]}} \quad (4.14)$$

from which it follows that

$$x = \frac{x_{st}}{\sqrt{\left[(1 - r^2)^2 + 4\xi^2 r^2\right]}}\sin(\omega t - \alpha) \quad (4.15)$$

or

$$x = x_{st}\text{MF}\sin(\omega t - \alpha) \quad (4.16)$$

where MF is the *dynamic magnification factor*

$$\text{MF} = \frac{1}{\sqrt{\left[(1 - r^2)^2 + 4\xi^2 r^2\right]}} \quad (4.17)$$

The value of the frequency ratio $r$ that yields the greatest response at steady-state vibration is found by differentiating eq. (4.17) with respect to $r$ and equating the result to zero. For real structures having damping ratios $\xi > \sqrt{2}$ the peak response frequency ratio is found to be at

$$r = \sqrt{(1 - \xi^2)} \quad (4.18)$$

when the structure is said to be in *resonance*. The corresponding maximum value of the MF is

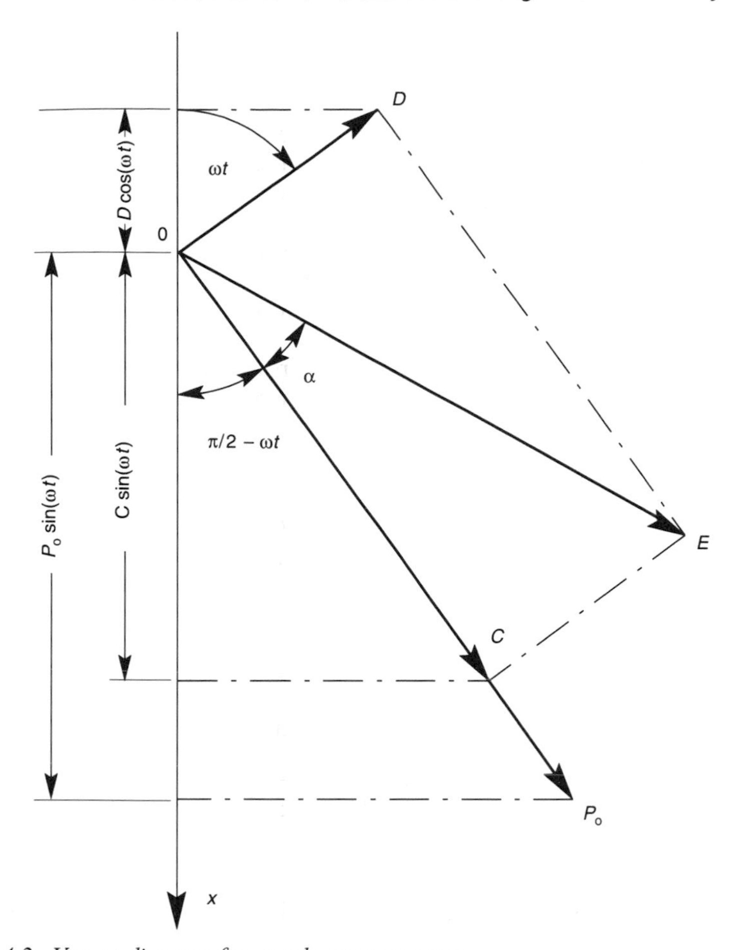


*Fig. 4.2. Vector diagram for steady-state response*

$$\mathrm{MF} = \frac{1}{2\xi\sqrt{(1-\xi^2)}} \tag{4.19}$$

In practice, values of the damping ratio $\xi$ are so small compared to unity that values of $\xi^2$ may be ignored. Thus the maximum response can be considered to occur when

$$r = 1$$
$$\mathrm{MF} = 1/2\xi$$
$$\alpha = \pi/2$$

The complete solution to eq. (4.4), an expression having been obtained for the particular integral, is therefore

$$x = \mathrm{e}^{-\xi\omega_n t}\{A\cos(\omega_d t) + B\sin(\omega_d t)\} + x_{st}\mathrm{MF}\sin(\omega t - \alpha) \tag{4.20}$$

## Response at resonance

At resonance, i.e. when $r = 1$, the response equation (eq. (4.20)) reduces to

$$x = e^{-\xi\omega_n t}\{A\cos(\omega_d t) + B\sin(\omega_d t)\} + \frac{x_{st}}{2\xi}\cos(\omega t) \qquad (4.21)$$

If it is assumed that $x = \dot{x} = 0$ when $t = 0$, it is found that

$$A = \frac{x_{st}}{2\xi}$$

$$B = \frac{x_{st}}{2\sqrt{(1-\xi^2)}}$$

and thus that

$$x = \frac{x_{st}}{2\xi}\left\{e^{-\xi\omega_n t}\left(\cos(\omega_d) + \frac{\xi}{\sqrt{(1-\xi^2)}}\sin(\omega_d t)\right) - \cos(\omega t)\right\} \qquad (4.22)$$

For the level of damping experienced in most structures, the contribution to the amplitude by the sine term is negligible. Equation (4.22) may therefore for practical purposes be written as

$$x = \frac{s_{st}}{2\xi}\{e^{-\xi\omega_n t}[\cos(\omega_d t) - \cos(\omega t)]\} \qquad (4.23)$$

For linear structures vibrated at resonance, $\omega = \omega_d$. Thus eq. (4.23) can be simplified and the expression of the response ratio at time $t$, $R(t) = x/x_{st}$, can be written as

$$R(t) = \frac{1}{2\xi}\{e^{-\xi\omega_n t} - 1\}\cos(\omega_d t) \qquad (4.24)$$

As $t$ increases, $e^{-\xi\omega t}$ approaches zero and eq. (4.24) reduces to

$$R(t) = \frac{1}{2\xi}\cos(\omega_d t) \qquad (4.25)$$

MF, as can be seen from eq. (4.17), is a function of the frequency ratio $r = \omega/\omega_n$ and the damping ratio $\xi$ and will therefore vary in magnitude with the exciting frequency $\omega$. The maximum value of MF occurs, as given in eq. (4.18), when $r = \sqrt{(1-\xi^2)}$. The variation of the dynamic magnification factor MF with the frequency ratio $r$ is shown in Fig. 4.3. Bearing in mind that a damping ratio of $\xi = 0{\cdot}1$, which is the same as a logarithmic decrement of damping of $\delta = 62{\cdot}4\%$, is much greater than that for normal structures, it will be realized that the maximum response can be assumed to occur when $\omega = \omega_n$. Fig. 4.3 also shows the significance of damping when a structure is being vibrated with an exciting frequency at or near its natural frequency.

The vector diagram shown in Fig. 4.2 shows that the force $P_0$ is in phase with the vector OC. From this and the expression for the response

$$x = e^{-\xi\omega_n t}\{A\cos(\omega_d t) + B\sin(\omega_d t)\} + E\sin(\omega t - \alpha) \qquad (4.26)$$

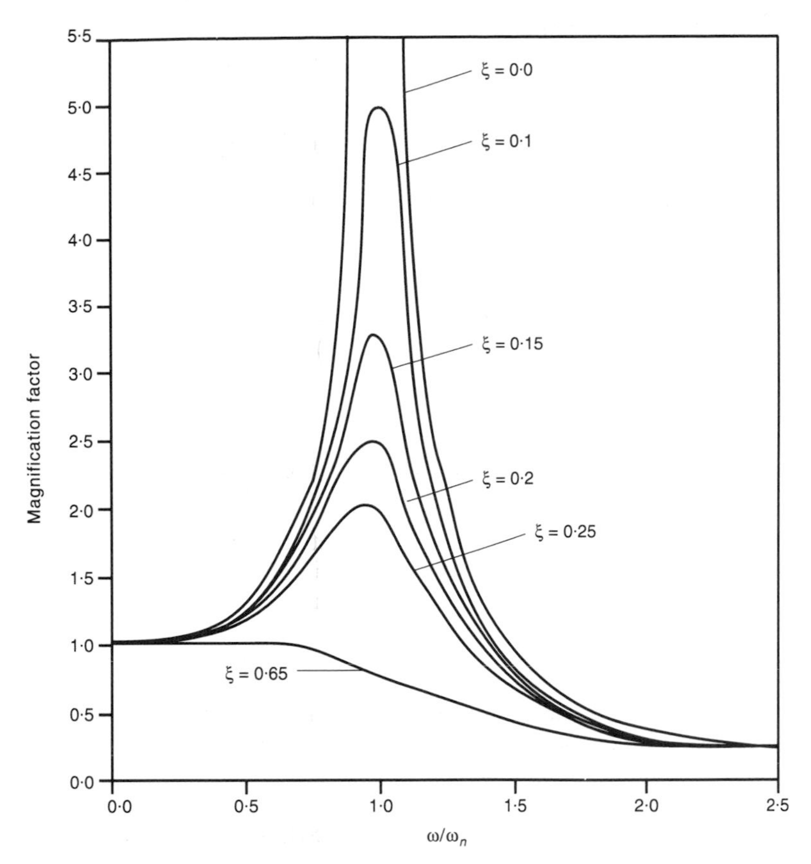


*Fig. 4.3. Variation of the dynamic magnification factor MF with the frequency ratio* $\omega/\omega_{\mathrm{n}}$

it follows that when $r < 1$ the response $x$ lags behind the disturbing force $P_0 \sin(\omega t)$ by a phase angle $\alpha$ and reaches its maximum $(\alpha/\omega)$ seconds after the maximum disturbing force $P_0$ has occurred. When $r > 1$ the response leads the disturbing force by a phase angle $\alpha$ and reaches its maximum $(\alpha/\omega)$ seconds before the maximum disturbing force occurs.

From eq. (4.12), the expression for $\alpha$ is given by

$$\alpha = \tan^{-1}\left\{\frac{2\xi r}{1 - r^2}\right\} \tag{4.27}$$

Thus when

$$\omega < \omega_{\mathrm{n}}, \ \alpha < \pi/2$$

when

$$\omega = \omega_{\mathrm{n}}, \ \alpha = \pi/2$$

when

$$\omega > \omega_{\mathrm{n}}, \ \alpha > \pi/2$$

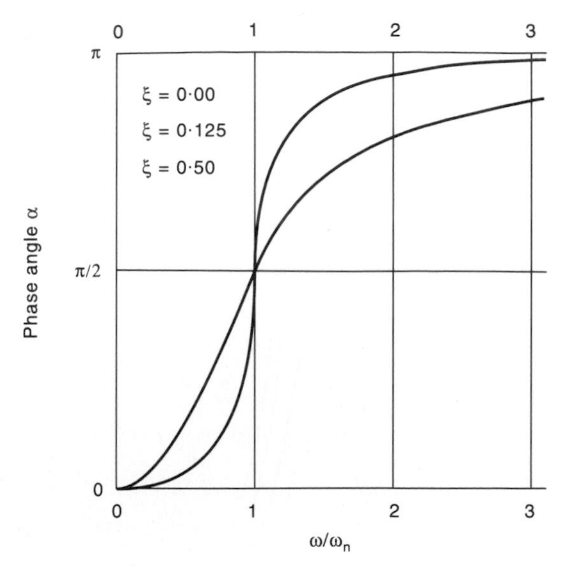


*Fig. 4.4. Variation of phase angle α with frequency ratio $r = \omega/\omega_n$ for different levels of damping*

In Fig. 4.4 values of $\alpha$ are plotted against the ratio $\omega/\omega_n$ for different levels of damping. The resulting curves show that in the region of resonance, where $\omega \approx \omega_n$, there is a sharp variation in the phase angle. In the limit when $\xi = 0$ the variation follows the broken line 01123.

What happens at resonance can be elucidated further by study of the equation of motion and the corresponding vector diagrams. From eq. (4.16), the amplitude at any time $t$ is given by

$$x = x_0 \sin(\omega t - \alpha) \tag{4.28}$$

where

$$x_0 = x_{st}\mathrm{MF}$$

Hence

$$\dot{x} = x_0 \omega \cos(\omega t - \alpha) \tag{4.29a}$$

$$\ddot{x} = -x_0 \omega^2 \sin(\omega t - \alpha) \tag{4.29b}$$

Substitution of the expressions for $x$, $\dot{x}$ and $\ddot{x}$ into eq. (4.3), the equation of motion, yields

$$\begin{aligned} &-Mx_0\omega^2 \sin(\omega t - \alpha) + Cx_0\omega \cos(\omega t - \alpha) + Kx_0 \sin(\omega t - \alpha) \\ &- P_0 \sin(\omega t) = 0 \end{aligned} \tag{4.30}$$

At resonance, when $\alpha = \pi/2$, eq. (4.30) reduces to

$$Mx_0\omega_n{}^2 \cos(\omega_n t) + Cx_0\omega_n \sin(\omega_n t) - Kx_0 \cos(\omega_n t) - P_0 \sin \omega_n t) = 0 \tag{4.31}$$

From eq. (4.31), if the coefficients of the sine and cosine terms are equated to zero,

$$\begin{aligned} P_0 &= Cx_0\omega_n \\ kx_0 &= Mx_0\omega_n^2 \end{aligned} \tag{4.32}$$

Thus the work done by the exciting force at resonance is only used to maintain the amplitude of response by replacing the energy lost through the structural damping mechanism. Equations (4.3) and (4.31) are represented in Fig. 4.5.

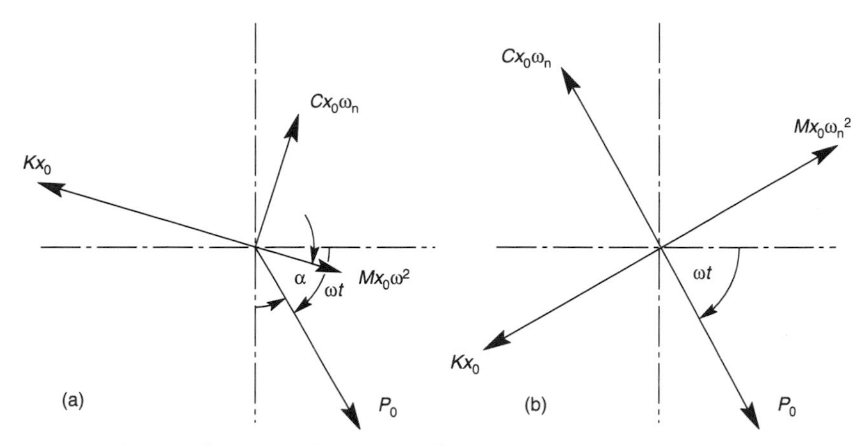


*Fig. 4.5. Vector diagrams of dynamic forces acting on a one DOF system (a) before and (b) at resonance*

**Example 4.1** A beam supports at its centre a machine weighing 71·5 kN. The beam is simply supported, has a span L = 3·5 m and a cross-sectional moment of inertia $I = 5{\cdot}3444 \times 10^7\ \text{mm}^4$, and weighs 18·2 kN/m. The motor runs at 300 rev/min, and its rotor is out of balance to the extent of 180 N at an eccentricity of 25 cm. What is the amplitude at steady-state response if the equivalent viscous damping ratio is 10% of $\xi_{critical}$? Determine also the phase angle $\alpha$ of response relative to that of the unbalanced force.

The equivalent stiffness of the beam is

$$\begin{aligned} K &= 48EI/L^3 = 48 \times 200 \times 5{\cdot}3444 \times 10^7/3{\cdot}5^3 \times 10^9 \\ &= 11{\cdot}96647\ \text{kN/mm} \end{aligned}$$

The equivalent weight of the beam at midspan is given by eq. (2.80). Thus

$$W = P + (17/35)wL = 71{\cdot}5 + (17/35) \times 18{\cdot}2 \times 3{\cdot}5 = 102{\cdot}44\ \text{kN}$$

The natural frequency of the beam is therefore

$$\omega_n = \sqrt{(Kg/M)} = \sqrt{(11{\cdot}96647 \times 1000 \times 1000 \times 9{\cdot}81/102{\cdot}44 \times 1000)}$$
$$= 33{\cdot}851876 \text{ rad/s}$$

The angular forcing frequency of the motor is

$$\omega = 300 \times 2\pi/60 = 31{\cdot}415927 \text{ rad/s}$$

Thus the unbalanced force when the motor is running at 300 rev/min is

$$P_0 = me\omega^2 = 180 \times 0{\cdot}25 \times 31{\cdot}415927^2/9{\cdot}81 \times 1000 = 4{\cdot}5273416 \text{ kN}$$

The expression for steady-state response is given by eq. (4.15), which yields the maximum response when $\sin(\omega t - \alpha) = 0$. Thus

$$x_{max} = \frac{P_0}{K}\text{MF} = \frac{me\omega^2}{K}\frac{1}{\sqrt{\left[(1-r^2)^2+(2\xi r)^2\right]}}$$

where

$$r = \omega/\omega_n = 31{\cdot}415927/33{\cdot}851176 = 0{\cdot}9280409$$
$$\xi = 0{\cdot}1$$

Substitution of the values for $r$ and $\xi$ into the expression for $x_{max}$ yields

$$\text{MF} = 4{\cdot}3153508$$

and

$$x_{max} = 4{\cdot}5273416 \times 4{\cdot}3153505/11{\cdot}96647 = \underline{1{\cdot}633 \text{ mm}}$$

The phase angle $\alpha$ is given by eq. (4.27) as

$$\alpha = \tan^{-1}\left\{\frac{2\xi r}{1-r^2}\right\} = \tan^{-1}\left\{\frac{2 \times 0{\cdot}1 \times 0{\cdot}9280409}{1-0{\cdot}9280409^2}\right\} = \underline{53{\cdot}222^\circ}$$

## Forces transmitted to the foundation by rotating unbalance in machines and motors

Consider a machine of mass $M$ mounted firmly on to the foundation as shown in Fig. 4.6(a). Let the machine have an unbalanced rotating mass $m$ at an eccentricity $e$. At an angular velocity of $\omega$ rad/s this mass will give rise to an unbalanced rotating force

$$P_0 = me\omega^2 \tag{4.33}$$

with vertical and horizontal components $P_0 \sin(\omega t)$ and $P_0 \cos(\omega t)$ respectively.

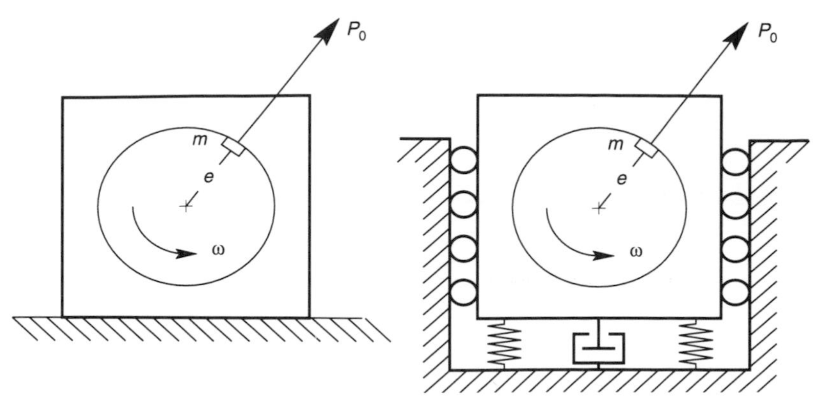


*Fig. 4.6. Motors with rotating unbalance mounted (a) directly to the foundation and (b) on springs and dampers*

In order to reduce the effect of the vertical pulsating force $P_0 \sin(\omega t)$, the machine may be mounted on springs with total stiffness $K$ and on dampers with a resultant damping coefficient $C$ as shown in Fig. 4.6(b). The differential equation of motion for the vertical vibration caused by $P_0 \sin(\omega t)$ for this case is

$$M\ddot{x} + C\dot{x} + Kx = P_0 \sin(\omega t) \tag{4.34}$$

The solution to eq. (4.34) is given by eq. (4.15). Thus

$$x = \frac{P_0}{K} = \frac{1}{\sqrt{\left[(1 - r^2)^2 + (2\xi r)^2\right]}} \sin(\omega t - \alpha) \tag{4.35}$$

where

$$\alpha = \tan^{-1} \frac{2\xi r}{1 - r^2} \tag{4.36}$$

The forces transmitted to the foundation by the springs and the dampers are $Kx$ and $C\dot{x}$ respectively. Thus the total force transmitted to the foundation is

$$F = Kx + C\dot{x} \tag{4.37}$$

or

$$F = P_0 \mathrm{MF} \sin(\omega t - \alpha) + P_0 C\omega / K \mathrm{MF} \cos(\omega t - \alpha) \tag{4.38}$$

With reference to the vector diagram shown in Fig. 4.7, it can be shown that the expression for the transmitted force also can be written as

$$F = P_0 \sqrt{\left[1 + (2\xi r)^2\right]} \mathrm{MF} \sin(\omega t - \alpha - \beta) \tag{4.39}$$

where

$$\beta = \tan^{-1}(C\omega / K) = \tan^{-1}(2\xi r) \tag{4.40}$$

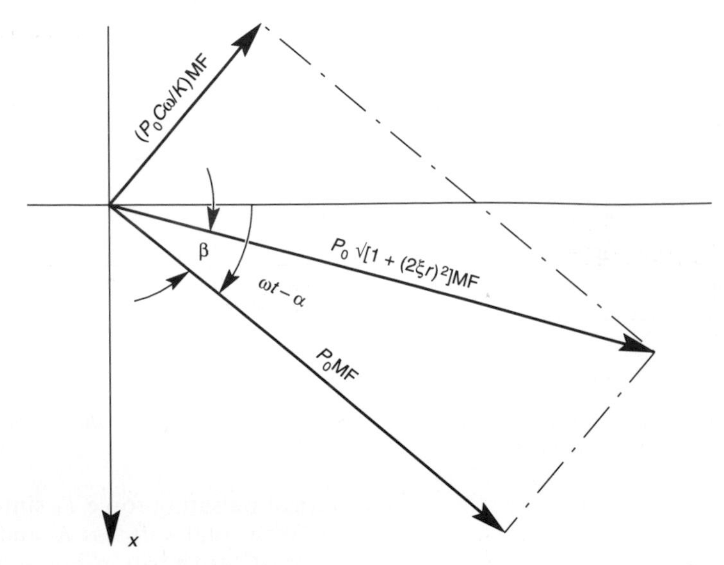


*Fig. 4.7. Vector diagram for forces in rotating unbalanced machines and motors*

Therefore the maximum force transmitted to the foundation is

$$F_0 = P_0 \frac{\sqrt{\left[1 + (2\xi r)^2\right]}}{\sqrt{\left[(1 - r^2)^2 + (2\xi r)^2\right]}} \tag{4.41}$$

The ratio

$$T = \frac{F_0}{P_0} = \frac{\sqrt{\left[1 + (2\xi r)^2\right]}}{\sqrt{\left[(1 - r^2)^2 + (2\xi r)^2\right]}} \tag{4.42}$$

is referred to as the *transmissibility* of the system, and $T$ is the *transmissibility factor*. From eq. (4.42) it can be shown that when

$$r < \sqrt{2},\ T > 1{\cdot}0$$

when

$$r = \sqrt{2},\ T = 1{\cdot}0$$

when

$$r > \sqrt{2},\ T < 1{\cdot}0$$

Transmissibility curves for different levels of damping are shown in Fig. 4.8. From these it can be seen that the transmissibility decreases with decreasing damping ratios when the frequency ratio is greater than $\sqrt{2}$, and increases with decreasing damping ratios when the frequency ratio is less than $\sqrt{2}$.

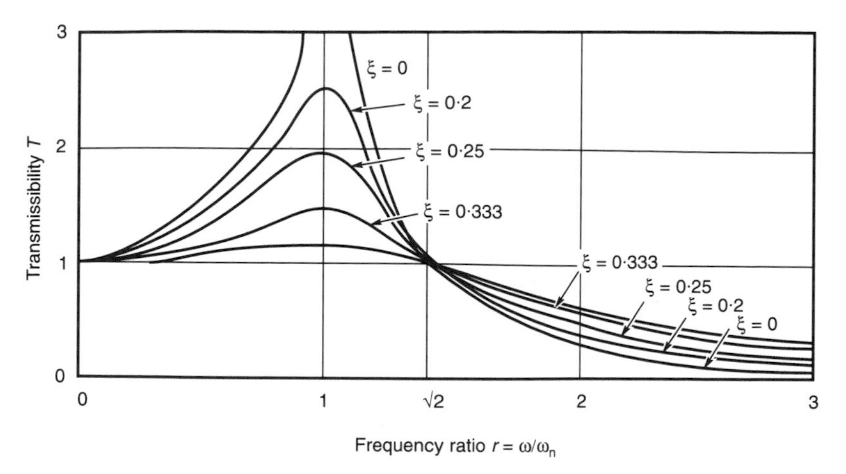


*Fig. 4.8. Transmissibility versus frequency ratios for vibration isolation*

**Example 4.2** A machine, which weighs 2·5 kN and is to be operated at frequencies of 1000 and 4000 rev/min, is to be installed in a factory and mounted on isolators with a combined damping ratio of 10% of critical. The machine has a total unbalanced mass of 0·01 kg at an eccentricity of 100 mm. Calculate the force transmitted and the required spring stiffness of the isolators if the maximum pushing force transmitted to the floor at the operating frequencies is to be reduced by 75%. Calculate also the maximum pulsating force transmitted to the floor when the speed of the machine increases from 0 to 4000 rev/min.

The maximum operating pulsating force transmitted occurs when the machine is running at 4000 rev/min. Thus

$$P_{4000} = me\omega^2 = 0{\cdot}01 \times 0{\cdot}1 \times (4000 \times 2\pi/60)^2 = 175{\cdot}45963\ \text{N}$$

Hence the maximum permissible operational force is

$$F_{4000} = 0{\cdot}25 \times 175{\cdot}45963 = \underline{43{\cdot}864908\ \text{N}}$$

The required frequency ratio to reduce the pulsating force by 75% can be found by use of eq. (4.42) which, when the values for $T$ and $\xi$ are substituted can be written as

$$0{\cdot}25 = \frac{\sqrt{\left[1 + (2 \times 0{\cdot}1 \times r)^2\right]}}{\sqrt{\left[(1 - r^2)^2 + (2 \times 0{\cdot}1 \times r)^2\right]}}$$

which yields

$$r = 1{\cdot}6689337$$

Thus the natural frequency of the machine mounted on isolators is

$$\omega_n = \frac{\omega}{r} = \frac{4000 \times 2\pi}{60 \times 1{\cdot}6689337} = 250{\cdot}98602\,\text{rad/s}$$

Hence the required spring stiffness is

$$K = \omega_n{}^2 M = 250{\cdot}98602^2 \times 2500/9{\cdot}81$$

$$= 16{\cdot}053512 \times 10^6\,\text{N/m}$$

$$= \underline{16{\cdot}053512\,\text{kN/mm}}$$

When the machine runs at 1000 rev/min the maximum pulsating force is

$$P_{1000} = 0{\cdot}01 \times 0{\cdot}1 \times (1000 \times 2\pi/60)^2 = 10{\cdot}966227\,\text{N}$$

The frequency ratio when the machine runs at 1000 rev/min is

$$r = (1000 \times 2\pi/60)/(250{\cdot}98602 = 0{\cdot}4172334$$

Thus the maximum force transmitted to the floor when the machine is running at this speed is

$$F_{1000} = \frac{P_{1000}\sqrt{\left[1 + (2\xi r)^2\right]}}{\sqrt{\left[(1 - r^2)^2 + (2\xi r)^2\right]}}$$

$$= \frac{10{\cdot}966227\sqrt{\left[1 + (2 \times 0{\cdot}1 \times 0{\cdot}4172334)^2\right]}}{\sqrt{\left[(1 - 0{\cdot}4172334^2)^2 + (2 \times 0{\cdot}1 \times 0{\cdot}4172334)^2\right]}}$$

$$= \underline{15{\cdot}913816\,\text{N}}$$

This force is less than the force transmitted when the machine runs at 4000 rev/min. The calculated spring stiffness is therefore satisfactory.

When $r = 1$ the maximum pulsating force due to the unbalanced mass is

$$P_n = me\omega_n{}^2 = 0{\cdot}01 \times 0{\cdot}1 \times 250{\cdot}98602^2 = 62{\cdot}993982\,\text{N}$$

The corresponding force transmitted to the floor in this case is given by

$$F_{max} = \frac{P_n\sqrt{\left[1 + (2\xi r)^2\right]}}{2\xi} = \frac{62{\cdot}993982\sqrt{\left[1 + (2 \times 0{\cdot}1 \times 1{\cdot}0)^2\right]}}{2 \times 0{\cdot}1}$$

$$= \underline{321{\cdot}20754\,\text{N}}$$

## Response to support motion

In the previous section we examined the transmission of harmonic forces, such as caused by unbalanced machinery, to the supports. In the following sections the reverse is undertaken: the response of structures to harmonic excitation of the supports themselves is studied. The shaking of supports or foundations is notably caused by earthquakes—even minor earthquakes can be particularly devastating if the depth of the soil above the bedrock is such that one of the dominant frequencies of the ground coincides with the frequency in which the structure is likely to respond. Other sources of ground motion are traffic and explosions. In the case of the latter the travel of shock waves through the ground is followed by pressure waves through the air, both of whose effects have to be considered if there are buildings in the vicinity. In the study of the response to ground vibration, the response relative to a fixed point, the absolute response of a structure, and the response relative to the support are considered. The first is important when studying the likely effect of ground motion on, say, sensitive electronic equipment in a building, and the second when assessing the strength of the building itself.

### *Response relative to a fixed point*

Consider the mass–spring system shown in Fig. 4.9, where the support subjects the left-hand end of the spring to a periodic displacement

$$x_g = x_{g0} \sin(\omega t) \tag{4.43}$$

where $x_{g0}$ is the maximum amplitude of the support motion and $x_g$ and $x$ are absolute displacements. The equation of motion for this case is given by

$$M\ddot{x} + C(\dot{x} - \dot{x}_g) + K(x - x_g) = 0 \tag{4.44}$$

or

$$M\ddot{x} + C\dot{x} + Kx = Kx_{g0} \sin(\omega_g t) + Cx_{g0}\omega_g \cos(\omega_g t) \tag{4.45}$$

With reference to the vector diagram shown in Fig. 4.10, it can be seen that eq. (4.45) may be written as

$$Mx + C\dot{x} + Kx = F_0 \sin(\omega_g t - \beta) \tag{4.46}$$

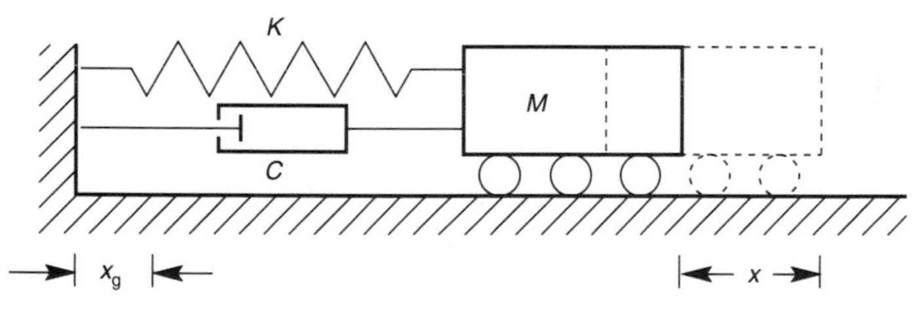


*Fig. 4.9. Mass–spring system with left-hand end of spring being subjected to a harmonic displacement $x_g$*

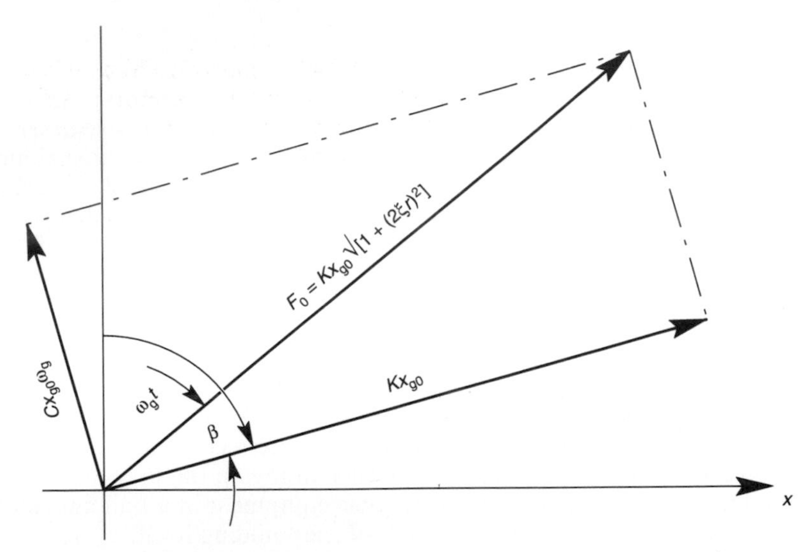


*Fig. 4.10. Vector diagram for forces arising from harmonic excitation of support of mass–spring system in Fig. 4.9*

where

$$F_0 = \sqrt{\left[(Cx_{g0}\omega_g)^2 + (Kx_{g0})^2\right]} = Kx_{g0}\sqrt{\left[1 + (2\xi r)^2\right]} \tag{4.47}$$

$$\beta = \tan^{-1} C\omega_g/K = \tan^{-1} 2\xi t \tag{4.48}$$

The solution to eq. (4.46) is given by eq. (4.15). Thus

$$x = \frac{x_{g0}\sqrt{\left[1 + (2\xi r)^2\right]}}{\sqrt{\left[(1 - r^2)^2 + (2\xi r)^2\right]}} \sin(\omega_g t - \alpha - \beta) \tag{4.49}$$

where from eq. (4.12)

$$\alpha = \tan^{-1}\frac{2\xi r}{1 - r^2} \tag{4.50}$$

The ratio

$$T = \frac{x_0}{x_{g0}} = \frac{\sqrt{\left[1 + (2\xi r)^2\right]}}{\sqrt{\left[(1 - r^2)^2 + (2\xi r)^2\right]}} \tag{4.51}$$

is the *absolute transmissibility* or the *transmissibility relative to a fixed point*, and is a measure of the extent to which the motion of the support is either magnified or reduced by the structure. It should be noted that the right-hand side of eq. (4.51) is identical to the right-hand side of eq. (4.42). Thus the transmissibility curves shown in Fig. 4.8 are also valid for structures subjected to harmonic excitation of the supports.

**Example 4.3** A delicate instrument, which weighs 450 N, is to be spring-mounted to the floor of a test laboratory, which occasionally vibrates with a frequency of 10 Hz and a maximum amplitude of 2·5 mm. Determine the stiffness of the isolation springs required to reduce the vertical amplitude of the instrument to 0·025 mm if the instrument is isolated with dampers with damping ratios (*a*) 2·0% and (*b*) 20·0% of critical.

(*a*) When $\xi = 0{\cdot}02$, the transmissibility, from eq. (4.51) is given by

$$T = \frac{0{\cdot}025}{2{\cdot}5} = \frac{\sqrt{\left[1 + (2 \times 0{\cdot}02 \times r)^2\right]}}{\sqrt{\left[(1 - r^2)^2 + (2 \times 0{\cdot}02 \times r)^2\right]}}$$

This yields

$$r = 10{\cdot}459366$$

Hence

$$\omega_n = \frac{\omega}{r} = \frac{2\pi \times 10}{10{\cdot}459366} = 6{\cdot}0072336 \text{ rad/s}$$

Thus the required spring stiffness when $\xi = 0{\cdot}02$ is

$$K = \omega^2 M = 6{\cdot}0072336^2 \times 450/9{\cdot}81 = \underline{1655{\cdot}3603 \text{ N/m}}$$

(*b*) When $\xi = 0{\cdot}2$, the transmissibility, again from eq. (4.51), is

$$T = \frac{0{\cdot}025}{2{\cdot}5} = \frac{\sqrt{\left[1 + (2 \times 0{\cdot}2 \times r)^2\right]}}{\sqrt{\left[(1 - r^2)^2 + (2 \times 0{\cdot}2 \times r)^2\right]}}$$

This yields

$$r = 40{\cdot}100599$$

Hence

$$\omega_n = \frac{\omega}{r} = \frac{2\pi \times 10}{40{\cdot}100599} = 1{\cdot}5666451 \text{ rad/s}$$

Thus the required spring stiffness when $\xi = 0{\cdot}2$ is

$$K = \omega_n^2 M = 1{\cdot}5666451^2 \times 450/9{\cdot}81 = \underline{112{\cdot}58609 \text{ N/m}}$$

Comparison of the calculated spring stiffnesses shows that, as expected, the stiffness reduces as the value of the damping ratio increases.

### Response relative to the support

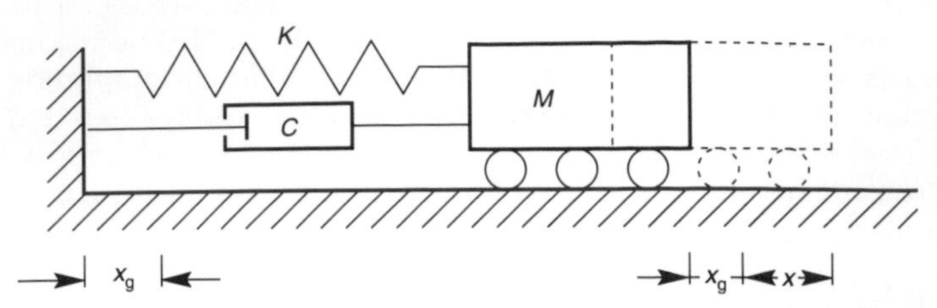


*Fig. 4.11. Mass–spring system with left-hand end of spring being subjected to a harmonic displacement*

If the movement of the mass is measured relative to the movement of the spring support, as shown in Fig. 4.11, and the movement of the support is given by eq. (4.43) as in the previous case, then the equation of motion may be written as

$$M(\ddot{x} + \ddot{x}_g) + C\dot{x} + Kx = 0 \tag{4.52}$$

or

$$M\ddot{x} + C\dot{x} + Kx = -Mx_{g0}\omega_g{}^2 \sin(\omega_g t) \tag{4.53}$$

The negative sign in eq. (4.53) is clearly irrelevant and may be ignored. The solution to eq. (4.53) is again given by eq. (4.15). The response relative to the support motion is therefore

$$x = \frac{Mx_{g0}\omega_g{}^2}{K} \frac{1}{\sqrt{\left[(1 - r^2) + (2\xi r)^2\right]}} \sin(\omega_g t - \alpha) \tag{4.54}$$

Since $\omega_n = K/M$, eq. (4.54) may be written as

$$x = \frac{x_{g0} r^2}{\sqrt{\left[(1 - r^2)^2 + (2\xi r)^2\right]}} \sin(\omega_g t - \alpha) \tag{4.55}$$

The transmissibility factor for the response relative to the support is therefore given by

$$T = \frac{x_0}{x_{g0}} = \frac{r^2}{\sqrt{\left[(1 - r^2)^2 + (2\xi r)^2\right]}} \tag{4.56}$$

Relative transmissibility curves constructed by substitution of increasing values for the frequency ratio $r$ and the damping ratio $\xi$ into eq. (4.56) are shown in Fig. 4.12.

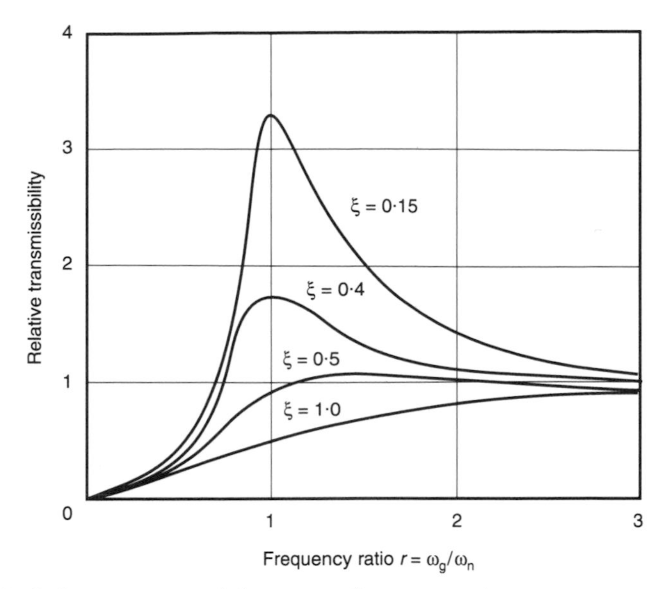


*Fig. 4.12. Relative transmissibility versus frequency ratio*

**Example 4.4** A rigid jointed rectangular steel portal frame has a span of 20 m. Each column is 4·0 m tall and pinned at the base. The weight of the horizontal beam, which may be assumed to be rigid, is 4·0 kN/m. The second moment of area and the section modulus of each of the columns, which may be considered to be weightless, are $3200\,\text{cm}^4$ and $286\,\text{cm}^3$ respectively. The damping ratio for the structure may be assumed to be 2·0% critical. Young's modulus $E = 100\,\text{kN/mm}^2$. If the frame is subjected to a sinusoidal ground motion $x_g = 8.0\sin(11.5t)$ mm, determine: (*a*) the transmissibility of the motion to the girder; (*b*) the maximum shear force in each column; (*c*) the maximum bending stress in each column.

The shear stiffness of the frame is

$$K = \frac{2 \times 3EI}{L^3} = \frac{2 \times 3 \times 200 \times 3200 \times 10^4}{4{\cdot}0 \times 10^9} = 0{\cdot}6\,\text{kN/mm}$$

Hence the natural frequency of the frame is

$$\omega_n = \sqrt{\left(\frac{K}{M}\right)} = \sqrt{\left(\frac{600 \times 10^3 \times 9{\cdot}81}{20 \times 4000}\right)} = 8{\cdot}5775871\,\text{rad/s}$$

Thus the frequency ratio is given by

$$r = \frac{\omega}{\omega_n} = \frac{11{\cdot}5}{8{\cdot}5775871} = 1{\cdot}3407034$$

From eq. (4.56) the relative transmissibility, i.e. the motion of the crossbeam relative to the ground, is

$$T = \frac{x}{x_g} = \frac{x}{8{\cdot}0} = \frac{1{\cdot}3407034^2}{\sqrt{\left[(1 - 1{\cdot}3407034^2)^2 + (2 \times 0{\cdot}02 \times 1{\cdot}3407034)^2\right]}}$$

$$= 2{\cdot}2488622$$

Hence the maximum horizontal motion of the beam is

$$x = 2{\cdot}2488622 \times 8{\cdot}0 = \underline{17{\cdot}990898\,\text{mm}}$$

The maximum shear force in each column is given by

$$\text{SF} = \tfrac{1}{2}Kx = \tfrac{1}{2} \times 0{\cdot}6 \times 17{\cdot}990898 = \underline{5{\cdot}3972694\,\text{kN}}$$

Therefore the maximum bending stresses in the columns are

$$\sigma_M = \frac{\text{SF} \times H}{Z} = \frac{5{\cdot}3972694 \times 4{\cdot}0 \times 10^3}{286 \times 10^3} = \underline{0{\cdot}0754862\,\text{kN/mm}^2}$$

## Seismographs

Movements of the ground due to earthquakes or other forms of disturbances can be recorded by the use of seismographs (Fig. 4.13). These instruments, which for arbitrarily chosen damping ratios will measure the relative displacement between the spring-supported mass and the base of the instrument, can be designed to measure either the displacement or the acceleration of the base support. From eq. (4.55), the relative response of the mass in the seismograph due to a base movement of $x_g$ is given by

$$x_g = x_{g0} \sin(\omega_g t) \tag{4.57}$$

$$x = \frac{x_{g0} r^2}{\sqrt{\left[(1 - r^2)^2 + (2\xi r)^2\right]}} \sin(\omega_g t - \alpha) \tag{4.58}$$

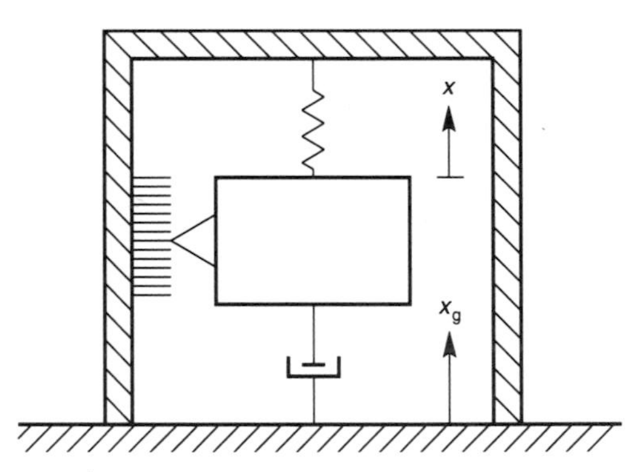


*Fig. 4.13. Seismograph*

From Fig. 4.12 the relative transmissibility is approximately equal to 1 when $r > 1$ and $\xi \approx 0{\cdot}5$. When this is the case the movement of the mass relative to the base is given by

$$x = x_{g0} \sin(\omega_g t - \alpha) \tag{4.59}$$

and the seismograph will therefore record the movement of the support.

From eq. (4.54), the relative movement of the mass may be written as

$$x = \frac{M x_{g0} \omega_g^2}{K} \frac{1}{\sqrt{\left[(1 - r^2)^2 + (2\xi r)^2\right]}} \sin(\omega_g - \alpha) \tag{4.60}$$

Since

$$\ddot{x}_g = -x_{g0}\omega_g^2 \sin(\omega_g t) = -\ddot{x}_{g0} \sin(\omega_g t) \tag{4.61}$$

it follows that eq. (4.60) may be written as

$$x = \frac{M \ddot{x}_{g0}}{K} \frac{1}{\sqrt{\left[(1 - r^2)^2 + (2\xi r)^2\right]}} \sin(\omega_g t - \alpha) \tag{4.62}$$

From Fig. 4.3, when $0 < r < 0{\cdot}6$ and $\xi \approx 0{\cdot}65$ the MF is approximately equal to 1. When this is the case the movement of the mass relative to the base is given by

$$x = \frac{M \ddot{x}_{g0}}{K} \sin(\omega_g t - \alpha) \tag{4.63}$$

Thus the recording of seismographs, designed with the above values for $r$ and $\xi$, will be proportional to the acceleration of the support. In both of the above cases the range of the seismograph may be increased by varying the spring stiffness or the size of the mass.

## Rotational response of one DOF systems with viscous damping to harmonic excitation

Consider the motion of the damped mass–spring system shown in Fig. 4.14 when subjected to the harmonic exciting moment

$$T(t) = T_0 \sin(\omega t) = P_0 e \sin(\omega t) \tag{4.64}$$

If d'Alembert's principle is applied

$$I_p \ddot{\theta} = -K_t \theta - C_t \dot{\theta} + P_0 e \sin(\omega t) \tag{4.65}$$

Hence

$$I_p \ddot{\theta} + K_t \theta + C_t \dot{\theta} = P_0 e \sin(\omega t) \tag{4.66}$$

where $I_p$ is the polar moment of inertia of the lumped mass, $K_t$ is the torsional stiffness of the spring, and $C_t$ is the equivalent torsional viscous damping coefficient, and from eq. (3.58)

$$C_t = 2\xi_t \omega_n I_p \tag{4.67}$$

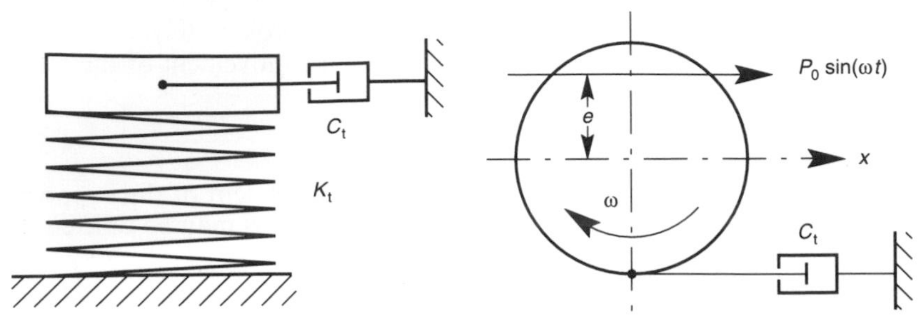


*Fig. 4.14. Elevation and plan view of damped mass–spring system subjected to harmonic exciting moment* $T(t) = P_0 e \sin(\omega t)$

Following exactly the same steps as when finding the solution to the rectilinear equation of motion, it can be shown that the steady-state response to torsional excitation is

$$\theta = \frac{P_0 e}{K_t} \times \frac{1}{\sqrt{\left[(1 - r^2)^2 + (2\xi r)^2\right]}} \sin(\omega t - \alpha) \tag{4.68}$$

where

$$\alpha = \tan^{-1} \frac{2\xi_t r}{1 - r^2} \tag{4.69}$$

**Example 4.5** The natural frequency of the translational motion of the structure shown in Fig. 4.15 is 0·8533 Hz. The corresponding equivalent mass and spring stiffness are respectively $4{\cdot}722 \times 10^6$ kg and $135{\cdot}748 \times 10^3$ kN/m. The natural rotational frequency about the vertical axis is 0·9199 Hz. The corresponding polar moment of inertia and torsional spring stiffness are respectively $1\,361{\cdot}2421 \times 10^6$ kg m$^2$ and $45\,475{\cdot}523 \times 10^3$ kN/rad. For the purpose of design it is assumed that the centre of gravity of the equivalent mass of the structure is located 1·0 m above the $x$-axis. Calculate the maximum translational and rotational response to a horizontal support motion $x_g t = 0{\cdot}02 \sin(6{\cdot}0t)$ m if the damping in both the translational and rotational modes is 2·0% of critical.

The expression for the translational response of one DOF systems is given by eq. (4.54). Thus

$$x_{max} = \frac{M x_{g0} \omega_g^{\,2}}{K} \times \frac{1}{\sqrt{\left[(1 - r^2)^2 + (2\xi r)^2\right]}}$$

where

$$r = 6{\cdot}0/2\pi \times 0{\cdot}8533 = 1{\cdot}1191019$$

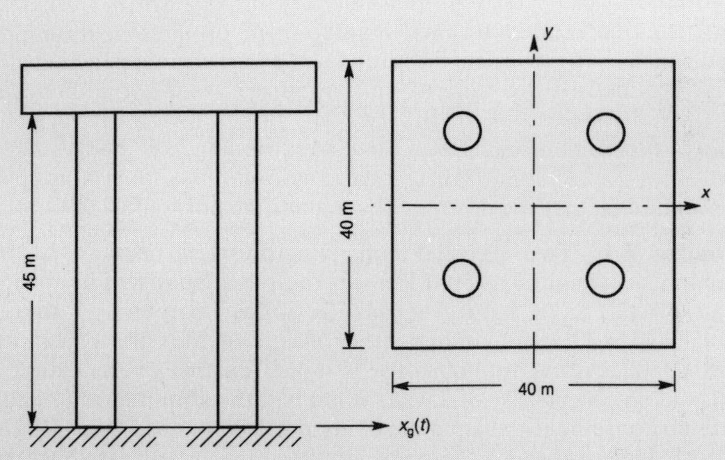


*Fig. 4.15. Platform structure with assumed nonsymmetric mass distribution subjected to harmonic support motion*

Hence

$$x_{\max} = \frac{4\cdot722 \times 10^6 \times 0\cdot02 \times 6\cdot0^2}{135\cdot748 \times 10^3 \times 10^3} \times \frac{1}{\sqrt{\left[(1 - 1\cdot1191019^2)^2 + (2 \times 0\cdot02 \times 1\cdot1191019)^2\right]}}$$

or

$$\underline{x_{\max} = 0\cdot3812\,\text{m}}$$

The expression of the rotational response of a one DOF system is given by eq. (4.68). Thus

$$\theta_{\max} = \frac{M\ddot{x}_g e \omega_g{}^2}{K_t} \times \frac{1}{\sqrt{\left[(1 - r^2)^2 + (2\xi r)^2\right]}}$$

where

$$r = 6\cdot0/2\pi \times 0\cdot9199 = 1\cdot0380799$$

Hence

$$\theta_{\max} = \frac{4\cdot722 \times 10^6 \times 0\cdot02 \times 6\cdot0^2 \times 1\cdot0}{45\,475\cdot523 \times 10^3 \times 10^3} \times \frac{1}{\sqrt{\left[(1 - 1\cdot0380799^2)^2 + (2 \times 0\cdot02 \times 1\cdot0380799)^2\right]}}$$

or

$$\underline{\theta_{\max} = 9\cdot64985 \times 10^{-3}\,\text{rad} = 0\cdot553^\circ}$$

Thus the maximum rotational displacement of each corner of the platform is

$$\alpha = \pm[\sqrt{[(20^2 + 20^2)]}] \times 9{\cdot}64985 \times 10^{-3} = \underline{\pm 0{\cdot}2729\,\text{m}}$$

**Problem 4.1** Two parallel simply supported beams support a machine weighing 150·0 kN at their midspans. The beams span 3·4 m, have a total cross-sectional moment of inertia $I = 5{\cdot}3444 \times 10^7\,\text{mm}^4$, and together weigh 18·2 kN/m. The motor runs at 800 rev/min, and its rotor is out of balance to the extent of 150 N at an eccentricity of 25 cm. What will the amplitude of steady-state response be if the equivalent viscous damping ratio is 10% of critical? Determine also the phase angle of response relative to that of the unbalanced force $E = 200\,\text{kN/mm}^2$.

**Problem 4.2** Determine the force transmitted by the machine to the supports of the beam whose data are given in problem 4.1. Calculate also the force transmitted if the motor runs at a speed equal to the natural frequency of the system. Start by developing the expression for the appropriate transmissibility factor and give the values for this factor for the two running speeds in question.

**Problem 4.3** Calculate the response of the top floor of the shear structure shown in Fig. 2.14 to the ground motion $x = 11{\cdot}0 \sin(18{\cdot}85t)$ mm. The weight of each floor is 20·0 kN/m, and the flexural rigidity of each of the columns is $EI = 87\,311{\cdot}477\,\text{kN}\,\text{m}^2$. Assume the response of the building to be the same as for an equivalent one DOF mass–spring system.

## References

Clough, R. W. & Penzien, J. *Dynamics of structures*. McGraw-Hill, London, 1975.

Craig, R. R. Jr. *Structural dynamics*. Wiley, Chichester, 1981.

Harris, C. M. *Shock vibration*, 3rd edn. McGraw-Hill, London, 1988.

Irvine, H. M. *Structural dynamics for the practising engineer*. Allen & Unwin, London, 1986.

Paz, M. *Structural dynamics*. Van Nostrand Reinhold, New York, 1980.

Stroud, K. A. *Engineering mathematics*. Macmillan, London, 1970.

# 5. Evaluation of equivalent viscous damping coefficients by harmonic excitation

## Introduction

Chapter 3 shows how the logarithmic decrement $\delta$ of viscous damping of a one DOF system can be determined from the decay function of free vibration by use of eq. (3.36) and plotting of $\ln(x_0/x_n)$ against the number of oscillations $n$, and that the relationship between the logarithmic decrement of damping and damping ratio is given by

$$\xi = \frac{\delta}{\sqrt{(4\pi^2 + \delta^2)}} \tag{5.1}$$

which, since $\delta$ is usually very much smaller than $4\pi^2$, may be written

$$\xi = \frac{\delta}{2\pi} \tag{5.2}$$

This chapter discusses various methods by which the viscous and equivalent viscous damping coefficients can be determined through harmonic excitation of a structure. The following methods are presented for the evaluation of damping.

- amplification of the static response at resonance
- vibration at resonance: (*a*) balancing of the maximum input and damping forces; (*b*) measurement of energy loss per cycle
- frequency sweep to obtain response functions, using (*a*) the bandwidth method, (*b*) amplitude ratios to obtain values for damping ratios at various points along the response curve, or (*c*) equivalent linear viscous response functions to calculate stiffnesses and damping ratios.

## Evaluation of damping from amplification of static response at resonance

For weakly damped structures, whose stiffness is known and whose maximum response $x_{n0}$ at resonance occurs when the frequency ratio $r$ is approximately equal to 1, the damping ratio $\xi$ is most easily obtained by measuring the response amplitude at resonance. From eq. (4.14) it follows that

$$\xi = \frac{x_{st}}{2x_{n0}} = \frac{P_0}{2Kx_{n0}} \tag{5.3}$$

For structures that possess a higher level of damping, eq. (5.1) will lead to an underestimation of the damping ratio as the maximum amplitude of response will have been reached before the frequency ratio is equal to 1. This underestimate will increase with increasing degrees of damping, as can be seen from Fig. 4.3. However, as the level of damping encountered in most structures is relatively low, the use of eq. (5.1) will in most cases be satisfactory.

## Vibration at resonance

### *Evaluation of viscous damping by balancing the maximum input and damping forces*

Another method of determining the damping of a structure is to vibrate it at resonance and then equate the maximum exciting and damping forces. The equivalent viscous damping coefficient can then be calculated by use of the first relationship given in eq. (4.32). This yields

$$C = \frac{P_0}{x_{n0}\omega_n} \tag{5.4}$$

$$\xi = \frac{P_0}{2Mx_{n0}\omega_n^2} \tag{5.5a}$$

or

$$\xi = \frac{P_0}{2Kx_{n0}} \tag{5.5b}$$

Equation (5.5b) may be more convenient if the equipment available is sufficiently sensitive to measure the response near zero frequency, as it permits a value for $K$ to be obtained without any static testing. The method requires that the instruments used be sufficiently sensitive to keep the phase angle $\alpha$ at steady-state response equal to $\pi/2$, and is accurate only when the damping is linearly viscous, in which case a graph of the exciting or input force plotted against the displacement will yield an ellipse—as shown by the broken line in Fig. 5.1, of area

$$A_n = \pi P_0 x_{n0} \tag{5.6}$$

### *Evaluation of equivalent viscous damping by measurement of energy loss per cycle*

If the damping mechanism does not possess a linear viscous characteristic, the graph resulting from plotting of the input force against the amplitude at steady-state response will be a curve similar to the solid line

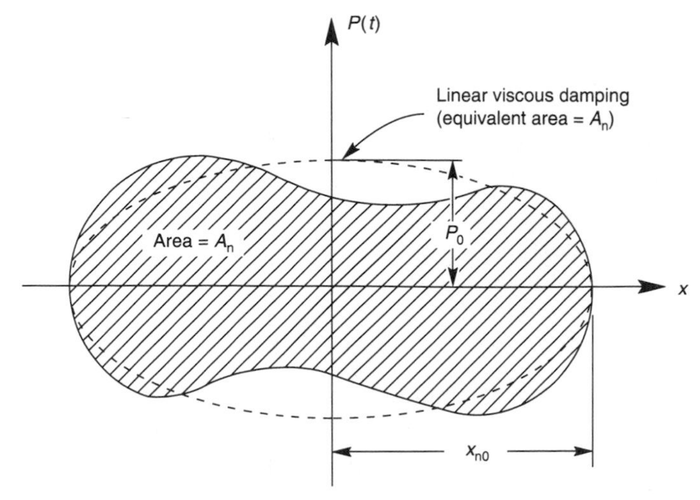


*Fig. 5.1. Input force plotted against displacement for linear (broken line) and nonlinear (soling line) viscous damping*

in Fig. 5.1. If the area enclosed by this curve is also denoted $A_n$, then the equivalent maximum force amplitude is, from eq. (5.6), given by

$$P_{e0} = \frac{A_n}{\pi x_{n0}} \tag{5.7}$$

Substitution of this value for the equivalent maximum input force into eq. (5.4) yields

$$C = \frac{A_n}{\pi x_{n0}{}^2 \omega_n} \tag{5.8}$$

$$\xi = \frac{A_n}{2\pi M x_{n0}{}^2 \omega_n{}^2} \tag{5.9a}$$

or

$$\xi = \frac{A_n}{2\pi K x_{n0}{}^2} \tag{5.9b}$$

## Evaluation of damping from response functions obtained by frequency sweeps

Another technique much used for measuring damping is based on frequency sweeps past the point where resonance occurs by construction of a frequency response curve, where each successive point is obtained from steady-state response after an incremental increase in the frequency of a vibrator. This procedure will lead to curves of the type shown in Fig. 4.3, where values of $x_{max}/x_{st} = \mathrm{MF}$ are plotted against the frequency ratio $r$. An examination of the curves in Fig. 4.3 shows that both the magnification factor and the general shape of the curves are functions of the level of damping in a structure. In particular it can be noted that the difference between two frequencies corresponding to a given magnification factor, referred to as the *bandwidth*, is a function of the degree of damping. Three different methods are now presented for determination of the damping of structures and structural elements from the response function, the first of which is based on a bandwidth corresponding to a specific magnification factor or response amplitude.

### Bandwidth method

In evaluation of damping by the bandwidth, method, it has been found convenient to measure the two frequencies and the distance between them at points where the amplitudes of the magnification factor or amplitudes of response, as shown in Fig. 5.1, are equal to $1/\sqrt{2}$ of the peak amplitude.

From eq. (4.15), the maximum responses when the frequency ratio $r$ is equal and not equal to 1 are respectively

$$x_{n0} = \frac{x_{st}}{2\xi} \tag{5.10}$$

$$x_0 = \frac{x_{st}}{\sqrt{\left[(1-r^2)^2 + (2\xi r^2)^2\right]}} \tag{5.11}$$

Thus the frequency ratio $r$ at which the amplitudes are equal to $1/\sqrt{2}$ of the amplitude when $r = 1$ can be determined by solving the equality

$$\frac{x_{st}}{\sqrt{\left[(1-r^2)^2 + (2\xi r)^2\right]}} = \frac{1}{\sqrt{2}}\frac{x_{st}}{2\xi} \tag{5.12}$$

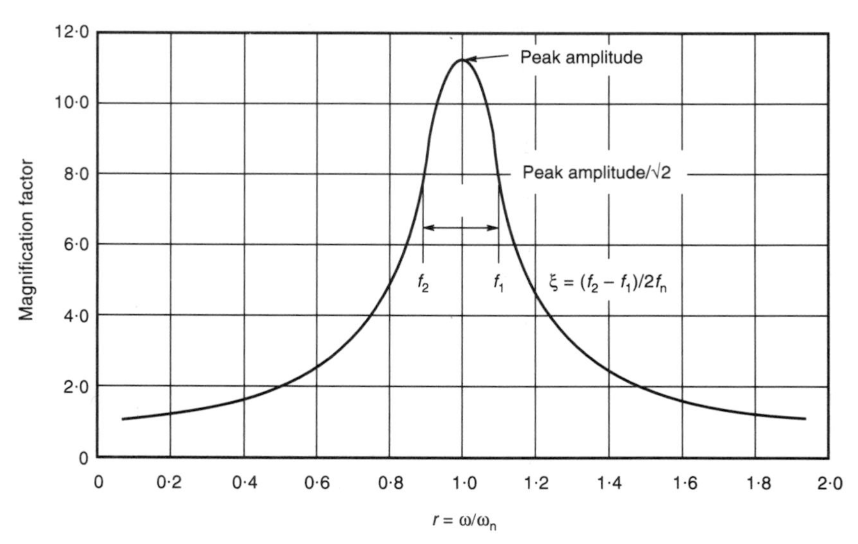


*Fig. 5.2. Frequency response curve showing the bandwidth at $1/\sqrt{2}$ of amplitude at $r = 1$*

Squaring both sides of eq. (5.12) and solving for $r^2$ yields

$$r^2 = 1 - 2\xi^2 \pm 2\xi\sqrt{(1 - \xi^2)} \tag{5.13}$$

Since usually $\xi^2 <<< 1{\cdot}0$, eq. (5.13) may be written as

$$r_1{}^2 = 1 - 2\xi^2 + 2\xi \tag{5.14a}$$

$$r_2{}^2 = 1 - 2\xi^2 - 2\xi \tag{5.14b}$$

Subtraction of eq. (5.14b) from eq. (5.14a) yields

$$4\xi = r_1{}^2 - r_2{}^2 = (r_1 + r_2)(r_1 - r_2) \tag{5.15}$$

or

$$4\xi\omega_n = (\omega_1 + \omega_2)(\omega_1 - \omega_2) \tag{5.16}$$

For weakly damped structures it may be assumed that

$$\omega_n = \tfrac{1}{2}(\omega_1 + \omega_2) \tag{5.17}$$

Substitution of this expression for $\omega_n$ into eq. (5.16) yields

$$\xi = \frac{\omega_1 - \omega_2}{2\omega_n} = \frac{\Delta\omega}{2\omega_n} \tag{5.18a}$$

or

$$\xi = \frac{f_1 - f_2}{2f_n} = \frac{\Delta f}{2f_n} \tag{5.18}$$

It should be noted that the size of the frequency step required in a frequency sweep in order to plot a steady-state response curve accurately, especially between $f_1$ and $f_2$, will depend on both the level of damping and the natural frequency of the structure. When, for example, $\xi = 0{\cdot}01$ and $f_n = 2{\cdot}0$ Hz, and the bandwidth is equal to

$$f_1 - f_2 = 2 \times 0{\cdot}01 \times 2{\cdot}0 = 0{\cdot}04\,\text{Hz}$$

a frequency step as small as, say, 0·004 Hz may be necessary to plot a satisfactory curve. If, however, the level of damping is higher and/or the natural frequency is greater, a larger step may be used. Thus, if $\xi 0{\cdot}05$ and $f_n = 20{\cdot}0$ Hz and hence the bandwidth is

$$f_1 - f_2 = 2 \times 0{\cdot}05 \times 20{\cdot}0 = 2{\cdot}0\,\text{Hz}$$

a frequency step of, say 0·2 Hz may be sufficient.

The expression for the damping ratio given by eq. (5.18a) or eq. (5.18b) assumes that the response curve, shown in Fig. 5.2, is obtained by vibrating a structure or structural element with a pulsating force $P(t) = P_0 \sin(\omega t)$, where $P_0$ is constant. This will not be the case if the vibrator consists of a motor with a rotating eccentric mass such as shown in Fig. 4.6. When this is the case eq. (5.12) must be modified and written as

$$\frac{me\omega^2}{K} = \frac{1}{\sqrt{\left[(1-r^2)^2 + (2\xi r)^2\right]}} = \frac{1}{\sqrt{2}}\frac{me\omega_n^2}{K}\frac{1}{2\xi} \qquad (5.19)$$

Simplification and rearrangement of eq. (5.19) yields

$$(1-8\xi^2)r^4 - (2-4\xi^2)r^2 + 1 = 0 \qquad (5.20)$$

which, solved with respect to $r^2$, yields

$$r^2 = \frac{1-2\xi^2 \pm 2\xi\sqrt{(1-\xi^2)}}{1-8\xi^2} \qquad (5.21)$$

If, as previously, it is assumed that

$$\sqrt{(1+\xi^2)} = 1$$

then

$$r_1^2 = \frac{1-2\xi^2+2\xi}{1-8\xi^2} \qquad (5.22a)$$

$$r_2^2 = \frac{1-2\xi^2-2\xi}{1-8\xi^2} \qquad (5.22b)$$

Subtraction of eq. (5.22b) from eq. (5.22a) yields

$$r_1^2 - r_2^2 = \frac{4\xi}{1-8\xi^2} \qquad (5.23)$$

which, solved with respect to $\xi$, gives

$$\xi = \frac{-1 \pm \sqrt{(1+8\Delta f/f_n)}}{8\Delta f/f_n} \qquad (5.24)$$

If the square root is expanded by the binomial theorem, ignoring the cubic and higher terms, and rejecting the negative sign in front of the square root for obvious reasons, then

$$\xi = \frac{\Delta f}{2f_n} + \frac{1}{8}\left(\frac{\Delta f}{f_n}\right)^2 \qquad (5.25)$$

The second term in eq. (5.25) results in increases of 0·25%, 1·25%, 2·5% and 5·0% in $2\xi$ when $\Delta f/f_n = 0{\cdot}01$, 0·05, 0·10 and 0·20 respectively.

### *Amplitude ratios*

Response curves similar to those shown in Figs 4.3 and 5.1 can sometimes be difficult to obtain because of limitations in the equipment available to perform the frequency sweeps. In such cases the damping ratios may be determined as follows.

Let the maximum response at resonance due to $P(t) = P_0 \sin(\omega_n t)$ be

$$x_n = \frac{P_0}{K}\frac{1}{2\xi} \qquad (5.26)$$

and that due to $P(t) = P_0 \sin(\omega t)$ be

$$x = \frac{P_0}{K}\frac{1}{\sqrt{\left[(1-r^2)^2 + (2\xi r)^2\right]}} \qquad (5.27)$$

Elimination of $P_0/K$ from eqs (5.26) and (5.27) leads to

$$\frac{x_{n0}}{x_0} = \frac{\sqrt{\left[(1-r^2)^2+(2\xi r)^2\right]}}{2\xi} \qquad (5.28)$$

which, when solved with respect to $\xi$, yields

$$\xi = \frac{1-r^2}{2\sqrt{(\gamma^2 - r^2)}} \qquad (5.29)$$

where

$$\gamma = \frac{x_n}{x}$$

For vibrators with eccentric masses, for which $P(t) = me\omega^2 \sin(\omega t)$, the expression for $\xi$ can be shown to be

$$\xi = \frac{1-r^2}{2r\sqrt{(\gamma^2 r^2 - 1)}} \qquad (5.30)$$

The percentage error caused by determination of the damping ratio using eq. (5.30) corresponds to the errors resulting from not including the second term in eq. (5.25).

### *Calculation of stiffness and damping ratios from an equivalent linear viscous response function*

The evaluation of damping ratios from decay and response functions will result in the same values only if the damping is linearly viscous. Generally this in not the case, and the methods described so far and based on such functions will usually yield different values, which often differ considerably. An alternative method for measuring nonlinear viscous damping, if the instruments available are not suitable for measuring the energy lost per cycle, is to undertake a frequency sweep and then establish an equivalent theoretical linear viscous response function as shown in Fig. 5.3.

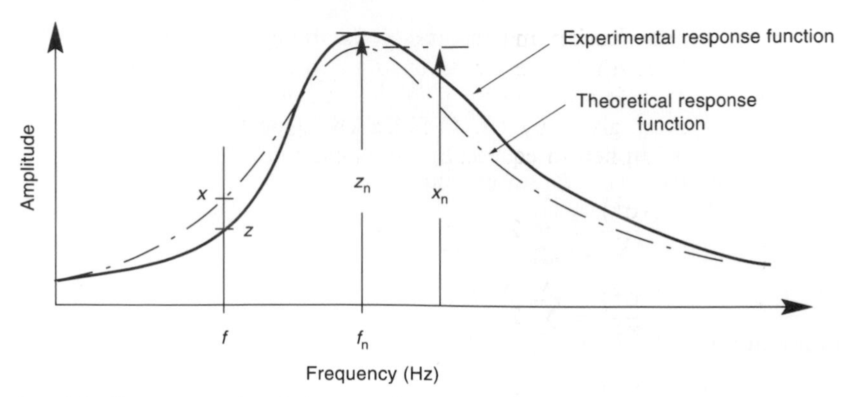


*Fig. 5.3. Experimental and theoretical response functions*

At a frequency $f$, let the experimental amplitude be $z$ and the theoretical one $x$. Thus the variance of the area between the two curves is given by

$$\sigma^2 = \sum_{n-1}^{N} (x - z)^2 \, \delta f \tag{5.31}$$

where $z$ is the experimental amplitude and $x$ is the theoretical response

$$x = \frac{P_0}{K} \frac{1}{\sqrt{\left[(1 - r^2)^2 + (\eta r)^2\right]}} \tag{5.32}$$

$$\eta = 2\xi \tag{5.33}$$

It is now assumed that the best equivalent theoretical linear response curve is the one for which the variance of the area between the two curves is a minimum with respect to both $K$ and $\eta$, i.e. when the gradient

$$g = \begin{bmatrix} \partial(\sigma^2)/\partial K \\ \partial(\sigma^2)/\partial \eta \end{bmatrix} = 0 \tag{5.34}$$

This condition can be achieved through the iterative process

$$\begin{bmatrix} K_{i+1} \\ \eta_{i+1} \end{bmatrix} = \begin{bmatrix} K_i + \delta K_i \\ \eta_i + \delta \eta_i \end{bmatrix} \tag{5.35}$$

by expanding the gradient vector at the $i$th iteration through a Taylor series, ignoring cubic and higher order terms and assuming that the gradient vector at the $(i + 1)$th iteration is zero. This yields

$$\begin{bmatrix} \delta K \\ \delta \eta \end{bmatrix}_i = - \begin{bmatrix} \partial^2(\sigma^2)/\partial k^2 & \partial^2(\sigma^2)/\partial K \partial \eta \\ \partial^2(\sigma^2)/\partial K \partial \eta & \partial^2(\sigma^2)/\partial \eta^2 \end{bmatrix}_i^{-1} \begin{bmatrix} \partial(\sigma^2)/\partial K \\ \partial(\sigma^2)/\partial \eta \end{bmatrix}_i \tag{5.36}$$

which may also be written as

$$\delta\{\boldsymbol{K}, \eta\} = \mathbf{H}_i^{-1} \boldsymbol{g}_i \tag{5.37}$$

where $\mathbf{H}_i$ is the Hessian matrix and $\boldsymbol{g}_i$ is the gradient vector of the variance $\sigma^2$ at the $i$th iterate, or point $(K_i, \eta_i)$ in optimization space.

The elements in $\mathbf{H}$ and $\boldsymbol{g}$ in eq. (5.37) are found by differentiating the expression for $\sigma^2$ given by eq. (5.31) with respect to $K$ and $\eta$. The differentiations implied in eq. (5.36), when carried out, yield

$$\frac{\partial(\sigma^2)}{\partial K} = \sum_{i=1}^{N} 2(x - z)\frac{\partial x}{\partial K}\,\delta f \tag{5.38a}$$

$$\frac{\partial^2(\sigma^2)}{\partial K^2} = \sum_{i=1}^{N} 2\left(\frac{\partial x}{\partial K}\frac{\partial x}{\partial K} + (x - z)\right)\frac{\partial^2 x}{\partial K^2}\,\delta f \tag{5.38b}$$

$$\frac{\partial(\sigma^2)}{\partial \eta} = \sum_{i=1}^{N} 2(x - z)\frac{\partial x}{\partial \eta}\,\delta f \tag{5.38c}$$

$$\frac{\partial^2(\sigma^2)}{\partial \eta^2} = \sum_{i=1}^{N} 2\left(\frac{\partial x}{\partial \eta}\frac{\partial x}{\partial \eta} + (x - z)\right)\frac{\partial^2 x}{\partial \eta^2}\,\delta f \tag{5.38d}$$

$$\frac{\partial^2(\sigma^2)}{\partial K\partial \eta} = \sum_{i=1}^{N} 2\left(\frac{\partial x}{\partial K}\frac{\partial x}{\partial \eta} + (x - z)\right)\frac{\partial^2 x}{\partial K\partial \eta}\,\delta f \tag{5.38e}$$

where the expressions for the partial differentials are obtained by differentiation of eq. (5.32) with respect to $K$ and $\eta$. Thus

$$\frac{\partial x}{\partial K} = -\frac{P_0}{K^2}\frac{1}{\sqrt{\left[(1 - r^2)^2 + (\eta r)^2\right]}} \tag{5.39a}$$

$$\frac{\partial^2 x}{\partial K^2} = \frac{2P_0}{K^3}\frac{1}{\sqrt{\left[(1 - r^2)^2 + (\eta r)^2\right]}} \tag{5.39b}$$

$$\frac{\partial x}{\partial \eta} = -\frac{P_0}{K}\frac{r^2\eta}{\left\{\sqrt{\left[(1 - r^2)^2 + (\eta r)^2\right]}\right\}^3} \tag{5.39c}$$

$$\frac{\partial^2 x}{\partial \eta^2} = \frac{3P_0}{K}\frac{r^4\eta^2}{\left\{\sqrt{\left[(1 - r^2)^2 + (\eta r)^2\right]}\right\}^5} - \frac{P_0}{K}\frac{r^2}{\left\{\sqrt{\left[(1 - r^2)^2 + (\eta r)^2\right]}\right\}^3} \tag{5.39d}$$

$$\frac{\partial^2 x}{\partial K\partial \eta} = \frac{P_0}{K^2}\frac{r^2\eta}{\left\{\sqrt{\left[(1 - r^2)^2 + (\eta r)^2\right]}\right\}^3} \tag{5.39e}$$

The iterative process described by eq. (5.35) has converged when $\boldsymbol{g}^{\mathrm{T}}\boldsymbol{g} = 0$, where the elements in $\boldsymbol{g}$ are found by substitution of the latest update of $K$ and $\eta$ in eqs (5.38a and 5.38c) and (5.39a and 5.39c).

The determination of an equivalent viscous response curve will in general lead to a value for $x_{n0}$ that is slightly different from $z_{n0}$. Thus, calculations using the value for the damping ratio $\xi = \eta/2$ to model the damping mechanism in the structure tested would lead to a different maximum amplitude for a given exciting force than the one that would be obtained experimentally. To overcome the problem it is suggested that the calculation of the damping ratio be modified as follows

$$\xi = \eta z_n / 2x_n \tag{5.40}$$

The justification for this is that at resonance eq. (5.32) would yield

$$z_{n0} = \frac{P_0}{K}\frac{1}{2\xi} \tag{5.41}$$

Hence

$$\frac{P_0}{2\xi} = K z_{n0} \tag{5.42}$$

Thus the calculated values for $K$ and $\xi$ can be verified by plotting $P_0/2\xi$ against $z_{n0}$ for different values of $P_0$ and comparing the resulting graph with one obtained from an ordinary static load test.

The method presented does not require instruments that can read or maintain the phase angle $\alpha$ at 90° at resonance, but it does require a computer than can store and analyse the data from frequency sweeps.

## Hysteretic damping

The expression for the viscous damping coefficient given by eq. (3.22) leads to the expression for the damping force

$$F_d = C\dot{x} = 2\xi\omega_n M\dot{x} \tag{5.43}$$

which shows that the damping force is a function not only of the mass and velocity of vibration, but also of the natural frequency of the structure. This contradicts a great deal of experimental evidence, which indicates that the damping often is very nearly independent of the mode frequency. A frequency-independent damping model is the *hysteretic* damping model, where the damping force is proportional to the stiffness and displacement of the structure, but in phase with the velocity. Mathematically the force may be expressed as

$$F_d = C_h K|x|\frac{\dot{x}}{|\dot{x}|} = C_h K\sqrt{\left[\left(\frac{x}{\dot{x}}\right)^2\right]}\dot{x} \tag{5.44}$$

The force displacement diagram for this form of damping for one cycle of vibration at resonance is shown in Fig. 5.4 in which the shaded area representing the energy lost per cycle is

$$A_n = 2C_h K x_{n0}{}^2 \tag{5.45}$$

Substitution of this expression for $A_n$ into eq. (5.9b) yields

$$C_h = \pi\xi \tag{5.46}$$

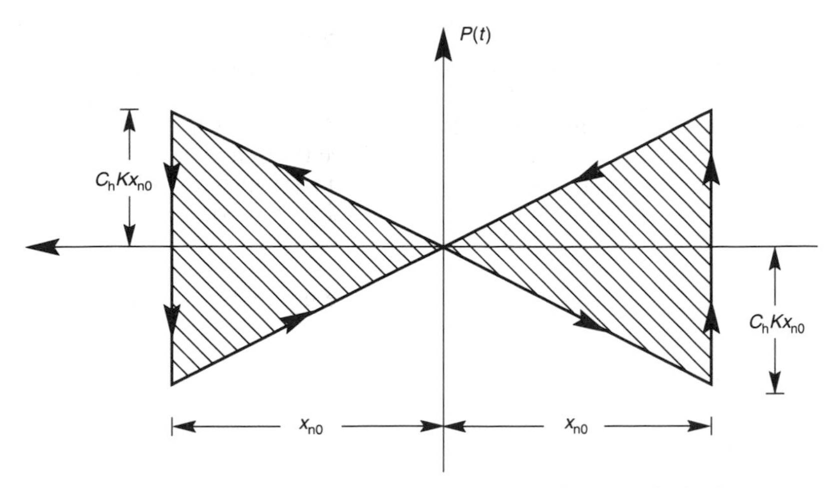


*Fig. 5.4. Assumed variation in hysteretic damping force with displacement at resonance*

which is independent of the mode frequency. The equation of motion for a one DOF system with damping independent of the frequency is therefore given by

$$M\ddot{x} + \pi\xi K\sqrt{\left[\left(\frac{x}{\dot{x}}\right)^2\right]}\dot{x} + Kx = P(t) \tag{5.47}$$

The solution of eq. (5.47) requires an iterative solution method, as the damping is a function of both displacement and velocity.

**Problem 5.1** The equivalent lumped mass of a shear structure is 100 t, the first natural frequency 3·0 Hz, and the damping 5·0% of critical. Plot the curves for the amplitude response and phase angle for a frequency sweep from 0·0 to 6·0 Hz when the building is vibrated by a variable speed motor with a mass of 5·0 kg at 50 cm eccentricity.

**Problem 5.2** Having plotted the response function for the structure in problem 5.1, verify the correctness of the curve plotted by calculating the damping using (*a*) the bandwidth method and (*b*) the amplitude ratio method. In the latter case calculate the two values for damping ratios by selecting one point on the curve corresponding to frequency of 2·0 Hz, and another corresponding to a frequency of 4·0 Hz. (*c*) What is the percentage error in the calculated values if the maximum pulsating force is assumed to be constant and equal to $me\omega_n$?

**Problem 5.3** Calculate the energy lost per cycle at resonance, if the damping is assumed to be linearly viscous, when the structure in problem 5.1 is vibrated by a pulsating force $P(t) = 1000\sin(\omega t)$ N. Calculate also the equivalent linear viscous damping ratio if a plot of the exciting force against the amplitude of response at resonance, during one cycle, is a rectangle and the maximum amplitude of response is 1·5 mm. Hence calculate the equivalent maximum damping force at resonance.

## References

Clough, R. W. & Penzien, J. *Dynamics of structures*. McGraw-Hill, London, 1975.

Harris, C. M. *Shock vibration*, 3rd edn. McGraw-Hill, London, 1988.

Stroud, K. A. *Engineering mathematics*. Macmillan, London, 1970.

# 6. Response of linear and nonlinear one degree of freedom systems to random loading: time domain analysis

## Introduction

The response of linear one DOF systems to harmonic excitation is presented in chapter 4 in terms of closed-form solutions. This chapter considers the response of the same type of equivalent mass–spring systems to random forms of loading, and extends the solution methods to include nonlinear structures. Examples of random types of loading are wind, waves and earthquakes, and examples of nonlinear structures are suspension bridges, cable-stayed footbridges and canopies, guyed masts, and cable and membrane roofs. Even structures that are regarded as linear may exhibit nonlinear characteristics if subjected to strong excitation. Nonlinear structures can be classified as either stiffening or softening. In the case of stiffening structures the rate of change of displacements will reduce with increasing deformation. In the case of stiffening structures the rate of change of displacements will reduce with increasing deformation. In the case of softening structures the reverse will be the case. Figure 6.1 shows typical load displacement curves for linear and nonlinear stiffening and softening one DOF structures.

When the stiffness of a structure varies with the amplitude of response, it follows that the natural frequencies will also vary. For this reason the closed-form solutions presented in chapter 4 are no longer valid. Figure 6.2 shows examples of frequency response curves for stiffening and softening one DOF systems to harmonic excitation.

The damping, in general, will also vary with the amplitude of vibration. Generally, however, structural damping is assumed to remain constant because of lack of information, and because the values given in codes of practice tend to be conservative. Even after extensive testing it is generally possible to produce only an approximate numerical model of a structural damping mechanism based on the damping ratios obtained from vibration in a few of the lower modes. The response of linear structures to random forms of loading such as wind and earthquakes may be carried out in the frequency domain using power spectra, as described

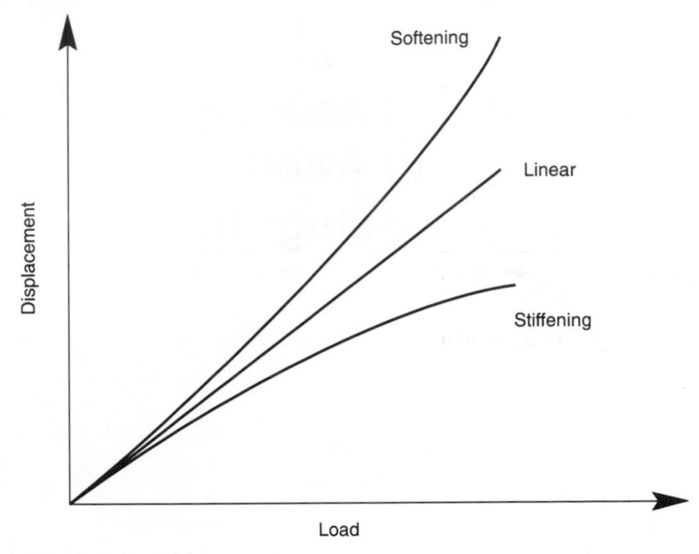


*Fig. 6.1. Load–displacement curves for linear and nonlinear one DOF structures*

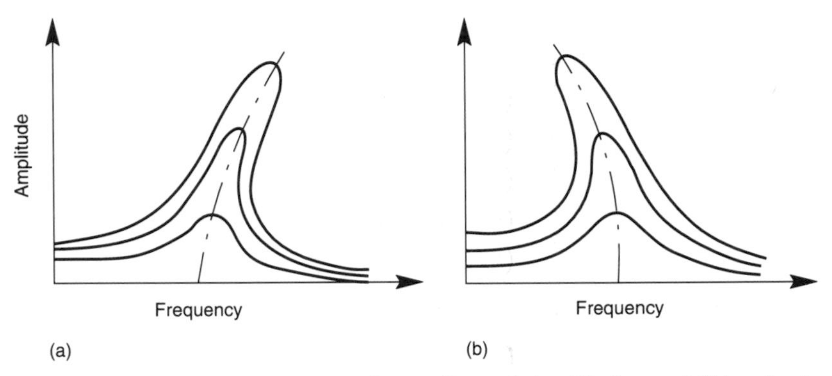


*Fig. 6.2. Frequency response curves for nonlinear (a) stiffening and (b) softening one DOF systems*

in chapters 10 and 12, that enable the use of closed-form solutions. For nonlinear structures this approach will underestimate the response in the case of softening structures and overestimate it in the case of stiffening ones. The general approach for predicting the behaviour of nonlinear structures to all types of dynamic loading is to predict the response by a forward integration in the time domain. Several such methods are presented and discussed in the following sections.

## Step by step integration methods

Let the force–time curve shown in Fig. 6.3(a) represent the variation of a random force $P(t)$ acting on a one DOF system, and the displacement–time curve in Fig. 6.3 represent the resulting dynamic response.

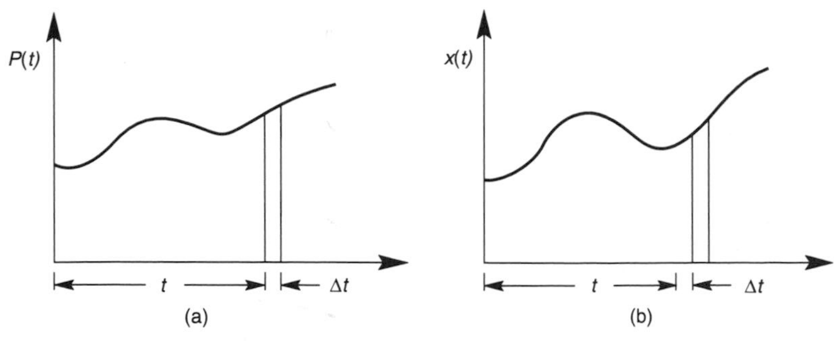


*Fig. 6.3. Variation of (a) random force P(t) and (b) response x(t) with time*

At time $t$ let

$$K(t) = K$$
$$P(t) = P$$
$$x(t) = x$$
$$\dot{x}(t) = \dot{x}$$
$$\ddot{x}(t) = \ddot{x}$$

and at time $(t + \Delta t)$ let

$$K(t + \Delta t) = K + \Delta K$$
$$P(t + \Delta t) = P + \Delta P$$
$$x(t + \Delta t) = x + \Delta x$$
$$\dot{x}(t + \Delta t) = \dot{x} + \Delta \dot{x}$$
$$\ddot{x}(t + \Delta t) = \ddot{x} + \Delta \ddot{x}$$

Thus at time $t$ the equation of motion is

$$M\ddot{x} + C\dot{x} + Kx = P \tag{6.1}$$

and at time $(t + \Delta t)$

$$M(\ddot{x} + \Delta\ddot{x}) + (C + \Delta C)(\dot{x} + \Delta\dot{x}) + (K + \Delta K)(x + \Delta x) = P + \Delta P \tag{6.2}$$

Subtraction of eq. (6.1) from eq. (6.2) yields

$$M\Delta\ddot{x} + C\Delta\dot{x} + \Delta C(\dot{x} + \Delta\dot{x}) + K\Delta x + \Delta K(x + \Delta x) = \Delta P \tag{6.3}$$

In practice it has been found sufficient to let the damping and the stiffness coefficients remain constant during each time step $\Delta t$, and update them only at the end of each step. Thus the terms $\Delta C(\dot{x} + \Delta\dot{x})$ and $\Delta K(x + \Delta x)$ may be ignored, and eq. (6.3) reduces to

$$M\Delta\ddot{x} + C\Delta\dot{x} + K\Delta x + \Delta P \tag{6.4}$$

which is referred to as the *incremental equation of motion* and can be solved only if there exists a relationship between $\Delta\ddot{x}$, $\Delta\dot{x}$ and $\Delta x$.

A number of such relationships are proposed in the literature. Here, only the three most commonly used are considered

- the linear change of acceleration method
- the Wilson θ-method
- the constant acceleration method.

### Linear change of acceleration method

In this method the acceleration during each time step is assumed to vary linearly as shown in Fig. 6.4, from which the slope at time $(t+\tau)$ is seen to be constant and can be written as

$$\dddot{x}(t+\tau)=\frac{\Delta\ddot{x}}{\tau}=A \tag{6.5}$$

$$\ddot{x}(t+\tau)=A\tau+B \tag{6.6}$$

$$\dot{x}(t+\tau)=\tfrac{1}{2}A\tau^2+B\tau+C \tag{6.7}$$

$$x(t+\tau)=\frac{1}{6}A\tau^3+\tfrac{1}{2}B\tau^2+C\tau+D \tag{6.8}$$

The constant $A$ is given by eq. (6.5), and the constants $B$, $C$ and $D$ may be determined by the condition that when $\tau=0$

$$\ddot{x}(t+\tau)=\ddot{x}(t)=\ddot{x}$$

$$\dot{x}(t+\tau)=\dot{x}(t)=\dot{x}$$

$$x(t+\tau)=x(t)=x$$

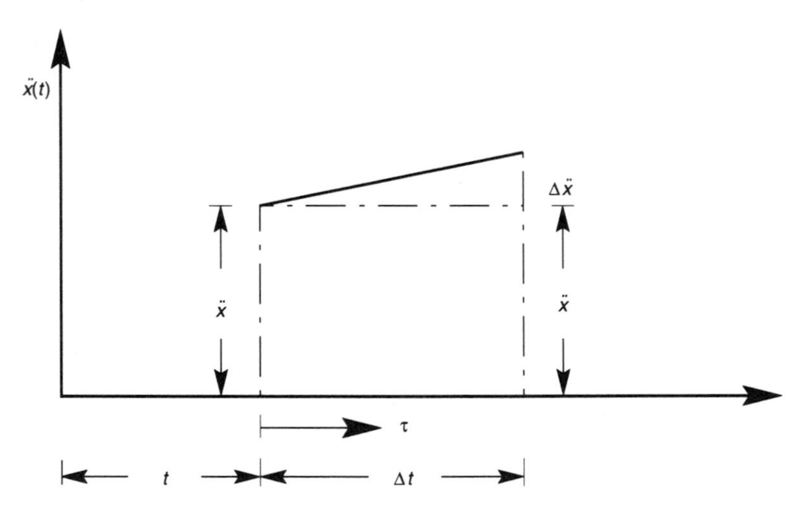


*Fig. 6.4. Assumed change in acceleration during a time step Δt in the linear acceleration method*

Hence

$$B = \ddot{x}$$
$$C = \dot{x}$$
$$D = x$$

Thus at time $(t + \Delta t)$

$$\dot{x} + \Delta\dot{x} = \tfrac{1}{2}\Delta\ddot{x}\Delta t + \ddot{\Delta} t + \dot{x} \tag{6.9}$$

$$x + \Delta x = \tfrac{1}{6}\Delta\ddot{x}\Delta t^3 + \tfrac{1}{2}\ddot{x}\Delta t^2 + \dot{x}\Delta t + x \tag{6.10}$$

from which the following expression is obtained

$$\Delta\ddot{x} = \frac{6}{\Delta t^2}\Delta x - \frac{6}{\Delta t}\dot{x} - 3\ddot{x} \tag{6.11}$$

Substitution of this expression for $\Delta\ddot{x}$ into eq. (6.9) yields

$$\Delta\dot{x} = \frac{3}{\Delta t}\Delta x - 3\dot{x} - \frac{1}{2}\ddot{x}\Delta t \tag{6.12}$$

Substitution of the expressions for $\Delta\dot{x}$ and $\Delta\ddot{x}$ given by eqs (6.11) and (6.12) into eq. (6.4) leads to the following formulation of the incremental equation of motion for a one DOF system

$$\left(K + \frac{3}{\Delta t}C + \frac{6}{\Delta t^2}M\right)\Delta x$$
$$= \Delta P + C\left(3\dot{x} + \frac{\Delta t}{2}\ddot{x}\right) + M\left(\frac{6}{\Delta t}\dot{x} + 3\ddot{x}\right) \tag{6.13}$$

or

$$\Delta x = K_{\mathrm{d}}^{-1}\Delta P_{\mathrm{d}} \tag{6.14}$$

where the *dynamic stiffness* $K_{\mathrm{d}}$ and the *equivalent dynamic load* $P_{\mathrm{d}}$ are given by

$$K_{\mathrm{d}} = K + \frac{3}{\Delta t}C + \frac{6}{\Delta t^2}M \tag{6.15}$$

$$\Delta P_{\mathrm{d}} = \Delta P + C\left(3\dot{x} + \frac{\Delta t}{2}\ddot{x}\right) + M\left(\frac{6}{\Delta t}\dot{x} + 3\ddot{x}\right) \tag{6.16}$$

Once $\Delta x$ has been calculated using eq. (6.14), the values for displacement, velocity and acceleration to be used at the beginning of the next time step are

$$x(t + \Delta t) = x + \Delta x \tag{6.17}$$

$$\dot{x}(t + \Delta t) = \frac{3}{\Delta t}\Delta x - 2\dot{x} - \frac{\Delta t}{2}\ddot{x} \tag{6.18}$$

$$\ddot{x}(t + \Delta t) = \frac{6}{\Delta t^2}\Delta x - \frac{6}{\Delta t}\dot{x} - 2\ddot{x} \tag{6.19}$$

The linear acceleration method tends to become unstable if $\Delta t > T/2$, where $T = 1/f$ is the period of natural vibration. Instability is, however, not usually a problem in the case of one DOF systems, where $\Delta t$ needs to be less than, say, $T/10$ to ensure sufficient accuracy in the predicted response.

### *Wilson θ-method*

In the Wilson θ-method the acceleration is assumed to vary linearly during a prolonged time step $\theta\Delta t$, where $> 1{\cdot}0$ as shown in Fig. 6.5. Each new step, however, is started from time $(t + \Delta t)$ and not $(t + \theta\Delta t)$.

From eqs (6.11) and (6.12) it follows that at time $(t + \theta\Delta t)$

$$\Delta\ddot{x} = \frac{6}{\theta^2\Delta t^2}\Delta x - \frac{6}{\theta\Delta t}\dot{x} - 3\ddot{x} \tag{6.20}$$

$$\Delta\dot{x} = \frac{3}{\theta\Delta t}\Delta x - 3\dot{x} - \frac{1}{2}\theta\Delta t\ddot{x} \tag{6.21}$$

Substitution of these expressions for $\Delta\ddot{x}$ and $\Delta\dot{x}$ into eq. (6.4) yields

$$\begin{aligned}&\left(K + \frac{3}{\theta\Delta t}C + \frac{6}{\theta^2\Delta t^2}M\right)\Delta x\\ &= \Delta P + C\left(3\dot{x} + \frac{\theta\Delta t}{2}\ddot{x}\right) + M\left(\frac{6}{\theta\Delta t}\dot{x} + 3\ddot{x}\right)\end{aligned} \tag{6.22}$$

or

$$\Delta x = K_{\mathrm{d}}^{-1}\Delta P_{\mathrm{d}} \tag{6.23}$$

where $\Delta x$ is the incremental displacement at the end of the time step $\theta\Delta t$, and

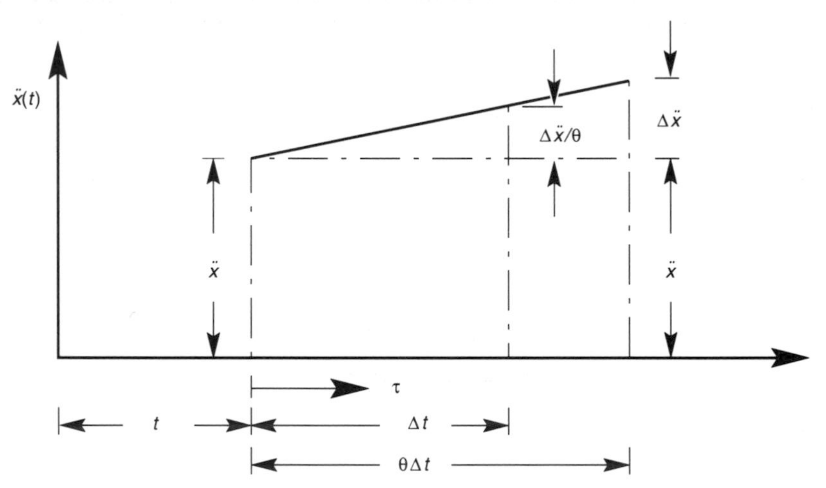


*Fig. 6.5. Assumed linear acceleration during time step $(t + \theta\Delta t)$ in the Wilson θ-method.*

$$K_{\mathrm{d}} = K + \frac{3}{\theta \Delta t} C + \frac{6}{\theta^2 \Delta t^2} M \tag{6.24}$$

$$\Delta P_{\mathrm{d}} = \Delta P + C\left(3\dot{x} + \frac{\theta \Delta t}{2} \ddot{x}\right) + M\left(\frac{6}{\theta \Delta t} \dot{x} + \ddot{x}\right) \tag{6.25}$$

The acceleration, velocity and displacement at the beginning of the next time step, at time $(t + \Delta t)$, are found by inspection of Fig. 6.5, from which it can be deduced that at time $(t + \Delta t)$ the change in acceleration during the time interval $\Delta x$ is

$$\frac{\Delta \ddot{x}}{\theta} = \frac{6}{\theta^3 \Delta t^2} \Delta x - \frac{6}{\theta^2 \Delta t} \dot{x} - \frac{3}{\theta} \ddot{x} \tag{6.26}$$

Hence the acceleration at time $(t + \Delta t)$ is given by

$$\ddot{x}(t + \Delta t) = \frac{6}{\theta^3 \Delta t^2} \Delta x - \frac{6}{\theta^2 \Delta t} \dot{x} + \left(1 - \frac{3}{\theta}\right) \ddot{x} \tag{6.27}$$

The expressions for the velocity and acceleration at time $(t + \Delta t)$ are then found by substitution of the expression for $\Delta \ddot{x}/\theta$ given by eq. (6.26) into eqs (6.9) and (6.10) respectively. This yields

$$\dot{x}(t + \Delta t) = \frac{3}{\theta^3 \Delta t} \Delta + \left(1 + \frac{3}{\theta^2}\right) \dot{x} + \frac{3\Delta t}{2} \left(1 - \frac{1}{\theta}\right) \ddot{x} \tag{6.28}$$

$$x(t + \Delta t) = \frac{1}{\theta^2} \Delta x + x + \Delta t \left(1 - \frac{1}{\theta^2}\right) \dot{x} + \frac{\Delta t^2}{6} \left(4 - \frac{3}{\theta}\right) \ddot{x} \tag{6.29}$$

For linear structures the method is stable when $\theta \geq 1{\cdot}37$. In general a value of $\theta = 1{\cdot}4$ appears to be satisfactory. Values of $\theta$ much in excess of 1·4 result in an increasing overestimation of the predicted amplitude of response, combined with an increasing phase lag relative to the dynamic force.

### *Constant acceleration method*

In this method the acceleration is assumed to remain constant during the time step $\Delta t$ and equal to $(\ddot{x} + \frac{1}{2}\Delta \ddot{x})$, as shown in Fig. 6.6, where it is compared with the assumption of linear acceleration. From Fig. 6.6, the acceleration at time $(t + \tau)$ is

$$\ddot{x}(t + \tau) = \ddot{x} + \tfrac{1}{2} \Delta \ddot{x} \tag{6.30}$$

$$\dot{x}(t + \tau) = \ddot{x}\tau + \tfrac{1}{2} \Delta \ddot{x} \tau + A \tag{6.31}$$

$$x(t + \tau) = \tfrac{1}{2} \ddot{x} \tau^2 + \tfrac{1}{4} \Delta \ddot{x} \tau^2 + A\tau + B \tag{6.32}$$

When $\tau = 0$

$$\dot{x}(t + \tau) = \dot{x}(t) = \dot{x}$$

$$x(t + \tau) = x(t) = x$$

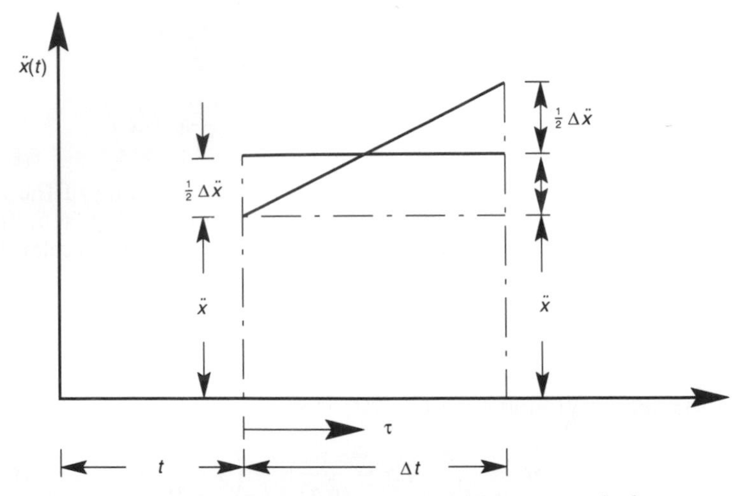


*Fig. 6.6. Assumed acceleration in the constant acceleration method*

Hence

$$A = \dot{x}$$
$$B = x$$

When $\tau = \Delta t$

$$\dot{x}(t+\tau) = \dot{x}(t+\Delta t) = \dot{x} + \Delta\dot{x}$$
$$x(t+\tau) = x(t+\Delta t) = x + \Delta x$$

Hence

$$\Delta\dot{x} = \Delta t\ddot{x} + \tfrac{1}{2}\Delta t\Delta\ddot{x} \tag{6.33}$$
$$\Delta x = \Delta t\dot{x} + \tfrac{1}{2}\Delta t^2\ddot{x} + \tfrac{1}{4}\Delta t^2\Delta\ddot{x} \tag{6.34}$$

from which

$$\Delta\ddot{x} = \frac{4}{\Delta t^2}\Delta x - \frac{4}{\Delta t}\dot{x} - 2\ddot{x} \tag{6.35}$$

Substitution of eq. (6.35) into eq. (6.33)

$$\Delta\ddot{x} = \frac{4}{\Delta t^2}\Delta x - 2\dot{x} \tag{6.36}$$

Substitution of the expressions for $\Delta\ddot{x}$ and $\Delta\dot{x}$ given by eqs (6.35) and (6.36) respectively into eq. (6.4) yields

$$\left(K + \frac{2}{\Delta t}C + \frac{4}{\Delta t^2}M\right)\Delta x = \Delta P + 2C\dot{x} + M\left(\frac{4}{\Delta t}\dot{x} + 2\ddot{x}\right) \tag{6.37}$$

or

$$\Delta x = K_{\mathrm{d}}^{-1}\Delta P_{\mathrm{d}} \tag{6.38}$$

where

$$K_d = K + \frac{2}{\Delta t}C + \frac{4}{\Delta t^2}M \tag{6.39}$$

$$\Delta P_d = \Delta P + 2C\dot{x} + M\left(\frac{4}{\Delta t}\dot{x} + 2\ddot{x}\right) \tag{6.40}$$

Having calculated the incremental displacement $\Delta x$ using eq. (6.38), the displacement, velocity and acceleration at the end of the time step at time $(t + \Delta t)$ can be calculated from

$$x(t + \Delta t) = x + \Delta x \tag{6.41}$$

$$\dot{x}(t + \Delta t) = \frac{2}{\Delta t}\Delta x - \dot{x} \tag{6.42}$$

$$\ddot{x}(t + \Delta t) = \frac{4}{\Delta t^2}\Delta x - \frac{4}{\Delta t}\dot{x} - \ddot{x} \tag{6.43}$$

The accuracy with which the linear acceleration, constant acceleration and Wilson θ-methods predict the response for a given load history depends on the size of the time step. This must be small enough to enable all the significant harmonic components in the load history to be taken into account, and should not be greater than 0·05 times the period of the one DOF system analysed. Experience indicates that with the same time step the constant acceleration and Wilson θ-methods, with $\theta = 1{\cdot}4$, yield similar results. The former may thus be preferable since less computational effort is required.

*The Newmark β-method*

Anybody who begins to study time domain methods for the dynamic analysis of linear and nonlinear structures will sooner or later come across references to the Newmark β-method. Newmark proposed the following expressions for the velocity and displacement at time $(t + \Delta t)$

$$\dot{x}(t + \Delta t) = \dot{x} + \Delta t(1 - \gamma)\ddot{x} + \Delta t\gamma(\ddot{x} + \Delta\ddot{x}) \tag{6.44}$$

$$x(t + \Delta t) = x + \Delta t\dot{x} + \Delta t^2\left(\tfrac{1}{2} - \beta\right)\ddot{x} + \Delta t^2\beta(\ddot{x} + \Delta\ddot{x}) \tag{6.45}$$

where $\gamma$ and $\beta$ are variable constants. Usually it is assumed that $\gamma = \frac{1}{2}$ while $\beta$ may be assigned different values. With $\gamma = \frac{1}{2}$ and $\beta = \frac{1}{6}$ or $\beta = \frac{1}{4}$ it can be shown that the Newmark β-method is identical to the linear acceleration or constant acceleration method respectively.

## Dynamic response to turbulent wind

Let the drag force due to wind per unit of velocity of the wind relative to that of the structure be $F_d$, and the corresponding wind velocities and velocities of the structure at times $t$ and $(t + \Delta t)$ be $V$, $V + \Delta V$, $\dot{x}$ and $\dot{x} + \Delta\dot{x}$ respectively. If it is assumed that the force due to wind is proportional to the square of the relative velocity of the wind to that of the structure, then the force exerted on the structure by the wind is

$$P = F_d(V - \dot{x})^2 \tag{6.46}$$

at time $t$, and

$$P + \Delta P = F_{\mathrm{d}}[(V + \Delta V) - (\dot{x} + \Delta \dot{x})]^2 \tag{6.47}$$

at time $(t + \Delta t)$. Subtraction of eq. (6.46) from eq. (6.47), ignoring the terms of second order of smallness, yields

$$\Delta P = 2F_{\mathrm{d}}[(V - \dot{x})\Delta V - (V - \dot{x})\Delta \dot{x}] \tag{6.48}$$

By using the constant acceleration method, from eq. (6.36)

$$\Delta \dot{x} = \frac{2}{\Delta t}\Delta x - 2\dot{x} \tag{6.49}$$

Hence

$$\Delta P = 2F_{\mathrm{d}}\left[(V - \dot{x})\Delta V + 2(v - \dot{x})\dot{x} - \frac{2}{\Delta t}(V - \dot{x})\Delta x\right] \tag{6.50}$$

Substitution of this expression for $\Delta P$ into eq. (6.37) yields the dynamic equation

$$\left[K + \frac{2}{\Delta t}C + \frac{4}{\Delta t^2}M + \frac{4}{\Delta t}F_{\mathrm{d}}(V - \dot{x})\right]\Delta x$$
$$= 2F_{\mathrm{d}}(V - \dot{x})(\Delta V + 2\dot{x}) + 2C\dot{x} + M\left(\frac{4}{\Delta t}\dot{x} + 2\ddot{x}\right) \tag{6.51}$$

which reveals that the wind, as well as exciting a structure, also increases its dynamic stiffness, which alternatively may be written as

$$K_{\mathrm{d}} = K + \frac{2}{\Delta t}[C + 2F_{\mathrm{d}}(V - \dot{x})] + \frac{4}{\Delta t^2}M \tag{6.52}$$

When the dynamic stiffness is expressed in this form it can be seen that the wind not only excites a structure, but also increases the damping coefficient by the term $F_{\mathrm{d}}(V - \dot{x})$. For certain types of structure, such as guyed masts, the *aerodynamic* damping may be more significant than the damping caused by friction in joints and hysteresis losses in the structure itself.

## Dynamic response to earthquakes

Let the acceleration of the ground at the support of a one DOF structure at times $t$ and $(t + \Delta t)$ be $\ddot{x}_{\mathrm{g}}$ and $\ddot{x}_{\mathrm{g}} + \Delta \ddot{x}_{\mathrm{g}}$ respectively. From eq. (4.53), the inertia force acting on the mass of the structure is $M\ddot{x}_{\mathrm{g}}(t)$. Thus, if $\ddot{x}_{\mathrm{g}}$ is the acceleration at time $t$ and $\ddot{x}_{\mathrm{g}} + \Delta \ddot{x}_{\mathrm{g}}$ is the acceleration at time $(t + \Delta t)$, the change in dynamic force during the time step $\Delta t$ is

$$\Delta P = M(\ddot{x}_{\mathrm{g}} + \Delta \ddot{x}_{\mathrm{g}}) - M\ddot{x}_{\mathrm{g}} = M\Delta \ddot{x}_{\mathrm{g}} \tag{6.53}$$

Substitution of this expression for $\Delta P$ into eq. (6.37) yields the response equation for earthquake excitation

$$\left(K + \frac{2}{\Delta t}C + \frac{4}{\Delta t^2}M\right) = M\left(\Delta\ddot{x}_g + \frac{4}{\Delta t}\dot{x} + 2\ddot{x}\right) + 2C\dot{x} \tag{6.54}$$

where the displacements $x$ and $\Delta x$, velocity $\dot{x}$ and acceleration $\Delta\ddot{x}$ are relative to the support.

Response analysis in the time domain to determine the effects of turbulent wind and earthquakes is normally carried out only for nonlinear structures. For linear structures such analysis is, as mentioned above, undertaken in the frequency domain.

## Dynamic response to impacts caused by falling loads

Consider the case when a weight of mass $m$ drops from a height $H$ on to a floor having an equivalent mass $M$ and an equivalent spring stiffness $K$. If the floor is assumed to respond linearly, the maximum response $x_0$ can most easily be determined by equating the initial maximum potential energy of the weight to the maximum strain energy stored in the floor, if the loss of energy at impact is ignored. Thus

$$mg(H + x_0) = \tfrac{1}{2}Kx_0^2 \tag{6.55}$$

Hence

$$x_0 = x_{st} \pm \sqrt{(x_{st}^2 + 2Hx_{st})} \tag{6.56}$$

where $x_{st} = mg/K$. The negative sign in front of the square root has no meaning and can be ignored. Usually since $H >>> x_{st}$ the square term in eq. (6.56) may be neglected, in which case the expression for the response $x_0$ reduces to

$$x_0 = x_{st} + \sqrt{(2Hx_{st})} \tag{6.57}$$

Thus it is necessary to undertake dynamic response analysis only if the structure is expected to exhibit nonlinear behaviour. Such an analysis may be undertaken using one of the forward step-by-step integration processes presented above by ignoring energy losses at impact and assuming that the kinetic energy after impact is equal to the initial potential energy of the falling load. This may be written as

$$mgH = \tfrac{1}{2}(M + m)\dot{x}_0^2 \tag{6.58}$$

and yields

$$\dot{x}_0 = \sqrt{\left(\frac{2mgH}{M + m}\right)} \tag{6.59}$$

In this case $\Delta P$ is zero and eq. (6.37) reduces to

$$\left(K + \frac{2}{\Delta t}C + \frac{4}{\Delta t^2}M\right)\Delta x = 2C\dot{x} + M\left(\frac{4}{\Delta t}\dot{x} + 2\ddot{x}\right) \tag{6.60}$$

where the initial value for the velocity $\dot{x}$ is given by eq. (6.59), and the initial value for $\ddot{x}$ is taken as zero.

**Example 6.1** A portal frame is subjected to a horizontal impulse force $P(t)$ at beam level. The specifications for the portal frame and dynamic load are shown in Figs 6.7(a) and 6.7(b) respectively. Use the linear acceleration method to predict the response of the frame

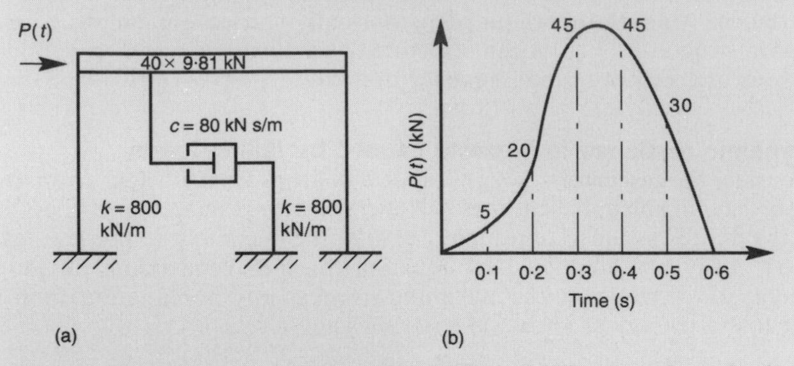


*Fig. 6.7. Portal frame: (a) subjected to shock load P(t); (b) at beam level*

The change of displacement during each time step is given by

$$\Delta x = K_{\mathrm{d}}^{-1} \Delta P_{\mathrm{d}}$$

where the expressions for $K_{\mathrm{d}}$ and $\Delta P_{\mathrm{d}}$ are given by eqs (6.15) and (6.16) respectively.

The periodic time of vibration of the frame is

$$T = 2\pi\sqrt{\left(\frac{M}{K}\right)} = 2\pi\sqrt{\left(\frac{40 \times 1000}{1600 \times 1000}\right)} = 0{\cdot}9934588\ \mathrm{s}$$

Hence a time step of $\Delta t = 0{\cdot}1$ s should suffice to describe the motion of response. With reference to the diagram of the force $P(t)$ in Fig. 6.7, a time step of 0·1 s may be slightly too large to describe the load history sufficiently accurately, but for the purpose of this example it is assumed that the chosen time step is small enough. By use of eqs (6.15), (6.16), (6.18) and (6.19), the following functions are obtained for $K_{\mathrm{d}}$, $\Delta P$, $\dot{x}$ and $\ddot{x}$

$$K_{\mathrm{d}} = 1600 + \frac{3}{0{\cdot}1} \times 80 + \frac{6}{0{\cdot}1^2} \times 40 = 28\,000\ \mathrm{kN/m}$$

$$\Delta P_{\mathrm{d}} = \Delta P + \left(3 \times 80 + \frac{6}{0{\cdot}1} \times 40\right)\dot{x} + \left(\frac{0{\cdot}1}{2} \times 80 + 3 \times 40\right)\ddot{x}$$

$$= \Delta P + 2640\dot{x} + 124\ddot{x}\ (\mathrm{kN})$$

$$\dot{x} = \frac{3}{0{\cdot}1}\Delta x - 2\dot{x} - \frac{0{\cdot}1}{2}\ddot{x} = 30\Delta x - 2\dot{x} - 0{\cdot}05\ddot{x}$$

$$\ddot{x} = \frac{6}{0{\cdot}1^2}\Delta x - \frac{6}{0{\cdot}1}\dot{x} - 2\ddot{x} = 600\Delta x - 60\dot{x} - 2\ddot{x}$$

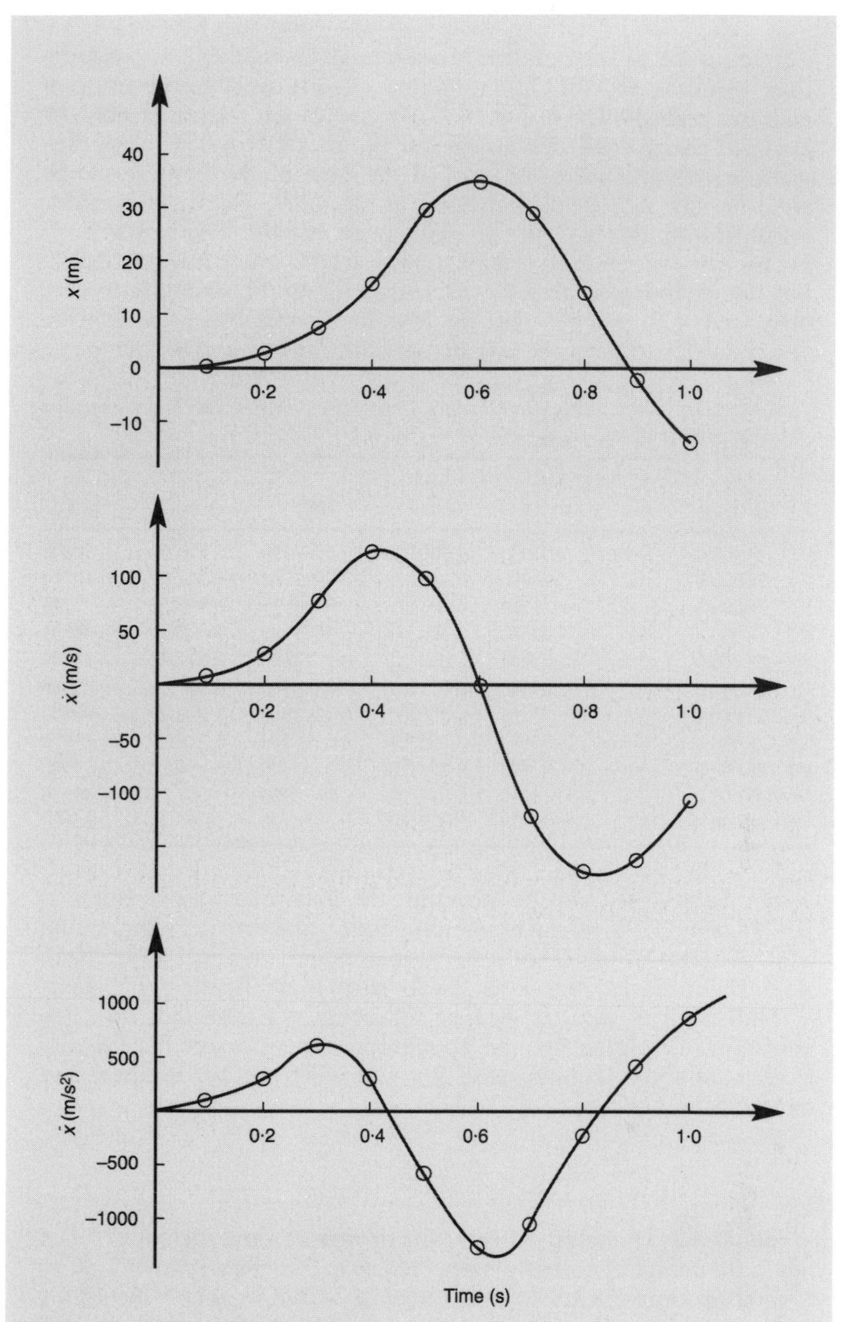


*Fig. 6.8. Displacement, velocity and acceleration histories for portal frame in example 6.1*

The sequence of the iterative process is given in Table 6.1 and the time histories for the displacement, velocity and acceleration of response are shown in Fig. 6.8. As a rule of thumb it may be assumed that the maximum response of one DOF systems with zero damping occurs when the ratio of the time of the impulse to the periodic time is $\leq 0{\cdot}8$ as indicated in Fig. 6.10, where the dynamic magnification factor is plotted against the impulse length ratio $\tau/T$. In this case the ratio is $0{\cdot}6/0{\cdot}9934588 = 0{\cdot}604$, which is less than 0·8, but the damping is 22·36% of critical. Thus the damping is very high, and it is assumed that the maximum response occurs within 1·0 s in order to keep the number of iterations to a minimum.

*Table 6.1. Response calculations using linear acceleration method to predict the response of portal frame shown in Fig. 6.7*

| $t$ (s) | $K_d$ (kN/m) | $\Delta P$ (kN) | $\Delta P_d$ (kN) | $\Delta x$ (m) | $x$ (m) | $\dot{x}$ (m/s) | $\ddot{x}$ (m/s$^2$) |
|---|---|---|---|---|---|---|---|
| 0·0 | 28 000·0 | 0·0 | 0·00000 | 0·000000 | 0·0000000 | 0·000000 | 0·00000 |
| 0·1 | 28 000·0 | 5·0 | 5·00000 | 0·000179 | 0·0001786 | 0·005357 | 0·10714 |
| 0·2 | 28 000·0 | 15·0 | 42·42784 | 0·001515 | 0·0016939 | 0·029388 | 0·37346 |
| 0·3 | 28 000·0 | 25·0 | 148·89388 | 0·005318 | 0·0070115 | 0·082080 | 0·68038 |
| 0·4 | 28 000·0 | 0·0 | 301·05866 | 0·010752 | 0·0177636 | 0·124384 | 0·16569 |
| 0·5 | 28 000·0 | −15·0 | 333·91788 | 0·011926 | 0·0296900 | 0·100718 | −0·63900 |
| 0·6 | 28 000·0 | −30·0 | 156·65791 | 0·005695 | 0·0352850 | 0·001637 | −1·40809 |
| 0·7 | 28 000·0 | 0·0 | 178·92592 | −0·005390 | 0·0298950 | −0·118030 | −0·91970 |
| 0·8 | 28 000·0 | 0·0 | −425·63507 | −0·015201 | 0·0146940 | −0·174000 | 0·19900 |
| 0·9 | 28 000·0 | 0·0 | −484·11919 | −0·017290 | −0·0025960 | −0·160720 | 0·46533 |
| 1·0 | 28 000·0 | 0·0 | −366·59374 | −0·013093 | −0·0156190 | −0·094610 | 0·85682 |

From Table 6.1 it can be seen that the maximum displacement is 35·285 mm. If the portal frame had been subjected to a maximum static force of 45·0 kN the displacement would have been $45{\cdot}0/16{\cdot}0 = 28{\cdot}125$ mm. Thus the dynamic magnification is 25·46%.

The histories plotted in Fig. 6.8 show, as expected, that the maximum displacement and acceleration occur when the velocity is zero and the maximum velocity occurs when the displacement and acceleration are zero.

**Example 6.2** A weight of 1·0 kN is dropped from a height of 1·0 m on to the centre of a simply supported beam having a span of 10·0 m. The beam supports a distributed load of 3·0 kN/m, which includes a self-weight. The $EI$ value for the beam is 28 000 kN m$^2$. Ignoring the loss of energy at impact and losses due to structural damping, calculate the initial velocity and maximum displacement of the beam, and the corresponding dynamic magnification factor. Use the constant acceleration method to calculate the maximum central displacement, if the structural damping of the beam is 2·0% of critical. Use a time step equal to approximately $\frac{1}{20}$ of the natural period of the beam.

The equivalent lumped weight, treating the beam and 1·0 kN falling load as a mass–spring system, is

$$W_e = P + \frac{17}{35}wL = 1{\cdot}0 + \frac{17}{35} \times 3{\cdot}0 \times 10 = \underline{15{\cdot}571429\,\text{kN}}$$

The equivalent stiffness is given by

$$K_e = \frac{6144EI}{125L^3} = \frac{6144 \times 28\,000{\cdot}0}{125 \times 10^3} = \underline{1376{\cdot}256\,\text{kN/m}}$$

Thus the natural frequency of the beam plus load is

$$f = \frac{1}{2\pi}\sqrt{\left(\frac{K_e g}{W_e}\right)} = \frac{1}{2\pi}\sqrt{\left(\frac{1376{\cdot}256 \times 1000 \times 9{\cdot}81}{15{\cdot}571429 \times 1000}\right)} = 4{\cdot}6864071\,\text{Hz}$$

Hence the time step

$$\Delta t = \frac{T}{20} = \frac{1}{20f} = \frac{1}{20 \times 4{\cdot}6864071} = 0{\cdot}0106691\,\text{s, say } \underline{\Delta t = 0{\cdot}01\,\text{s}}$$

The damping coefficient is given by

$$C = \xi C_c = 2\xi\sqrt{(K_e M_e)} = 2 \times 0{\cdot}2\sqrt{\left(\frac{1376{\cdot}256 \times 1\,571\,429}{9{\cdot}81}\right)}$$

$$= \underline{1{\cdot}8695597\,\text{kN s/m}}$$

The initial velocity of the beam after impact is given by

$$\dot{x}_0 = \sqrt{\left(\frac{2mgH}{M+m}\right)} = \sqrt{\frac{2PgH}{W_e}} = \sqrt{\left(\frac{2 \times 1{\cdot}0 \times 9{\cdot}81 \times 1{\cdot}0}{15{\cdot}571429}\right)}$$

$$= \underline{1{\cdot}1224972\,\text{m/s}}$$

The maximum displacement of the beam is found by equating the maximum potential energy to the maximum strain energy. This yields

$$x_{max} = x_{st}[1 + \sqrt{(1 + 2H/x_{st})}]$$

where

$$x_{st} = P/K_e = 1{\cdot}0/1376{\cdot}256 = 0{\cdot}726609 \times 10^{-3}\,\text{m}$$

$$x_{max} = 0{\cdot}726609 \times 10^{-3}[1 + \sqrt{(1 + 2 \times 1{\cdot}0/0{\cdot}726609 \times 10^{-3})}]$$

$$= \underline{38{\cdot}85463 \times 10^{-3}\,\text{m}}$$

$$\left(K + \frac{2}{\Delta t}C + \frac{4}{\Delta t^2}M\right)\Delta x = 2C\dot{x} + M\left(\frac{4}{\Delta t}\dot{x} + 2\ddot{x}\right)$$

or

$$K_d \Delta x = \Delta P_d$$

where

$$\left(K + \frac{2}{\Delta t}C + \frac{4}{\Delta t^2}M\right)$$
$$= \left(1376{\cdot}256 + \frac{2}{0{\cdot}01}\,1{\cdot}8695597 + \frac{4}{0{\cdot}01^2}\,\frac{15{\cdot}571429}{9{\cdot}81}\right)$$
$$= \underline{65\,242{\cdot}233\ \text{kN/m}}$$

$$2C\dot{x} + M\left(\frac{4}{\Delta t}\dot{x} + 2\ddot{x}\right) = 2 \times 1{\cdot}8695597\dot{x} + \frac{15{\cdot}571429}{9{\cdot}81}\left(\frac{4}{0{\cdot}01}\dot{x} + 2\ddot{x}\right)$$
$$= \underline{638{\cdot}65977\dot{x} + 3{\cdot}1746033\ddot{x}}$$

Thus

$$\Delta P_0 = 638{\cdot}65977 \times 1{\cdot}1224972 + 3{\cdot}1746033 \times 0{\cdot}0 = \underline{716{\cdot}8938\ \text{kN}}$$

$$x(t + \Delta t) = \underline{x + \Delta x}$$

$$\dot{x}(t + \Delta t) = \frac{2}{\Delta t}\Delta x - \dot{x} = \frac{2}{0{\cdot}01}\Delta x - \dot{x} = \underline{200\Delta x - \dot{x}}$$

$$\ddot{x}(t + \Delta t) = \frac{4}{\Delta t^2}\Delta x - \frac{4}{\Delta t}\dot{x} - \ddot{x} = \frac{4}{0{\cdot}01^2}\Delta x - \frac{4}{0{\cdot}01}\dot{x} - \ddot{x}$$
$$= \underline{40\,000\Delta x - 400\dot{x} - \ddot{x}}$$

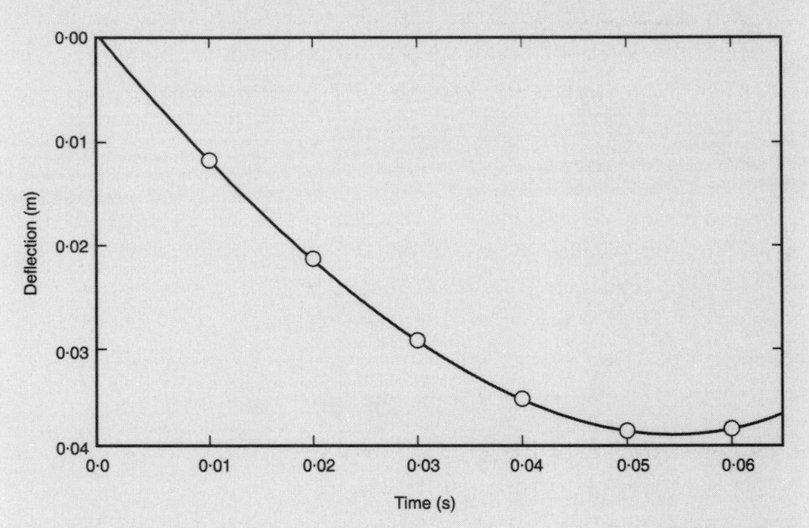


*Fig. 6.9. Variation of the central deflection of the beam in example 6.2 caused by a falling load with time*

*Table 6.2. Response calculations using the constant acceleration method to predict the response of the simply supported beam in example 6.2 to the impact caused by a falling load*

| $t$ (s) | $K_d$ (kN/m) | $\Delta x$ (m) | $x$ (m) | $\dot{x}$ (m/s) | $\ddot{x}$ (m/s$^2$) | $\Delta P_d$ (kN) |
|---|---|---|---|---|---|---|
| 0·00 | 65 242·233 | 0·00000000 | 0·0000000 | 1·1224972 | 0·00000 | 716·89380 |
| 0·01 | 65 242·233 | 0·01098810 | 0·0109881 | 1·0751228 | −9·47488 | 656·55869 |
| 0·02 | 65 242·233 | 0·01006340 | 0·0210515 | 0·9375572 | −18·03824 | 541·51581 |
| 0·03 | 65 242·233 | 0·00830008 | 0·0293515 | 0·7224588 | −24·98144 | 382·09921 |
| 0·04 | 65 242·233 | 0·00585662 | 0·0352081 | 0·4488652 | −29·74896 | 192·23100 |
| 0·05 | 65 242·233 | 0·00294641 | 0·0381545 | 0·1404168 | −31·94072 | −11·72055 |
| 0·06 | 65 242·233 | −0·00017965 | 0·0379748 | −0·1763460 | −31·41184 | −212·34139 |

## Response to impulse loading

Time domain methods may also be used to study how the response of one DOF systems varies with the duration of different forms of impulse, such as those shown in Fig. 6.10, where the ratio of the dynamic to the static response, the *dynamic magnification factor*, is plotted against the ratio of the duration $\tau$ of the impulse to that of the natural period $T$ of the oscillator.

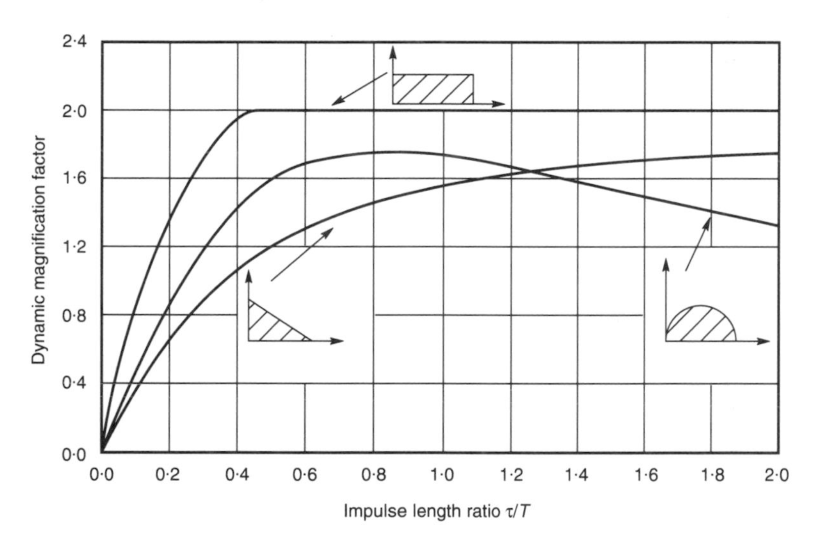


*Fig. 6.10. Dynamic magnification factor versus impulse length ratio $\tau/T$ for rectangular, triangular and half-sinusoidal impulses of $\tau$ s duration (after Clough and Penzien, 1974)*

### Incremental equations of motion for multi-DOF systems

The method for predicting the response of linear and nonlinear one DOF systems for random loading may be extended to multi-DOF systems by writing the incremental equations of motion in matrix form. Thus eq. (6.37), in which the acceleration is assumed to remain constant during the time step $\Delta t$, may be written as

$$\left[\mathbf{K}+\frac{2}{\Delta t}\mathbf{C}+\frac{4}{\Delta t^2}\mathbf{M}\right]\Delta \boldsymbol{x}=\Delta \boldsymbol{P}+2\mathbf{C}\dot{\boldsymbol{x}}+\mathbf{M}\left[\frac{4}{\Delta t}\dot{x}+2\ddot{x}\right] \tag{6.61}$$

where $\mathbf{K}$, $\mathbf{C}$ and $\mathbf{M}$ are the stiffness, damping and mass matrices for a multi-DOF structure, $\Delta\boldsymbol{x}$ is the incremental displacement vector, $\boldsymbol{x}$, $\dot{\boldsymbol{x}}$ and $\ddot{\boldsymbol{x}}$ are the displacement, velocity and acceleration vectors at time $t$, and $\Delta P$ is the incremental load vector.

The *ith* elements in vectors $\boldsymbol{x}$, $\dot{\boldsymbol{x}}$ and $\ddot{\boldsymbol{x}}$ at time $t+\Delta t$ are given by eqs (6.41)–(6.43). Thus

$$x_i(t+\Delta t)=x_i+\Delta x_i \tag{6.62}$$

$$\dot{x}_i(t+\Delta t)=\frac{2}{\Delta t}\Delta x_i-\dot{x}_i \tag{6.63}$$

$$\ddot{x}_i(t+\Delta t)\frac{4}{\Delta t^2}\Delta x_i-\frac{4}{\Delta t}\dot{x}_i-\ddot{x}_i \tag{6.64}$$

From eq. (6.61) it follows that the dynamic stiffness matrix and incremental dynamic load vector are

$$\mathbf{K}_\mathrm{d}=\left[\mathbf{K}+\frac{2}{\Delta t}\mathbf{C}+\frac{4}{\Delta t^2}\mathbf{M}\right] \tag{6.65}$$

$$\Delta\boldsymbol{P}_\mathrm{d}=\Delta\boldsymbol{P}+2\mathbf{C}\dot{\boldsymbol{x}}+\mathbf{M}\left[\frac{4}{\Delta t}\dot{\boldsymbol{x}}+2\ddot{\boldsymbol{x}}\right] \tag{6.66}$$

Hence

$$\Delta\boldsymbol{x}=\mathbf{K}_\mathrm{d}{}^{-1}\Delta\boldsymbol{P} \tag{6.67}$$

Similarly, the equations of motion for structures subjected to turbulent wind $\boldsymbol{V}(\boldsymbol{t})$ may be written as

$$\begin{aligned}&\left[\mathbf{K}+\frac{2}{\Delta t}\mathbf{C}+\frac{4}{\Delta t^2}\mathbf{M}+\frac{4}{\Delta t}\boldsymbol{F}_\mathrm{d}(\boldsymbol{V}-\dot{\boldsymbol{x}})\right]\Delta\boldsymbol{x}\\&=2\boldsymbol{F}_\mathrm{d}(\boldsymbol{V}-\dot{\boldsymbol{x}})(\Delta\boldsymbol{V}+2\dot{\boldsymbol{x}})+2\mathbf{C}\dot{\boldsymbol{x}}+\mathbf{M}\left(\frac{4}{\Delta t}\dot{\boldsymbol{x}}+2\ddot{\boldsymbol{x}}\right)\end{aligned} \tag{6.68}$$

and for structures subjected to ground acceleration $\ddot{x}_\mathrm{g}(t)$ as

$$\left(\mathbf{K}+\frac{2}{\Delta t}\mathbf{C}+\frac{4}{\Delta t^2}\mathbf{M}\right)\Delta\boldsymbol{x}=\mathbf{M}\left(\Delta\ddot{\boldsymbol{x}}_\mathrm{g}+\frac{4}{\Delta t}\dot{\boldsymbol{x}}+2\ddot{\boldsymbol{x}}\right)+2\mathbf{C}\dot{\boldsymbol{x}} \tag{6.69}$$

Equations (6.61), (6.68) and (6.69) require the assembly of stiffness, mass and structural damping matrices. The construction of stiffness matrices and mass matrices is considered in chapters 7 and 8, and in most modern books on structural analysis, such as Coates *et al.* (1972). The construction of damping matrices is dealt with in chapter 9, which shows how damping matrices can be constructed from modal damping ratios, whose values need to be obtained from codes of practice or published papers, or by dynamic testing.

**Problem 6.1** Use the constant acceleration method to predict the first 1 s response of the portal frame in example 6.1. Plot the time histories of displacement, velocity and acceleration response and compare them with those shown in Fig. 6.8.

**Problem 6.2** A steel ball of mass 0·76 kg is dropped in turn from heights of 100 mm and 200 mm on to the centre of a prestressed concrete plate. The equivalent mass of the plate is 230 kg and the equivalent stiffness 3·77 kN/mm. The damping is 1·6% of critical. Use the constant acceleration method to calculate the maximum response for each drop. The size of a suitable time step may be assumed to be approximately $\frac{1}{20}$ of the period of the plate. Compare the values obtained with those obtained by equating the initial potential energy of the ball to the strain energy stored in the plate.

## References

Buchholdt, H. A. *Introduction to cable roof structures.* Cambridge University Press, Cambridge, 1985.

Clough, R. W. & Penzien, J. *Dynamics of structures.* McGraw-Hill, London, 1975.

Coates, R. C., Coutie, M. G. & Kong, F. K. *Structural analysis.* Nelson, London, 1972.

Wood, W. L. *Practical time-stepping schemes.* Oxford Applied Mathematics, Oxford, 1990.

# 7. Free vibration of multi-degrees of freedom systems

## Introduction

In chapter 1 it is mentioned that structures in general have an infinite number of degrees of freedom, which usually are approximated to $N$ DOF systems by replacing the distributed mass of structures with an equivalent system of lumped masses, and assuming the elastic members to be weightless. A general preliminary introduction to the free and forced vibration of multi-DOF systems entails an excessive amount of algebra and requires the use of computers to study their behaviour if there are more than three DOF. In order to avoid these difficulties, the dynamic analysis of large systems is introduced by studying structures with only two and three degrees of freedom. This is quite feasible since the procedure for determining free vibration, as well as response to dynamic loads, is exactly the same as for structures with $N$ degrees of freedom. At this stage it is merely noted that an $N$ DOF structure has $N$ eigenvalues and eigenvectors associated with the system of equations that defines its motion, and that the square roots of the eigenvalues are equal to the natural angular frequencies of the structure. The eigenvectors corresponding to the natural frequencies represent the natural modes or modeshapes in which the structure can vibrate. The determination of eigenvalues of eigenvectors is of fundamental importance to the frequency domain method of analysis, in which the distribution of energy of random forces such as wind, waves and earthquakes are given as functions of their frequency content in terms of power spectra. Structural damping is usually not included when one is formulating the eigenvalue problem, as it increases the numerical effort considerably and has only a second order effect on the calculated frequencies.

## Eigenvalues and eigenvectors

The mathematical concept arises from the solution of a set of $N$ homogeneous equations where two $N \times N$ matrices **A** and **B** are related by a set of vectors $V$ and scalars $\lambda$ such that the relationship

$$\mathbf{A}X - \lambda \mathbf{B}X = \mathbf{0} \tag{7.1}$$

or

$$(\mathbf{A} - \lambda \mathbf{B})X = \mathbf{0} \tag{7.2}$$

is valid for non-zero values of $\boldsymbol{X}$. For a set of homogeneous equations represented by eq. (7.1) or (7.2) to have a non-trivial solution, the determinant of the matrix $(\mathbf{A} - \lambda\mathbf{B})$ must be zero, i.e.

$$|\mathbf{A} - \lambda\mathbf{B}| = 0 \tag{7.3}$$

The matrix $(\mathbf{A} - \lambda\mathbf{B})$ is called the *characteristic matrix of the system*, and its determinant is called the *characteristic function*, while $|\mathbf{A} - \lambda\mathbf{B}|$ is the *characteristic equation*. For a structure with $N$ DOF the characteristic equation is a polynomial of degree $N$ in $\lambda$. The equation therefore has $N$ roots—$\lambda_1, \lambda_2, \lambda_3, \ldots, \lambda_N$—which are real if the *system matrix*

$$\mathbf{S} = \mathbf{B}^{-1}\mathbf{A} \tag{7.4}$$

is symmetric. The roots of the characteristic equation are called the *characteristic* or *latent roots* or eigenvalues of the matrix $\mathbf{S}$. In structural engineering, as mentioned above, the eigenvalues are associated with more than one DOF and associated matrices of order greater than 3, the numerical work involved in solving the eigenvalue problem is too great and too time-consuming to be carried out by hand, and digital computers are needed for its solution. A number of methods can be used for the hand calculations of small problems; there are also approximate methods for the calculation of the first few eigenvalues of larger problems. There are three basic approaches for solving the eigenvalue problem

- direct solution of the characteristic polynomial
- iterative optimization of eigenvectors
- transformation of the system matrix.

The first two methods can be used relatively easily to determine the eigenvalues and eigenvectors for structures with up to three DOF, while the third approach requires the use of computers. In the following the first two methods are applied to the solution of two and three DOF mass–spring systems.

## Determination of free normal mode vibration by solution of the characteristic equation

Consider the two DOF mass–spring system shown in Fig. 7.1 which, for example, could be considered as the mass–spring model of a column with the mass lumped together at two points along its length. From the free body diagram the equations of motion for the two masses are

$$M_1\ddot{x}_1 = -K_1x_1 - K_cx_1 + K_cx_2 \qquad +P_1(t) \tag{7.5a}$$

$$M_2\ddot{x}_2 = \qquad +K_cx_1 - K_cx_2 - K_2x_2 + P_2(t) \tag{7.5b}$$

Equations (7.5a) and (7.5b) may be written in matrix form as

$$\begin{bmatrix} M_1 & 0 \\ 0 & M_2 \end{bmatrix}\begin{bmatrix} \ddot{x}_1 \\ \ddot{x}_2 \end{bmatrix} + \begin{bmatrix} (K_1 + K_c) & -K_c \\ -K_c & (K_2 + K_c) \end{bmatrix}\begin{bmatrix} x_1 \\ x_2 \end{bmatrix} = \begin{bmatrix} P_1(t) \\ P_2(t) \end{bmatrix} \tag{7.6}$$

In order to determine the natural frequencies and modeshapes of vibration, put $P_1(t) = P_2(t) = 0$. This yields

$$\begin{bmatrix} M_1 & 0 \\ 0 & M_2 \end{bmatrix} \begin{bmatrix} \ddot{x}_1 \\ \ddot{x}_2 \end{bmatrix} + \begin{bmatrix} (K_1 + K_c & -K_c \\ -K_c & (K_2 + K_c \end{bmatrix} \begin{bmatrix} x_1 \\ x_2 \end{bmatrix} = \begin{bmatrix} 0 \\ 0 \end{bmatrix} \tag{7.7}$$

It it is assumed that the motion of each mass in free vibration is simple harmonic, then

$$x_1 = X_1 \sin(\omega t) \tag{7.8a}$$

$$x_2 = X_2 \sin(\omega t) \tag{7.8b}$$

$$\ddot{x}_1 = -X_1 \omega^2 \sin(\omega t) \tag{7.9a}$$

$$\ddot{x}_2 = -X_2 \omega^2 \sin(\omega t) \tag{7.9b}$$

Substitution of the expressions for $x$ and $\ddot{x}$ into eq. (7.7) yields

$$\begin{bmatrix} (K_1 + K_c & -K_c \\ -K_c & (K_2 + K_c) \end{bmatrix} \begin{bmatrix} X_1 \\ X_2 \end{bmatrix} - \omega^2 \begin{bmatrix} M_1 & 0 \\ 0 & M_2 \end{bmatrix} \begin{bmatrix} X_1 \\ X_2 \end{bmatrix} = \begin{bmatrix} 0 \\ 0 \end{bmatrix} \tag{7.10}$$

or

$$\begin{bmatrix} (K_1 + K_c - \omega^2 M_1) & -K_c \\ -K_c & (K_2 + K_c - \omega^2 M_2) \end{bmatrix} \begin{bmatrix} X_1 \\ X_2 \end{bmatrix} = \begin{bmatrix} 0 \\ 0 \end{bmatrix} \tag{7.11}$$

Equation (7.11) is satisfied only if the determinant

$$\begin{vmatrix} (K_1 + K_c - \omega^2 M_1) & -K_c \\ -K_c & (K_2 + K_c - \omega^2 M_2) \end{vmatrix} = 0 \tag{7.12}$$

Expansion of the above determinant yields the characteristic equation

$$\begin{aligned} &\omega^4 - [(K_1 + K_c)/M_1 + (K_2 + K_c)/M_2]\omega^2 + \\ &[K_1 K_2 + (K_1 + K_2) K_c] M_1 M_2 = 0 \end{aligned} \tag{7.13}$$

from which the two angular frequencies $\omega_1$ and $\omega_2$ can be determined. Substitution in turn of the calculated values for $\omega_1$ and $\omega_2$ into eq. (7.10) yields

$$\left(\frac{X_1}{X_2}\right)_1 = \frac{K_c}{K_1 + K_c - \omega_1{}^2 M_1} = \frac{K_2 + K_c - \omega_1{}^2 M_2}{K_c} \tag{7.14a}$$

$$\left(\frac{X_1}{X_2}\right)_2 = \frac{K_c}{K_1 + K_c - \omega_2{}^2 M_1} = \frac{K_2 + K_c - \omega_2{}^2 M_2}{K_c} \tag{7.14b}$$

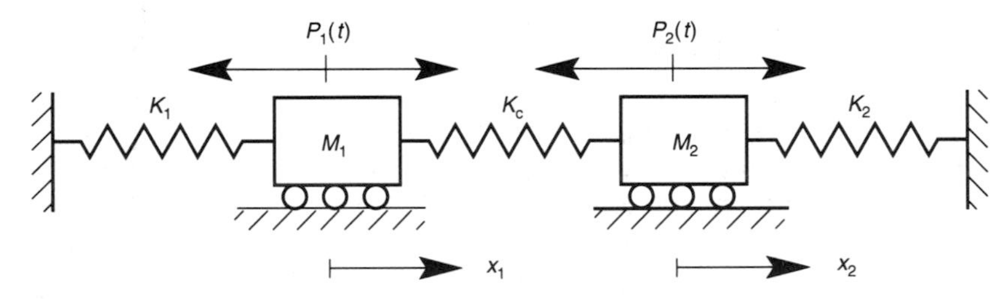


*Fig. 7.1. Two DOF mass–spring system*

With no forces applied to the system the amplitudes of vibration will have no absolute values, and only the amplitude ratios that determine the modeshapes can be determined. Thus the first and second modeshapes for the system shown in Fig. 7.1 are determined by first calculating the ratios $(X_1/X_2)_1$ and $(X_2/X_2)_2$ corresponding respectively to $\omega_1$ and $\omega_2$ from eqs (7.14a) and (7.14b), and then assigning a value of, say, 1 to either $X_1$ or $X_2$.

**Example 7.1** Determine the natural frequencies and corresponding modeshapes of vibration for the two DOF mass–spring system shown in Fig. 7.1 if $K_1 = K_c + K_2 = K$, and $M_1 = M_2 = M$.

Substitution for $M_1$, $M_2$, $K_1$, $K_c$ and $K_2$ in eq. (7.12) yields

$$\omega^4 - \frac{4K}{M}\omega^2 + \frac{3K^2}{M^2} = 0$$

Hence

$$\omega^2 = \frac{K}{M}(2 \pm 1)$$

Therefore

$$\omega_1 = \sqrt{\left(\frac{K}{M}\right)}$$

$$\omega_2 = \sqrt{\left(\frac{3K}{M}\right)}$$

Substitution of the expressions for $\omega_1$ and $\omega_2$ into eqs (7.13a) and (7.13b) yields the amplitude ratios

$$\left(\frac{X_1}{X_2}\right)_1 = \frac{2K - M(K/M)}{K} = 1$$

$$\left(\frac{X_1}{X_2}\right)_2 = \frac{2K - M(3K/M)}{K} = -1$$

The above values for the amplitude ratios imply that in the first mode the two masses move in the same direction as if connected by a solid rod, and in the second mode they move in the opposite direction such that the midpoint of the central spring is at rest at all times, as shown in Figs 7.2(a) and 7.2(b) respectively.

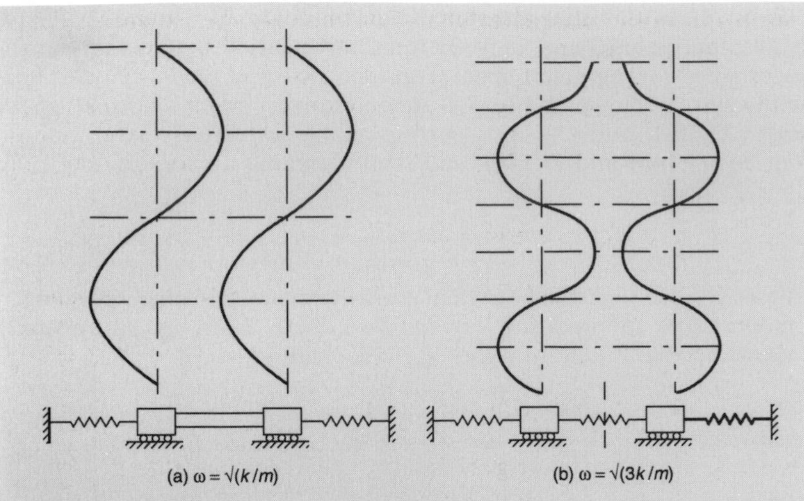


*Fig. 7.2. (a) First and (b) second modes of vibration of the mass–spring system in example 7.1*

**Example 7.2** Write down the equations for free vibration of the structure in example 2.5 (Fig. 2.14), and hence establish the characteristic equation for the structure.

The equations of motion are given by

$$\begin{bmatrix} 3M & 0 & 0 \\ 0 & 2M & 0 \\ 0 & 0 & M \end{bmatrix} \begin{bmatrix} \ddot{x}_1 \\ \ddot{x}_2 \\ \ddot{x}_3 \end{bmatrix} + \begin{bmatrix} 7K & -3K & 0 \\ -3K & 5K & -2K \\ 0 & -2K & 2K \end{bmatrix} \begin{bmatrix} x_1 \\ x_2 \\ x_3 \end{bmatrix} = \begin{bmatrix} 0 \\ 0 \\ 0 \end{bmatrix}$$

If SHM is assumed, the corresponding eigenvalue equation is

$$\begin{bmatrix} 7K & -3K & 0 \\ -3K & 5K & -2K \\ 0 & -2K & 2K \end{bmatrix} \begin{bmatrix} X_1 \\ X_2 \\ X_3 \end{bmatrix} - \lambda \begin{bmatrix} 3M & 0 & 0 \\ 0 & 2M & 0 \\ 0 & 0 & M \end{bmatrix} \begin{bmatrix} X_1 \\ X_2 \\ X_3 \end{bmatrix} = \begin{bmatrix} 0 \\ 0 \\ 0 \end{bmatrix}$$

The above eigenvalue equation will have a non-trivial solution only if

$$\begin{vmatrix} (7K - 3M\lambda) & -3K & 0 \\ -3K & (5K - 2M\lambda) & -2 \\ 0 & -2K & (2K - M\lambda) \end{vmatrix} = 0$$

Evaluation of this determinant leads to the characteristic equation

$$\phi(\lambda) = 6\lambda^3 - 41\alpha\lambda^2 + 72\alpha^2\lambda - 24\alpha^3 = 0$$

where

$$\alpha = K/M$$

### Solution of cubic characteristic equations by the Newton approximation method

In the Newton approximation method, successive estimates of $\lambda$ are achieved through the iterative procedure

$$\lambda_{i+1} = \lambda_1 - \frac{\phi\{\lambda_i\}}{\phi'\{\lambda_i\}} \tag{7.15}$$

where $\phi\{\lambda\}$ is the characteristic polynomial

$$\phi\{\lambda\} = \lambda^3 + a\lambda^2 + b\lambda + c = 0 \tag{7.16}$$

Thus

$$\lambda_{i+1} = \lambda_i - \frac{\lambda_i^{\,3} + a\lambda_i^{\,2} + b\lambda_i + c}{3\lambda_i^{\,2} + 2a\lambda_i + b} \tag{7.17}$$

When calculating the first eigenvalue let the initial value of $\lambda$ be $\lambda_{i=1} = 0$, and when calculating the third eigenvalue set the initial value of $\lambda$ equal to the trace of $\mathbf{M}^{-1}\mathbf{K}$. The second eigenvalue can then be calculated by applying theorem 7.1 below.

### Solution of cubic characteristic equations by the direct method

For the general characteristic equation given by eq. (7.16), let

$$Q = \frac{a^2 - 3b}{9} \qquad R = \frac{2a^3 - 9ab + 27c}{54} \tag{7.18}$$

If $Q^3 - R^2 < 0$ the characteristic equation has only one root, but if $Q^3 - R^2 \geq 0$ the equation has three real roots, which can be found by first calculating

$$\theta = \cos^{-1}\left(R/\sqrt{Q^3}\right) \tag{7.19}$$

The roots of the cubic equation are then found in terms of $\theta$ and are

$$\lambda_1 = -\sqrt{(Q)}\cos\frac{\theta}{3} - \frac{a}{3} \tag{7.20a}$$

$$\lambda_2 = -\sqrt{(Q)}\cos\frac{\theta + 2\pi}{3} - \frac{a}{3} \tag{7.20b}$$

$$\lambda_3 = -\sqrt{(Q)}\cos\frac{\theta + 4\pi}{3} - \frac{a}{3} \tag{7.20c}$$

### Two eigenvalue and eigenvector theorems

The following two theorems from matrix theory are useful for checking calculated eigenvalues and eigenvectors and, in the case of two and three DOF systems, for reducing the amount of calculations.

*Theorem 7.1*
The sum of the elements along the leading diagonal of the system matrix $\mathbf{S} = \mathbf{M}^{-1}\mathbf{K}$, referred to as the *trace* of $\mathbf{S}$ is equal to the sum of its eigenvalues.

From this theorem it follows that

$$\sum_{i=1}^{N} S_{ii} = \sum_{i=1}^{N} \lambda_i = \sum_{i=1}^{N} \omega_i^2 \tag{7.21}$$

*Theorem 7.2*
If $\boldsymbol{X}_i$ and $\boldsymbol{X}_j$, $i \neq j$, are two of the eigenvectors of the eigenvalue equation $\mathbf{K}\boldsymbol{X} - \lambda\mathbf{M}\boldsymbol{X} = 0$, then $\boldsymbol{X}_i^{\mathrm{T}}\mathbf{M}\boldsymbol{X}_j = 0$. If $\mathbf{M}^{-1}\mathbf{K}$ is symmetric, then $\boldsymbol{X}_i^{-1}\boldsymbol{X}_j = 0$.

In the case of structures with three DOF the second angular frequency and modeshape vector may be found by applying the two theorems from matrix algebra stated above.

**Example 7.3** Use the Newton approximation method to determine the eigenvalues $\lambda$ for the characteristic equation established in example 7.2. Hence determine also the natural frequencies and modeshape vectors for the three-storey shear structure in example 2.5 (Fig. 2.14) in terms of the flexural rigidity $EI$ of the columns and the weight $w$ per metre span of the floors.

The Newton approximation formula for the characteristic equation developed in example 7.2 is

$$\lambda_{i+1} = \lambda_i - \frac{\phi(\lambda)}{\phi'(\lambda)} = \lambda_i - \frac{6\lambda_i^3 - 41\alpha\lambda_i^2 + 72\alpha^2\lambda_i - 24\alpha^3}{18\lambda_i^2 - 82\alpha\lambda_i + 72\alpha^2}$$

The above function will always converge towards the nearest root. In order to determine the first eigenvalue it is therefore convenient to assume that $\lambda_{i=1} = 0{\cdot}0\alpha$. To four decimal places, this yields

$$\lambda_{i=2} = 0{\cdot}0000\alpha + 0{\cdot}3333\alpha = 0{\cdot}3333\alpha$$
$$\lambda_{i=3} = 0{\cdot}3333\alpha + 0{\cdot}0929\alpha = 0{\cdot}4262\alpha$$
$$\lambda_{i=4} = 0{\cdot}4262\alpha + 0{\cdot}0074\alpha = 0{\cdot}4336\alpha$$
$$\lambda_{i=5} = 0{\cdot}4336\alpha + 0{\cdot}0000\alpha = 0{\cdot}4336\alpha$$

Hence

$$\lambda_1 = \omega_1^2 = 0{\cdot}4336K/M$$

In order to determine the highest eigenvalue, bearing in mind that the Newton method converges towards the nearest root, the initial value for $\lambda$ is assumed to be equal to the trace of the system matrix.

Thus

$$\lambda_{i=1} = \frac{7}{3}\alpha + \frac{5}{2}\alpha + 2\alpha = \frac{41}{6}\alpha = 6{\cdot}3333\alpha$$

This yields

$$\lambda_{i=2} = 6{\cdot}8333\alpha - 1{\cdot}3289\alpha = 5{\cdot}5044\alpha$$
$$\lambda_{i=3} = 5{\cdot}5044\alpha - 0{\cdot}7875\alpha = 4{\cdot}7169\alpha$$
$$\lambda_{i=4} = 4{\cdot}7169\alpha - 0{\cdot}3860\alpha = 4{\cdot}3309\alpha$$
$$\lambda_{i=5} = 4{\cdot}3309\alpha - 0{\cdot}1138\alpha = 4{\cdot}2171\alpha$$
$$\lambda_{i=6} = 4{\cdot}2171\alpha - 0{\cdot}0101\alpha = 4{\cdot}2070\alpha$$
$$\lambda_{i=7} = 4{\cdot}2070\alpha - 0{\cdot}0001\alpha = 4{\cdot}2069\alpha$$
$$\lambda_{i=8} = 4{\cdot}2069\alpha - 0{\cdot}0000\alpha = 4{\cdot}2069\alpha$$

Hence

$$\lambda_3 = \omega_3{}^2 = 4{\cdot}2069K/M$$

The second eigenvalue can now easily be determined by applying theorem 7.1, which states that the sum of the eigenvalues is equal to the trace of the system matrix. Thus

$$\lambda_1 + \lambda_2 + \lambda_3 = \frac{41}{6}\frac{K}{M}$$

Substitution of the values for $\lambda_1$ and $\lambda_3$ into the above equation yields

$$\lambda_2 = \omega_2{}^2 = 2{\cdot}1928K/M$$

The eigenvectors can now be determined by assuming that, say, $X_1 = 1$, and substituting the different values for $\lambda$ one at a time into the eigenvalue equation. This yields

$$\boldsymbol{X}_1 = \{1{\cdot}0000 \quad 1{\cdot}8997 \quad 2{\cdot}4256\}$$
$$\boldsymbol{X}_2 = \{1{\cdot}0000 \quad 0{\cdot}1405 \quad -1{\cdot}4578\}$$
$$\boldsymbol{X}_3 = \{1{\cdot}0000 \quad -1{\cdot}8736 \quad 1{\cdot}6979\}$$

Finally, substitution of the values for $K$ and $M$ used in example 2.4 gives the following values for the eigenvalues, natural angular frequencies and frequencies

$$\omega_1{}^2 = 8{\cdot}1300 \times 10^{-3}\frac{EIg}{w} \qquad \omega_1 = 0{\cdot}0901665\sqrt{\left(\frac{EIg}{w}\right)}$$

$$f_1 = 0{\cdot}0143504\sqrt{\left(\frac{EIg}{w}\right)}$$

$$\omega_2{}^2 = 41{\cdot}1150 \times 10^{-3}\frac{EIg}{w} \qquad \omega_2 = 0{\cdot}2027683\sqrt{\left(\frac{EIg}{w}\right)}$$

$$f_2 = 0{\cdot}0322715\sqrt{\left(\frac{EIg}{w}\right)}$$

$$\omega_3{}^2 = 78{\cdot}8793 \times 10^{-3}\frac{EIg}{w} \qquad \omega_3 = 0{\cdot}2808547\sqrt{\left(\frac{EIg}{w}\right)}$$

$$f_3 = 0{\cdot}0446994\sqrt{\left(\frac{EIg}{w}\right)}$$

**Example 7.4** Determine the roots of the characteristic equation developed in example 7.2 by the direct method.

The characteristic equation may be written as

$$\lambda^3 - \frac{41}{6}\alpha\lambda^2 + 12\alpha^2\lambda - 4\alpha^3 = 0$$

Hence

$$Q = \frac{a^2 - 3b}{9} = \frac{(-41/6)^2 - 3 \times 12}{9}\alpha^2 = 1{\cdot}1882716\alpha^2$$

$$R = \frac{2a^3 - 9ab + 27c}{54}$$

$$= \frac{2\times(-41/6)^3 - 9\times(-41/6)\times 12 + 27\times(-4)}{54}\alpha^3 = -0{\cdot}1510631\alpha^3$$

$$Q^3 - R^2 = \left\{1{\cdot}1882716^3 - (-0{\cdot}1510631)^2\right\}\alpha^6 = 1{\cdot}700647\alpha^6 > 0$$

Therefore the characteristic equation has three real roots

$$\theta = \cos^{-1}(R/\surd Q^3) = \cos^{-1}\left(-0{\cdot}1510631/\surd 1{\cdot}1882716^3\right)$$

$$= 96{\cdot}697255$$

$$\lambda_1 = -2\surd(Q)\cos\left(\frac{\theta}{3}\right) - \frac{a}{3}$$

$$= \left\{-2\surd(1{\cdot}1882716)\cos\left(\frac{96.697255}{3}\right) - \frac{-41/6}{3}\right\}\alpha = 0{\cdot}4336011\alpha$$

$$\lambda_2 = -2\surd(Q)\cos\left(\frac{\theta + 2\pi}{3}\right) - \frac{a}{3}$$

$$= \left\{-2\surd(1{\cdot}1882716)\cos\left(\frac{96{\cdot}697255 + 360}{3}\right) - \frac{-41/6}{3}\right\}\alpha$$

$$= 4{\cdot}2068786\alpha$$

$$\lambda_3 = -2\sqrt{(Q)}\cos\left(\frac{\theta + 4\pi}{3} - \frac{a}{3}\right)$$

$$= \left\{-2\sqrt{(1{\cdot}1882716)}\cos\left(\frac{96{\cdot}697255 + 7620}{3}\right) - \frac{-41/6}{3}\right\}\alpha$$

$$= 2{\cdot}1928537\alpha$$

The direct method therefore yields the same values for the eigenvalues, to four decimal places, as the Newton approximation method. However, the former does not necessarily calculate the roots in ascending order, as is apparent from the above results.

### Iterative optimization of eigenvectors

Equation (7.10) may be written in general matrix notation as

$$\mathbf{K}X - \omega^2\mathbf{M}X = 0 \tag{7.22}$$

where **K** is the stiffness matrix and **M** is the mass matrix for the structure. When the mass of a structure is lumped together at nodes, which is usually the case in hand calculations, the mass matrix is diagonal.

In the iterative method the eigenvalues $\omega^2$ and eigenvectors $X$ are determined by optimizing an assumed modeshape vector through an iterative procedure on either

$$\omega^2 X = \mathbf{M}^{-1}\mathbf{K}X \tag{7.23}$$

or

$$X/\omega^2 = \mathbf{K}^{-1}\mathbf{M} \tag{7.24}$$

Iterations on eq. (7.23) will cause the assumed eigenvector to converge towards the mode corresponding to the highest eigenvector and hence the highest frequency; iterations on eq. (7.24) will cause the assumed vector to converge towards the eigenvector corresponding to the lowest frequency. Equation (7.23) involves the inversion of the mass matrix **M**, which when the matrix is diagonal is achieved by simply inverting each of the elements on the leading diagonal. The calculation of the lowest eigenvalue using eq. (7.24) requires the inversion of the stiffness matrix **K**. Because the stiffness matrix is banded, the inversion process takes more time than the inversion of the mass matrix. The inversion of the stiffness matrix can, however, be avoided by calculating the lowest eigenvalue and eigenvector as follows. Let

$$\mathbf{B}X_i = \left[\alpha\mathbf{I} - \mathbf{M}^{-1}\mathbf{K}\right]X_i \tag{7.25}$$

where $\alpha$ is a constant larger than the highest eigenvalue, **I** is a unit matrix and **B** is a square matrix of the same order as **M** and **K**. From eq. (7.23) it follows that

$$\mathbf{M}^{-1}\mathbf{K}X_i = \omega_i^2\mathbf{I}X_i \tag{7.26}$$

Substitution of the expression for $\mathbf{M}^{-1}\mathbf{K}X_i$ given in eq. (7.26) into eq. (7.25) yields

$$\mathbf{B}X_i = [\alpha - \omega_i^2]\mathbf{I}X_i = [\alpha - \omega_i^2]X_i \tag{7.27}$$

Assuming an initial vector $X_i$, iterations on eq. (7.27) will yield the highest value of $[\alpha - \omega_i^2]$ and hence the lowest possible value for $\omega_i^2$. Thus

$$\begin{aligned} \omega_i^2 &= \omega_1^2 \\ X_i &= X_1 \end{aligned} \tag{7.28}$$

Iteration algorithms based on eqs (7.23) and (7.27) will yield the highest and lowest natural frequency and corresponding modeshapes for any structure. In the following the iterative method for determining the natural frequencies and modeshapes is demonstrated by solving first a two and then a three DOF system.

**Example 7.5** Use two iterative optimization procedures to determine the highest and lowest frequencies and modeshapes for the mass–spring system specified in example 7.1.

The eigenvalue equation for the mass–spring system is given by

$$\begin{bmatrix} 2K & -K \\ -K & 2K \end{bmatrix}\begin{bmatrix} X_1 \\ X_2 \end{bmatrix} - \omega^2\begin{bmatrix} M & 0 \\ 0 & M \end{bmatrix}\begin{bmatrix} X_1 \\ X_2 \end{bmatrix} = \begin{bmatrix} 0 \\ 0 \end{bmatrix}$$

Hence

$$\mathbf{K} = K\begin{bmatrix} 2 & -1 \\ -1 & 2 \end{bmatrix}$$

$$\mathbf{M} = M\begin{bmatrix} 1 & 0 \\ 0 & 1 \end{bmatrix}$$

Thus the equation that will yield the highest natural angular frequency is

$$\omega_2^2 X_2 = \frac{K}{M}\begin{bmatrix} 2 & -1 \\ -1 & 2 \end{bmatrix}\begin{bmatrix} X_{12} \\ X_{22} \end{bmatrix}$$

Assume the vector for starting the iterative process to be

$$X_2 = \begin{bmatrix} X_{12} \\ X_{22} \end{bmatrix} = \begin{bmatrix} 1{\cdot}0 \\ 1{\cdot}0 \end{bmatrix}$$

The iterative process then proceeds as follows

$$\text{1st iteration:} \quad \omega_2^2\begin{bmatrix} X_{12} \\ X_{22} \end{bmatrix} = \frac{K}{M}\begin{bmatrix} 2 & -1 \\ -1 & 2 \end{bmatrix}\begin{bmatrix} 1{\cdot}000 \\ 0{\cdot}000 \end{bmatrix}$$

$$= \frac{2{\cdot}000 \times K}{M}\begin{bmatrix} 1{\cdot}000 \\ -0{\cdot}500 \end{bmatrix}$$

$$\text{2nd iteration:} \quad \omega_2^2 \begin{bmatrix} X_{12} \\ X_{22} \end{bmatrix} = \frac{K}{M} \begin{bmatrix} 2 & -1 \\ -1 & 2 \end{bmatrix} \begin{bmatrix} 1{\cdot}000 \\ -0{\cdot}500 \end{bmatrix}$$

$$= \frac{2{\cdot}500 \times K}{M} \begin{bmatrix} 1{\cdot}000 \\ -0{\cdot}800 \end{bmatrix}$$

$$\text{3rd iteration:} \quad \omega_2^2 \begin{bmatrix} X_{12} \\ X_{22} \end{bmatrix} = \frac{K}{M} \begin{bmatrix} 2 & -1 \\ -1 & 2 \end{bmatrix} \begin{bmatrix} 1{\cdot}000 \\ 0{\cdot}800 \end{bmatrix}$$

$$= \frac{2{\cdot}800 \times K}{M} \begin{bmatrix} 1{\cdot}000 \\ -0{\cdot}929 \end{bmatrix}$$

$$\text{4th iteration:} \quad \omega_2^2 \begin{bmatrix} X_{12} \\ X_{22} \end{bmatrix} = \frac{K}{M} \begin{bmatrix} 2 & -1 \\ -1 & 2 \end{bmatrix} \begin{bmatrix} 1{\cdot}000 \\ 0{\cdot}929 \end{bmatrix}$$

$$= \frac{2{\cdot}927 \times K}{M} \begin{bmatrix} 1{\cdot}000 \\ -0{\cdot}976 \end{bmatrix}$$

As the iterative process proceeds, the values for the product $\omega_2{}^2 \boldsymbol{X}_2$ will converge to

$$\omega_2{}^2 \begin{bmatrix} X_{12} \\ X_{22} \end{bmatrix} = \frac{3{\cdot}0 \times K}{M} \begin{bmatrix} 1{\cdot}0 \\ -1{\cdot}0 \end{bmatrix}$$

Thus

$$\omega_2 = \sqrt{\left(\frac{3K}{M}\right)}$$

$$\boldsymbol{X}_2 = \{1{\cdot}0 \quad -1{\cdot}0\}$$

The first eigenvalue is now found by applying theorem 7.1, which yields

$$\omega_1{}^2 + \omega_2{}^2 = \frac{K}{M}(2 + 2)$$

Substitution of the expression for $\omega_2$ into the above equation is determined by applying theorem 7.2. If one sets $X_{11} = 1{\cdot}0$, then

$$[1{\cdot}0 \quad X_{21}] \begin{bmatrix} 1{\cdot}0 \\ -1{\cdot}0 \end{bmatrix} = 0$$

which yields

$$X_{21} = 1{\cdot}0$$

and hence

$$\boldsymbol{X}_i = \{1{\cdot}0 \quad 1{\cdot}0\}$$

**Example 7.6** Use the iterative optimization method to determine the first natural frequency and modeshape vector for the structure given in example 2.5 and shown in Fig. 2.14.

Let the shear stiffness $12EI/L^3$ of each column be $K$ and the mass per span of each floor be $M$. With this notation, the matrix formulation of the equation of motion is

$$M\begin{bmatrix} 3 & 0 & 0 \\ 0 & 2 & 0 \\ 0 & 0 & 1 \end{bmatrix}\begin{bmatrix} \ddot{x}_1 \\ \ddot{x}_2 \\ \ddot{x}_3 \end{bmatrix} + K\begin{bmatrix} 7 & -3 & 0 \\ -3 & 5 & -2 \\ 0 & -2 & 2 \end{bmatrix}\begin{bmatrix} x_1 \\ x_2 \\ x_3 \end{bmatrix} = \begin{bmatrix} 0 \\ 0 \\ 0 \end{bmatrix}$$

Assuming SHM, substitution for $x$ and $\ddot{x}$ yields

$$K\begin{bmatrix} 7 & -3 & 0 \\ -3 & 5 & -2 \\ 0 & -2 & 2 \end{bmatrix}\begin{bmatrix} X_1 \\ X_2 \\ X_3 \end{bmatrix} - \omega^2 M\begin{bmatrix} 3 & 0 & 0 \\ 0 & 2 & 0 \\ 0 & 0 & 1 \end{bmatrix}\begin{bmatrix} X_1 \\ X_2 \\ X_3 \end{bmatrix} = \begin{bmatrix} 0 \\ 0 \\ 0 \end{bmatrix}$$

To determine the lowest eigenvalue, one iterates on eq. (7.27) which requires that the matrix

$$\mathbf{B} = [\alpha\mathbf{I} - \mathbf{M}^{-1}\mathbf{K}]$$

be established. This in turn requires that one first establish the system matrix $\mathbf{M}^{-1}\mathbf{K}$ and then choose a value for $\alpha$

$$\mathbf{M}^{-1}\mathbf{K} = \frac{K}{6M}\begin{bmatrix} 14 & -6 & 0 \\ -9 & 15 & -6 \\ 0 & -12 & 12 \end{bmatrix}$$

The value of $\alpha$ must be greater than the highest eigenvalue. Theorem 7.1 states that the trace of the system matrix is equal to the sum of the eigenvalues. Therefore a value for $\alpha$ equal to the trace is satisfactory, i.e.

$$\alpha = \frac{K}{6K}(14 + 15 + 12) = \frac{41K}{6K}$$

Hence

$$\mathbf{B} = \left[\alpha\mathbf{I} - \mathbf{M}^{-1}\mathbf{K}\right] = \frac{K}{6M}\begin{bmatrix} 27 & 6 & 0 \\ 9 & 26 & 6 \\ 0 & 12 & 29 \end{bmatrix}$$

It remains to assume a suitable initial vector $\boldsymbol{X}_1$. A simple choice would be

$$\boldsymbol{X}_1 = \{1{\cdot}0 \quad 0{\cdot}0 \quad 0{\cdot}0\}$$

Alternatively one can choose the vector used in example 2.5, which assumes that the modeshape is similar to the deflected form caused by a horizontal force applied at each level, and proportional to the weight of the corresponding floor. If the latter is assumed

$$X_1 = \{1{\cdot}0 \quad 1{\cdot}67 \quad 2{\cdot}0\}$$

then from the

$$\text{1st iteration:}\quad \lambda_3 X_1 = \frac{K}{6M}\begin{bmatrix} 27 & 6 & 0 \\ 9 & 26 & 6 \\ 0 & 12 & 29 \end{bmatrix}\begin{bmatrix} 1{\cdot}000 \\ 1{\cdot}670 \\ 2{\cdot}000 \end{bmatrix}$$

$$= \frac{6{\cdot}170 \times K}{M}\begin{bmatrix} 1{\cdot}000 \\ 1{\cdot}740 \\ 2{\cdot}104 \end{bmatrix}$$

$$\text{2nd iteration:}\quad \lambda_3 X_1 = \frac{K}{6M}\begin{bmatrix} 27 & 6 & 0 \\ 9 & 26 & 6 \\ 0 & 12 & 29 \end{bmatrix}\begin{bmatrix} 1{\cdot}000 \\ 1{\cdot}740 \\ 2{\cdot}104 \end{bmatrix}$$

$$= \frac{6{\cdot}240 \times K}{M}\begin{bmatrix} 1{\cdot}000 \\ 1{\cdot}786 \\ 2{\cdot}187 \end{bmatrix}$$

$$\text{3rd iteration:}\quad \lambda_3 X_1 = \frac{K}{6M}\begin{bmatrix} 27 & 6 & 0 \\ 9 & 26 & 6 \\ 0 & 12 & 29 \end{bmatrix}\begin{bmatrix} 1{\cdot}000 \\ 1{\cdot}786 \\ 2{\cdot}187 \end{bmatrix}$$

$$= \frac{6{\cdot}286 \times K}{M}\begin{bmatrix} 1{\cdot}000 \\ 1{\cdot}818 \\ 2{\cdot}250 \end{bmatrix}$$

$$\text{4th iteration:}\quad \lambda_3 X_1 = \frac{K}{6M}\begin{bmatrix} 27 & 6 & 0 \\ 9 & 26 & 6 \\ 0 & 12 & 29 \end{bmatrix}\begin{bmatrix} 1{\cdot}000 \\ 1{\cdot}818 \\ 2{\cdot}250 \end{bmatrix}$$

$$= \frac{6{\cdot}318 \times K}{M}\begin{bmatrix} 1{\cdot}000 \\ 1{\cdot}840 \\ 2{\cdot}297 \end{bmatrix}$$

$$\text{5th iteration:}\quad \lambda_3 X_1 = \frac{K}{6M}\begin{bmatrix} 27 & 6 & 0 \\ 9 & 26 & 6 \\ 0 & 12 & 29 \end{bmatrix}\begin{bmatrix} 1{\cdot}000 \\ 1{\cdot}840 \\ 2{\cdot}297 \end{bmatrix}$$

$$= \frac{6{\cdot}340 \times K}{M}\begin{bmatrix} 1{\cdot}000 \\ 1{\cdot}857 \\ 2{\cdot}322 \end{bmatrix}$$

$$\text{6th iteration:} \quad \lambda_3 X_1 = \frac{K}{6M}\begin{bmatrix} 27 & 6 & 0 \\ 9 & 26 & 6 \\ 0 & 12 & 29 \end{bmatrix}\begin{bmatrix} 1{\cdot}000 \\ 1{\cdot}857 \\ 2{\cdot}322 \end{bmatrix}$$

$$= \frac{6{\cdot}357 \times K}{M}\begin{bmatrix} 1{\cdot}000 \\ 1{\cdot}867 \\ 2{\cdot}350 \end{bmatrix}$$

$$\text{7th iteration:} \quad \lambda_3 X_1 = \frac{K}{6M}\begin{bmatrix} 27 & 6 & 0 \\ 9 & 26 & 6 \\ 0 & 12 & 29 \end{bmatrix}\begin{bmatrix} 1{\cdot}000 \\ 1{\cdot}867 \\ 2{\cdot}350 \end{bmatrix}$$

$$= \frac{6{\cdot}367 \times K}{M}\begin{bmatrix} 1{\cdot}000 \\ 1{\cdot}875 \\ 2{\cdot}370 \end{bmatrix}$$

$$\text{8th iteration:} \quad \lambda_3 X_1 = \frac{K}{6M}\begin{bmatrix} 27 & 6 & 0 \\ 9 & 26 & 6 \\ 0 & 12 & 29 \end{bmatrix}\begin{bmatrix} 1{\cdot}000 \\ 1{\cdot}875 \\ 2{\cdot}370 \end{bmatrix}$$

$$= \frac{6{\cdot}375 \times K}{M}\begin{bmatrix} 1{\cdot}000 \\ 1{\cdot}882 \\ 2{\cdot}385 \end{bmatrix}$$

$$\text{9th iteration:} \quad \lambda_3 X_1 = \frac{K}{6M}\begin{bmatrix} 27 & 6 & 0 \\ 9 & 26 & 6 \\ 0 & 12 & 29 \end{bmatrix}\begin{bmatrix} 1{\cdot}000 \\ 1{\cdot}882 \\ 2{\cdot}385 \end{bmatrix}$$

$$= \frac{6{\cdot}382 \times K}{M}\begin{bmatrix} 1{\cdot}000 \\ 1{\cdot}887 \\ 2{\cdot}396 \end{bmatrix}$$

$$\text{10th iteration:} \quad \lambda_3 X_1 = \frac{K}{6M}\begin{bmatrix} 27 & 6 & 0 \\ 9 & 26 & 6 \\ 0 & 12 & 29 \end{bmatrix}\begin{bmatrix} 1{\cdot}000 \\ 1{\cdot}887 \\ 2{\cdot}396 \end{bmatrix}$$

$$= \frac{6{\cdot}387 \times K}{M}\begin{bmatrix} 1{\cdot}000 \\ 1{\cdot}890 \\ 2{\cdot}404 \end{bmatrix}$$

11th iteration: $$\lambda_3 X_1 = \frac{K}{6M}\begin{bmatrix} 27 & 6 & 0 \\ 9 & 26 & 6 \\ 0 & 12 & 29 \end{bmatrix}\begin{bmatrix} 1{\cdot}000 \\ 1{\cdot}890 \\ 2{\cdot}404 \end{bmatrix}$$

$$= \frac{6{\cdot}390 \times K}{M}\begin{bmatrix} 1{\cdot}000 \\ 1{\cdot}893 \\ 2{\cdot}410 \end{bmatrix}$$

12th iteration: $$\lambda_3 X_1 = \frac{K}{6M}\begin{bmatrix} 27 & 6 & 0 \\ 9 & 26 & 6 \\ 0 & 12 & 29 \end{bmatrix}\begin{bmatrix} 1{\cdot}000 \\ 1{\cdot}893 \\ 2{\cdot}410 \end{bmatrix}$$

$$= \frac{6{\cdot}393 \times K}{M}\begin{bmatrix} 1{\cdot}000 \\ 1{\cdot}895 \\ 2{\cdot}414 \end{bmatrix}$$

13th iteration: $$\lambda_3 X_1 = \frac{K}{6M}\begin{bmatrix} 27 & 6 & 0 \\ 9 & 26 & 6 \\ 0 & 12 & 29 \end{bmatrix}\begin{bmatrix} 1{\cdot}000 \\ 1{\cdot}895 \\ 2{\cdot}414 \end{bmatrix}$$

$$= \frac{6{\cdot}395 \times K}{M}\begin{bmatrix} 1{\cdot}000 \\ 1{\cdot}896 \\ 2{\cdot}417 \end{bmatrix}$$

14th iteration: $$\lambda_3 X_1 = \frac{K}{6M}\begin{bmatrix} 27 & 6 & 0 \\ 9 & 26 & 6 \\ 0 & 12 & 29 \end{bmatrix}\begin{bmatrix} 1{\cdot}000 \\ 1{\cdot}896 \\ 2{\cdot}417 \end{bmatrix}$$

$$= \frac{6{\cdot}396 \times K}{M}\begin{bmatrix} 1{\cdot}000 \\ 1{\cdot}897 \\ 2{\cdot}419 \end{bmatrix}$$

15th iteration: $$\lambda_3 X_1 = \frac{K}{6M}\begin{bmatrix} 27 & 6 & 0 \\ 9 & 26 & 6 \\ 0 & 12 & 29 \end{bmatrix}\begin{bmatrix} 1{\cdot}000 \\ 1{\cdot}897 \\ 2{\cdot}419 \end{bmatrix}$$

$$= \frac{6{\cdot}397 \times K}{M}\begin{bmatrix} 1{\cdot}000 \\ 1{\cdot}898 \\ 2{\cdot}421 \end{bmatrix}$$

From eq. (7.27)

$$\lambda_3 = \alpha - \omega_1{}^2$$

Hence

$$\omega_1{}^2 = \frac{K(6{\cdot}8333333 - 6{\cdot}397)}{M} = \frac{0{\cdot}4363333 \times K}{M}$$

where

$$K = \frac{12EI}{4{\cdot}0^3}$$

$$M = \frac{10w}{g}$$

Substitution of the expressions for $K$ and $M$ into the expression for $\omega_1{}^2$ yields

$$f_1 = 0{\cdot}0143956\sqrt{\left(\frac{EIg}{w}\right)}$$

This implies that the frequency, whose value is given in example 2.4, after 15 iterations has converged to within 0·23% of the correct value, with a corresponding modeshape of

$$\boldsymbol{X}_1 = \{\,1{\cdot}000 \quad 1{\cdot}898 \quad 2{\cdot}421\,\}$$

The highest natural frequency can be determined by iterations on eq. (7.23) and the second frequency by applying theorem 7.1. Alternatively, all the eigenvalues may be determined by setting up and solving the characteristic equation, which in this case is a cubic equation in $\lambda$. This can be solved either graphically or by application of the theory for solving cubic equations. Alternatively, having determined one eigenvalue the characteristic polynomial can be reduced by factorization, in which case the resulting quadratic characteristic equation can be solved by using the standard formula for determining the roots of such equations.

## The Rayleigh quotient

The eigenvalue equation for a general $N$ DOF system is given by eq. (7.22). Premultiplication of each term by $\boldsymbol{X}^{\mathrm{T}}$ yields

$$\boldsymbol{X}^{\mathrm{T}}\mathbf{K}\boldsymbol{X} - \omega^2\boldsymbol{X}^{\mathrm{T}}\mathbf{M}\boldsymbol{X} = 0 \tag{7.29}$$

Hence

$$\omega^2 = \frac{\boldsymbol{X}^{\mathrm{T}}\mathbf{K}\boldsymbol{X}}{\boldsymbol{X}^{\mathrm{T}}\mathbf{M}\boldsymbol{X}} \tag{7.30}$$

The expression for the square of the natural frequency given by eq. (7.30) is referred to as the Rayleigh quotient. It has the property that for even approximately correct values of the eigenvectors or modeshape vectors the values for the frequencies are reasonably correct, as demonstrated by examples 2.1, 2.2, 2.5 and 2.6 where the quotient is used without explicitly stating it. That this is the case can be seen simply by premultiplying each term in eq. (7.29) by one-half. This yields

$$\boldsymbol{X}^{\mathrm{T}}\mathbf{K}\boldsymbol{X} = \omega^2 \boldsymbol{X}^{\mathrm{T}}\mathbf{M}\boldsymbol{X} \tag{7.31}$$

which states that the maximum strain energy is equal to the maximum kinetic energy.

## Condensation of the stiffness matrix in lumped mass analysis

When the mass of a structure is assumed to be concentrated at the nodes, it is usual to consider only the inertia due to translational movements and to ignore that due to rotation. This assumes that the lumped masses are concentrated as point masses with radii of gyration equal to zero. Thus in the case of flexible structures where the joints rotate, the elements on the leading diagonal of the mass matrix corresponding to the rotational degrees of freedom will be zero. In such cases the mass matrix cannot be inverted. Therefore the elements related to rotation need to be eliminated by condensing the stiffness matrix. Condensation of the stiffness matrix may also be desirable to reduce the overall degree of freedom of structures with a very large number of DOF in order to reduce the numerical problem. Assume that the degrees of freedom to be reduced or condensed are the first $\theta$ unknown rotations, and carry out a Gauss–Jordan elimination of these coordinates. After this elimination process the stiffness equation may be arranged in partitioned form as follows

$$\begin{bmatrix} \mathbf{I} & -\tilde{\mathbf{T}} \\ \mathbf{0} & \bar{\mathbf{K}} \end{bmatrix} \begin{bmatrix} \boldsymbol{\theta} \\ \boldsymbol{x} \end{bmatrix} = \begin{bmatrix} \mathbf{0} \\ \mathbf{P} \end{bmatrix} \tag{7.32}$$

where $\boldsymbol{\theta}$ is the displacement vector corresponding to the $\theta$ degrees of freedom to be reduced and $\boldsymbol{x}$ is the vector corresponding to the remaining $x$ independent DOF. It should be noted that in eq. (7.32) it is assumed that at the dependent degrees of freedom $\theta$ the external forces are zero. Equation (7.32) is equivalent to the following two relationships

$$\boldsymbol{\theta} = \tilde{\mathbf{T}}\boldsymbol{x} \tag{7.33}$$

$$\tilde{\mathbf{K}}\boldsymbol{x} = \mathbf{P} \tag{7.34}$$

Equation (7.33), which expresses the relationship between the displacement vectors $\boldsymbol{x}$ and $\boldsymbol{\theta}$, may also be written as

$$\begin{bmatrix} \boldsymbol{\theta} \\ \boldsymbol{x} \end{bmatrix} = \begin{bmatrix} \tilde{\mathbf{T}} \\ \mathbf{I} \end{bmatrix} \begin{bmatrix} x \end{bmatrix} \tag{7.35}$$

In equation (7.34), which shows the relationship between the displacement vector $\boldsymbol{x}$ and the force vector $\boldsymbol{P}$, $\tilde{\mathbf{K}}$ is the reduced stiffness matrix. $\tilde{\mathbf{K}}$ may also be expressed by the following transformation of the system matrix

$$\tilde{\mathbf{K}} = \mathbf{T}^{\mathrm{T}}\mathbf{K}\mathbf{T} \tag{7.36}$$

where

$$\mathbf{T} = \begin{bmatrix} \tilde{\mathbf{T}} \\ \mathbf{I} \end{bmatrix} \tag{7.37}$$

Similarly, the mass and the damping matrix (the latter is introduced in chapter 8) may be reduced by the transformations

$$\tilde{\mathbf{M}} = \mathbf{T}^{\mathrm{T}}\mathbf{M}\mathbf{T} \tag{7.38}$$

$$\tilde{\mathbf{C}} = \mathbf{T}^{\mathrm{T}}\mathbf{C}\mathbf{T} \tag{7.39}$$

where the transformation matrix $\mathbf{T}$ is given by eq. (7.37).

**Example 7.7** Reduce the DOF of the three-storey shear structure in example 2.5 (Fig. 2.14) to a one DOF system by eliminating the translational displacements at the first and second floor levels. Hence calculate the natural frequency and compare the value obtained with those obtained previously.

A Gauss–Jordan elimination of the elements in the stiffness matrix corresponding to the displacements at the first and second floor levels results in the following transformation

$$\mathbf{K} = K\begin{bmatrix} 7 & -3 & 0 \\ -3 & 5 & -2 \\ 0 & -2 & 2 \end{bmatrix} \rightarrow K\begin{bmatrix} 1 & 0 & -\frac{3}{7} \\ 0 & 1 & -\frac{7}{13} \\ 0 & 0 & \frac{6}{13} \end{bmatrix}$$

Hence

$$\tilde{\mathbf{K}} = \frac{6}{13}K$$

The corresponding reduced mass matrix is found through the transformation

$$\tilde{\mathbf{M}} = \mathbf{T}^{\mathrm{T}}\mathbf{M}\mathbf{T} = M\begin{bmatrix} \frac{3}{7} & \frac{7}{13} & 1 \end{bmatrix}\begin{bmatrix} 3 & 0 & 0 \\ 0 & 2 & 0 \\ 0 & 0 & 1 \end{bmatrix}\begin{bmatrix} \frac{3}{7} \\ \frac{7}{13} \\ 1 \end{bmatrix} = \frac{9365}{8281}M$$

Hence

$$\omega^2 = \frac{6 \times 8281 \times K}{13 \times 9365 \times M} = \frac{0{\cdot}4081153 \times K}{M}$$

where as before

$$K = \frac{12EI}{L^3}$$
$$M = \frac{10w}{g}$$

Substitution of the expressions for $K$ and $M$ into the expression for $\omega^2$ yields

$$f = 0{\cdot}0139223\sqrt{\left(\frac{EIg}{w}\right)}$$

This represents an error of 3·96% as compared with the 4·28% error resulting from the much simpler method of reduction used in chapter 2. The above reduction can be checked by substitution of the values for the matrices **T** and **K** into eq. (7.36) and carrying out the implied matrix multiplications.

## Consistent mass matrices

The modelling of structural mass by lumped mass matrices usually leads to satisfactorily accurate values for the frequencies, and has the advantage of reducing the amount of computer storage and calculations involved in solving the eigenvalue problem. In the case of buildings the total mass will vary with the usage, and it is usually difficult to estimate the mass and mass distribution accurately. This further justifies the lumped mass approach. From a computational point of view, however, it is probably equally convenient to set up a mass matrix that takes account of the distribution of the mass in individual members by using *consistent* element mass matrices. For plane frames it can be shown that the relationship between the inertia force vector, mass matrix and acceleration vector for an uniform element of length $L$ and mass $m$ per unit length is given by

$$\begin{bmatrix} I_{x1} \\ I_{y1} \\ I_{\theta 1} \\ I_{x2} \\ I_{y2} \\ I_{\theta 2} \end{bmatrix} = \frac{mL}{420} \begin{bmatrix} 140 & 0 & 0 & 70 & 0 & 0 \\ 0 & 156 & 22L & 0 & 54 & -13L \\ 0 & 22L & 4L^2 & 0 & 13L & -3L^2 \\ 70 & 0 & 0 & 140 & 0 & 0 \\ 0 & 54 & 13L & 0 & 156 & -22L \\ 0 & -13L & -3L^2 & 0 & -22L & 4L^2 \end{bmatrix} \begin{bmatrix} \ddot{x} \\ \ddot{y} \\ \ddot{\theta} \\ \ddot{x} \\ \ddot{y} \\ \ddot{\theta} \end{bmatrix} \tag{7.40}$$

from which it can be seen that the consistent mass matrix has the same banded form as the stiffness matrix for the member. Similar 12 × 12 mass matrices can also be set up for three-dimensional structures. Consistent mass matrices for both two-dimensional and three-dimensional structures can be reduced through the transformation given by eq. (7.38) by, for example, eliminating the rotational coordinates.

**Example 7.8** Construct the stiffness and mass matrices for the stepped antenna-mast shown in Fig. 7.3, assuming that the flexural rigidity and mass of the lower half of the mast are $2EI$ and 2 m per unit length, and the flexural rigidity and mass of the top half of the mast are $EI$ and $m$ per unit length. Ignoring axial stiffness, condense the matrices to include horizontal translations only.

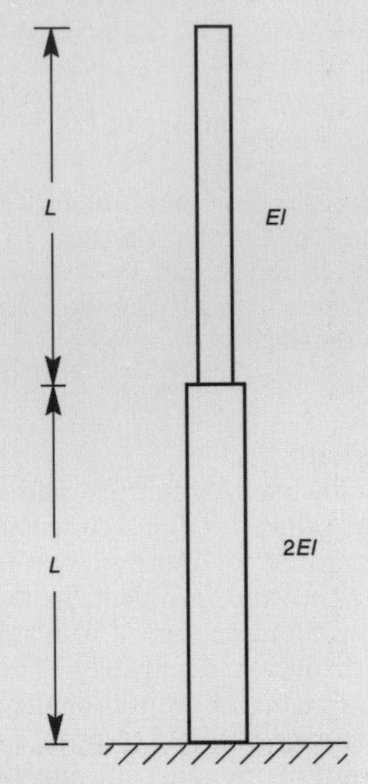


*Fig. 7.3. Stepped antenna-mast*

The general stiffness matrix for a plane frame member, ignoring the axial stiffness, is given by

$$\mathbf{K} = EI \begin{bmatrix} 12/L^3 & 6/L^2 & -12/L^3 & 6/L^2 \\ 6/L^2 & 4/L & -6/L^2 & 2/L \\ -12/L^3 & -6/L^2 & 12/L^3 & -6/L^2 \\ 6/L^2 & 2/L & -6/L^2 & 4/L \end{bmatrix}$$

The general mass matrix for a plane frame member, ignoring the axial inertia forces, is

$$\mathbf{M} = \frac{mL}{420} \begin{bmatrix} 156 & 22L & 54 & -13L \\ 22L & 4L^2 & 13L & -3L^2 \\ 54 & 13L & 156 & -22L \\ -13L & -3L^2 & -22L & 4L^2 \end{bmatrix}$$

The stiffness matrix for the mast is now constructed as indicated by

$$\mathbf{K} = \begin{bmatrix} \mathbf{K}_{22}^{(1)} + \mathbf{K}_{11}^{(2)} & \mathbf{K}_{12}^{(2)} \\ \mathbf{K}_{21}^{(2)} & \mathbf{K}_{22}^{(2)} \end{bmatrix}$$

Hence

$$\mathbf{K}\boldsymbol{x} = EI \begin{bmatrix} 36/L^3 & -6/L^2 & -12/L^3 & 6/L^2 \\ -6/L^2 & 12/L & -6/L^2 & 2/L \\ -12/L^3 & -6/L^2 & 12/L^3 & -6/L^2 \\ 6/L^2 & 2/L & -6/L^2 & 4/L \end{bmatrix} \begin{bmatrix} x_1 \\ \theta_1 \\ x_2 \\ \theta_2 \end{bmatrix}$$

Rearrangement of $\mathbf{K}$ in order to reduce it by eliminating $\theta_1$ and $\theta_2$ yields

$$\mathbf{K}\boldsymbol{x} = EI \begin{bmatrix} 12/L & 2/L & -6/L^2 & -6/L^2 \\ 2/L & 4/L & 6/L^2 & -6/L^2 \\ -6/L^2 & 6/L^2 & 36/L^3 & -12/L^3 \\ -6/L^2 & -6/L^2 & -12/L^3 & 12/L^3 \end{bmatrix} \begin{bmatrix} \theta_1 \\ \theta_2 \\ x_1 \\ x_2 \end{bmatrix}$$

Finally, reduction of the matrix by the Gauss–Jordan elimination process leads to

$$\mathbf{K}\boldsymbol{x} = EI \begin{bmatrix} 1 & 0 & -9/11L & -3/11L \\ 0 & 1 & 21/11L & -15/11L \\ 0 & 0 & 216/11L^3 & -60/11L^3 \\ 0 & 0 & -60/11L^3 & 54/11L^3 \end{bmatrix} \begin{bmatrix} \theta_1 \\ \theta_2 \\ x_1 \\ x_2 \end{bmatrix}$$

Hence the condensed stiffness matrix $\tilde{\mathbf{K}}$ and associated transformation matrix $\mathbf{T}$ are given by

$$\tilde{\mathbf{K}} = \frac{6EI}{11L^3} \begin{bmatrix} 36 & -10 \\ -10 & 9 \end{bmatrix}$$

$$\mathbf{T} = \begin{bmatrix} 9/11L & 3/11L \\ -21/11L & 15/11L \\ 1 & 0 \\ 0 & 1 \end{bmatrix}$$

The mass matrix for the mast is assembled in exactly the same way as the stiffness matrix. Hence

$$\mathbf{M}\ddot{\boldsymbol{x}} = \frac{mL}{420} \begin{bmatrix} 468 & -22L & 54 & -13L \\ -22L & 12L^2 & 13L & -3L^2 \\ 54 & 13L & 156 & -22L \\ -13L & -3L^2 & -22L & 4L^2 \end{bmatrix} \begin{bmatrix} \ddot{x}_1 \\ \ddot{\theta}_1 \\ \ddot{x}_2 \\ \ddot{\theta}_2 \end{bmatrix}$$

Transformation of the above mass matrix to conform with the Gauss–Jordan elimination of $\theta_1$ and $\theta_2$ yields

$$\mathbf{M}\ddot{x} = \frac{mL}{420}\begin{bmatrix} 12L^2 & -3L^2 & -22L & 13L \\ -3L^2 & 4L^2 & -13L & -22L \\ -22L & -13L & 468 & 54 \\ 13L & -22L & 54 & 156 \end{bmatrix}\begin{bmatrix} \theta_1 \\ \theta_2 \\ x_1 \\ x_2 \end{bmatrix}$$

The condensed mass matrix is now found through the transformation given by eq. (7.38) as

$$\tilde{\mathbf{M}} = \mathbf{T}^{\mathrm{T}}\mathbf{M}\mathbf{T}$$

$$\tilde{\mathbf{M}} = \frac{mL}{50\,820}\begin{bmatrix} 62\,148 & 8880 \\ 8880 & 13\,212 \end{bmatrix}$$

## Orthogonality and normalization of eigenvectors

Before proceeding to determine the dynamic response of multi-DOF structures it is necessary to consider what are known as the orthogonality properties of the eigenvectors. Let $\lambda_i$ and $\lambda_j$ be two of the eigenvalues corresponding to the eigenvectors or modeshape vectors $X_i$ and $X_j$, where $i \neq j$, of a multi-DOF system represented by the eigenvalue equation

$$\mathbf{K}X - \lambda\mathbf{M}X = 0 \tag{7.41}$$

Hence

$$\mathbf{K}X_i - \lambda_i\mathbf{M}X_i = 0 \tag{7.42}$$

$$\mathbf{K}X_j - \lambda_j\mathbf{M}X_j = 0 \tag{7.43}$$

Transposition of each term in eq. (7.42) yields

$$X_i{}^{\mathrm{T}}\mathbf{K}^{\mathrm{T}} - \lambda_i X_i{}^{\mathrm{T}}\mathbf{M}^{\mathrm{T}} = 0 \tag{7.44}$$

If both **K** and **M** are symmetric matrices, then

$$X_i{}^{\mathrm{T}}\mathbf{K} - \lambda_i X_i{}^{\mathrm{T}}\mathbf{M} = 0 \tag{7.45}$$

Postmultiplication of each term in eq. (7.45) by $X_i$ yields

$$X_i{}^{\mathrm{T}}\mathbf{K}X_j - \lambda_i X_i{}^{\mathrm{T}}\mathbf{M}X_j = 0 \tag{7.46}$$

Premultiplication of each term in eq. (7.43) by $X_i{}^{\mathrm{T}}$ yields

$$X_i{}^{\mathrm{T}}\mathbf{K}X_j - \lambda_j X_k{}^{\mathrm{T}}\mathbf{M}X_j = 0 \tag{7.47}$$

Finally, subtraction of eq. (7.47) from eq. (7.46) yields

$$(\lambda_j - \lambda_i)X_i{}^{\mathrm{T}}\mathbf{M}X_j = 0 \tag{7.48}$$

Since $\lambda_j \neq \lambda_i$, it follows that

$$X_i{}^{\mathrm{T}}\mathbf{M}X_j = 0 \tag{7.49}$$

If the zero value for $\boldsymbol{X}_i{}^{\mathrm{T}}\mathbf{M}\boldsymbol{X}_j$ is substituted into either eq. (7.46) or eq. (7.47), it follows also that

$$\boldsymbol{X}_i{}^{\mathrm{T}}\mathbf{K}\boldsymbol{X}_j = 0 \tag{7.50}$$

The relationships given by eqs (7.49) and (7.50) still apply if the eigenvectors are normalized. Let

$$\boldsymbol{X}_i{}^{\mathrm{T}}\mathbf{M}\boldsymbol{X}_i = \tilde{M}_i \tag{7.51a}$$

$$\boldsymbol{X}_j{}^{\mathrm{T}}\mathbf{M}\boldsymbol{X}_j = \tilde{M}_j \tag{7.51b}$$

and hence

$$\boldsymbol{Z}_i = \boldsymbol{X}_i/\sqrt{\tilde{M}_i} \tag{7.52a}$$

$$\boldsymbol{Z}_j = \boldsymbol{X}_j/\sqrt{\tilde{M}_j} \tag{7.52b}$$

where $\boldsymbol{Z}_i$ and $\boldsymbol{Z}_j$ are the normalized eigenvectors of vectors $\boldsymbol{X}_i$ and $\boldsymbol{X}_j$ with respect to $\mathbf{M}$. Hence

$$\boldsymbol{Z}_i{}^{\mathrm{T}}\mathbf{M}\boldsymbol{Z}_j = \boldsymbol{X}_i{}^{\mathrm{T}}\mathbf{M}\boldsymbol{X}_j/\sqrt{M_i}\sqrt{M_j} = 0 \tag{7.53}$$

$$\boldsymbol{Z}_i{}^{\mathrm{T}}\mathbf{M}\boldsymbol{Z}_i = \boldsymbol{X}_i{}^{\mathrm{T}}\mathbf{M}\boldsymbol{X}_i/\tilde{M}_i = 1 \tag{7.54a}$$

$$\boldsymbol{Z}_j{}^{\mathrm{T}}\mathbf{M}\boldsymbol{Z}_j = \boldsymbol{X}_j{}^{\mathrm{T}}\mathbf{M}\boldsymbol{X}_j/\tilde{M}_j = 1 \tag{7.54b}$$

Premultiplication of eq. (7.42) by $\boldsymbol{X}_i{}^{\mathrm{T}}$ and eq. (7.43) by $\boldsymbol{X}_j{}^{\mathrm{T}}$ and then substitution of $\sqrt{(M_i)}\boldsymbol{Z}_i$ for $\boldsymbol{X}_i$ and $\sqrt{(M_j)}\boldsymbol{Z}_j$ for $\boldsymbol{X}_j$ in the resulting equations yields

$$\boldsymbol{Z}_i{}^{\mathrm{T}}\mathbf{K}\boldsymbol{Z}_i - \lambda_i\boldsymbol{Z}_i{}^{\mathrm{T}}\mathbf{M}\boldsymbol{Z}_i = 0 \tag{7.55a}$$

$$\boldsymbol{Z}_j{}^{\mathrm{T}}\mathbf{K}\boldsymbol{Z}_j - \lambda_j\boldsymbol{Z}_j{}^{\mathrm{T}}\mathbf{M}\boldsymbol{Z}_j = 0 \tag{7.55b}$$

From eq. (7.53) it follows that

$$\boldsymbol{Z}_j{}^{\mathrm{T}}\mathbf{M}\boldsymbol{Z}_j = 0 \tag{7.56}$$

Since

$$\boldsymbol{Z}_j{}^{\mathrm{T}}\mathbf{M}\boldsymbol{Z}_j = \boldsymbol{Z}_j{}^{\mathrm{T}}\mathbf{M}\boldsymbol{Z}_j = 1$$

it follows that

$$\boldsymbol{Z}_j{}^{\mathrm{T}}\mathbf{K}\boldsymbol{Z}_i = \lambda_i = \omega_i{}^2 \tag{7.57a}$$

$$\boldsymbol{Z}_j{}^{\mathrm{T}}\mathbf{K}\boldsymbol{Z}_j = \lambda_j = \omega_j{}^2 \tag{7.57b}$$

The matrix $\mathbf{Z}$ in which the columns are the normalized eigenvectors

$$\boldsymbol{Z}_1, \boldsymbol{Z}_2, \ldots, \boldsymbol{Z}_i, \ldots, \boldsymbol{Z}_j, \ldots, \boldsymbol{Z}_N$$

is referred to as the *modal* or *modeshape matrix* of the dynamic matrix $\mathbf{M}^{-1}\mathbf{K}$. From eqs (7.55a), (7.55b), (7.56a) and (7.56b) it follows that

$$\boldsymbol{Z}^{\mathrm{T}}\mathbf{M}\boldsymbol{Z} = \mathbf{I} \tag{7.57}$$

$$\boldsymbol{Z}^{\mathrm{T}}\mathbf{K}\boldsymbol{Z} = \lambda \tag{7.58}$$

where $\mathbf{I}$ is the identity or unit matrix and $\lambda$ is the diagonal matrix

$$\lambda = \mathrm{diag}\{\lambda_1, \lambda_2, \ldots, \lambda_i, \ldots, \lambda_j, \ldots, \lambda_N\} \tag{7.59}$$

**Example 7.9** Normalize the eigenvectors calculated in example 7.3 with respect to the mass matrix, and write down the normalized modeshape matrix. The weight of the floors is 20·0 kN/m.

$\tilde{M}_1 = X_1^{T}\mathbf{M}X_1$

$$= [1{\cdot}0000\ 1{\cdot}8997\ 2{\cdot}4256]\begin{bmatrix} 3M & 0 & 0 \\ 0 & 2M & 0 \\ 0 & 0 & M \end{bmatrix}\begin{bmatrix} 1{\cdot}0000 \\ 1{\cdot}8997 \\ 2{\cdot}4256 \end{bmatrix} = 16{\cdot}101256M$$

$\tilde{M}_2 = X_2^{T}\mathbf{M}X_2$

$$= [1{\cdot}0000\ 0{\cdot}1405\ -1{\cdot}4578]\begin{bmatrix} 3M & 0 & 0 \\ 0 & 2M & 0 \\ 0 & 0 & M \end{bmatrix}\begin{bmatrix} 1{\cdot}0000 \\ 0{\cdot}1405 \\ -1{\cdot}4578 \end{bmatrix} = 5{\cdot}1646613M$$

$\tilde{M}_3 = X_3^{T}\mathbf{M}X_3$

$$= [1{\cdot}0000\ -1{\cdot}8736\ 1{\cdot}6979]\begin{bmatrix} 3M & 0 & 0 \\ 0 & 2M & 0 \\ 0 & 0 & M \end{bmatrix}\begin{bmatrix} 1{\cdot}0000 \\ -1{\cdot}8736 \\ 1{\cdot}6979 \end{bmatrix} = 12{\cdot}902600M$$

$$Z_1 = X_1/\sqrt{\tilde{M}_1}$$

$$Z_2 = X_2/\sqrt{\tilde{M}_2}$$

$$Z_3 = X_3/\sqrt{\tilde{M}_3}$$

Hence

$$\mathbf{Z} = \begin{bmatrix} 1{\cdot}7454 & 3{\cdot}0818 & 1{\cdot}9498 \\ 3{\cdot}3157 & 0{\cdot}4330 & -3{\cdot}6531 \\ 4{\cdot}2336 & -4{\cdot}4926 & 3{\cdot}3105 \end{bmatrix} \times 10^{-3}$$

**Problem 7.1** Formulate the equations of motion for the free vibration of the three-storey shear structure shown in Fig. 7.4. Assume the mass of each floor to be $M$, the shear stiffness of each column to be $K/2$, and the damping to be negligible. The weight of the columns may be ignored. Establish the characteristic equation for the building. Solve the equation by plotting the values of the characteristic polynomial versus increasing values of $\omega^2$. Hence determine the modeshapes of vibration by substitution of the obtained values for $\omega^2$ into the equations of motion. Normalize the modeshape vectors and give the resulting modeshape matrix.

**Problem 7.2** Use Newton's approximation method to solve the characteristic polynomial established for the structure in problem 7.1.

**Problem 7.3** Use iterative procedures to determine the first and third eigenvalues for the structure shown in Fig. 7.4. Hence determine the second eigenvalue, and the natural frequencies of the building. Finally, establish the eigenvectors and check the results by applying the orthogonality properties of eigenvectors.

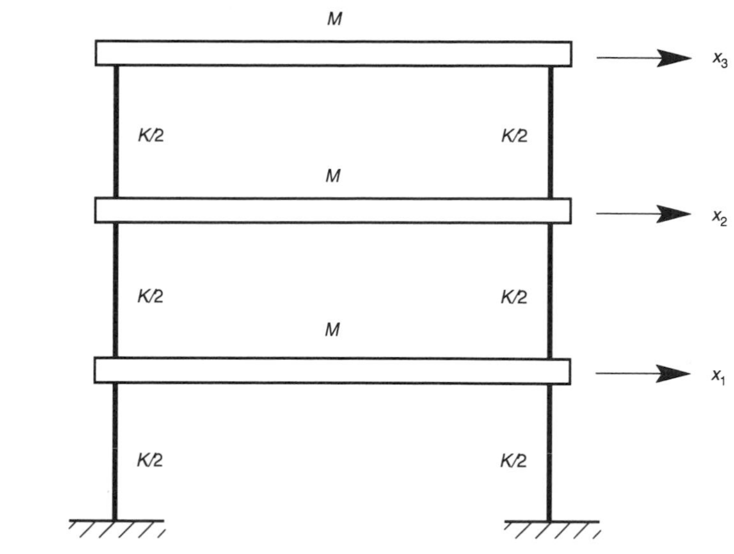


*Fig. 7.4. Three-storey shear structure*

**Problem 7.4** Determine the *EI* values for the two sections of the antenna-mast in example 7.8 if each section is 10·0 m long, the mast supports a disc of mass 500 kg at the top, and the first natural frequency has to be equal to or greater than 4·0 Hz. Assume the mass of the upper half of the mast to be 1600 kg/m and that of the lower half to be 3200 kg/m. Determine also the first and second modeshapes.

## References

Clough, R. W. & Penzien, J. *Dynamics of structures.* McGraw-Hill, London, 1975.
Coates, R. C., Coutie, M. G. & Kong, F. K. *Structural analysis.* Nelson, London, 1972.
Harris, C. M. *Shock vibration*, 3rd edn. McGraw-Hill, London, 1988.
Kreider, D. L. *et al. An introduction to linear analysis.* Addison-Wesley, London, 1966.
Paz, M. *Structural dynamics.* Van Nostrand Reinhold, New York, 1980.
Stroud, K. A. *Engineering mathematics.* Macmillan, London, 1970.

# 8. Forced harmonic vibration of multi-degrees of freedom systems

## Introduction

Structures when excited by random forces such as wind, waves and earthquakes will respond in a number of different modes, although most civil engineering structures respond mainly in the first mode. Particularly in the case of line-like structures such as towers and chimneys, response in higher modes will contribute to the maximum stresses and strain set up in the structure. Thus it is necessary to take account of the contribution from these modes. In the case of linear structures this can be done by calculating the response in the individual modes and then applying the principle of superimposition. In order to present the method of approach it is, as in the case of free vibration, necessary only to consider two and three DOF systems, as the principles applied for the solution of the equations of motion for these systems are equally applicable to structures with more DOF. This chapter presents not only methods for solving the equations of motion, but also methods for constructing suitable damping matrices that as far as possible will model the damping in the different structural modes correctly. Before taking damping into account it is convenient first to consider the problem of solving an undamped two DOF system subjected to harmonic excitation.

## Forced vibration of undamped two DOF systems

Consider the two DOF system shown in Fig. 8.1, where the two masses are acted upon by the two pulsating forces $P_1 \sin(\omega_1 t)$ and $P_2 \sin(\omega_2 t)$ as

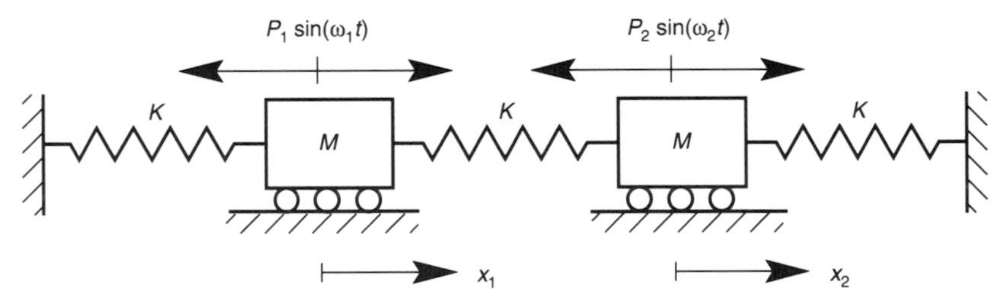


*Fig. 8.1. Two DOF lumped mass–spring system acted on by harmonic forces*

shown. The equation of motion for this system is given by

$$\begin{bmatrix} M & 0 \\ 0 & M \end{bmatrix} \begin{bmatrix} \ddot{x}_1 \\ \ddot{x}_2 \end{bmatrix} + \begin{bmatrix} 2K & -K \\ -K & 2K \end{bmatrix} \begin{bmatrix} x_1 \\ x_2 \end{bmatrix} = \begin{bmatrix} P_1 \sin(\omega_1 t) \\ P_2 \sin(\omega_2 t) \end{bmatrix} \tag{8.1}$$

or

$$\mathbf{M}\ddot{\boldsymbol{x}} + \mathbf{K}\boldsymbol{x} = P(\boldsymbol{t}) \tag{8.2}$$

Inspection of eq. (8.1) does not immediately indicate a straightforward method of solution. In the following it is therefore demonstrated how the use of the eigenvectors of the equations of motion for free vibrations can be used to reduce the two DOF system shown in Fig. 8.1 to two equivalent one DOF systems by decoupling the equations of motion. In example 7.1 it is shown that the frequencies and modeshapes for the mass–spring system are $\omega_1 = \sqrt{(K/M)}$ and $\omega_2 = \sqrt{(3K/M)}$, and $\mathbf{X}_1 = \{1, \ 1\}$ and $\mathbf{X}_2 = \{1, \ -1\}$. Thus the modeshape matrix $\mathbf{X}$ is given by

$$\mathbf{X} = \begin{bmatrix} 1 & 1 \\ 1 & -1 \end{bmatrix}$$

Let

$$\boldsymbol{x} = \mathbf{X}\boldsymbol{q}$$

and hence

$$\ddot{\boldsymbol{x}} - \mathbf{X}\ddot{\boldsymbol{q}} \tag{8.3}$$

Substitution of these expressions for $\ddot{\boldsymbol{x}}$ and $\boldsymbol{x}$ into eq. (8.2) yields

$$\mathbf{MX}\ddot{\boldsymbol{q}} + \mathbf{KX}\boldsymbol{q} = \mathbf{P}(\boldsymbol{t}) \tag{8.4}$$

Premultiplication of each term in eq. (8.4) by $\mathbf{X}^{\mathrm{T}}$ yields

$$\mathbf{X}^{\mathrm{T}}\mathbf{MX}\ddot{\boldsymbol{q}} + \mathbf{X}^{\mathrm{T}}\mathbf{KX}\boldsymbol{q} = \mathbf{X}^{\mathrm{T}}\mathbf{P}(\boldsymbol{t}) \tag{8.5}$$

Hence

$$\begin{aligned} &\begin{bmatrix} 1 & 1 \\ 1 & -1 \end{bmatrix} \begin{bmatrix} M & 0 \\ 0 & M \end{bmatrix} \begin{bmatrix} 1 & 1 \\ 1 & -1 \end{bmatrix} \begin{bmatrix} \ddot{q}_1 \\ \ddot{q}_2 \end{bmatrix} \\ &+ \begin{bmatrix} 1 & 1 \\ 1 & -1 \end{bmatrix} \begin{bmatrix} 2K & -K \\ -K & 2K \end{bmatrix} \begin{bmatrix} 1 & 1 \\ 1 & -1 \end{bmatrix} \begin{bmatrix} q_1 \\ q_2 \end{bmatrix} \\ &= \begin{bmatrix} 1 & 1 \\ 1 & -1 \end{bmatrix} \begin{bmatrix} P_1(t) \\ P_2(t) \end{bmatrix} \end{aligned} \tag{8.6}$$

The matrix multiplications yield

$$\begin{bmatrix} 2M & 0 \\ 0 & 2M \end{bmatrix} \begin{bmatrix} \ddot{q}_1 \\ \ddot{q}_2 \end{bmatrix} + \begin{bmatrix} 2K & 0 \\ 0 & 6K \end{bmatrix} \begin{bmatrix} q_1 \\ q_2 \end{bmatrix} = \begin{bmatrix} P_1(t) + P_2(t) \\ P_1(t) - P_2(t) \end{bmatrix} \tag{8.7}$$

Since the stiffness matrix as a result of this operation has been diagonalized

$$2M\ddot{q}_1 + 2Kq_1 = P_1(t) + P_2(t) \tag{8.8a}$$

$$2M\ddot{q}_2 + 6Kq_2 = P_1(t) - P_2(t) \tag{8.8b}$$

If

$$P_1 \sin(\omega_1 t) = P_2 \sin(\omega_2 t) = P \sin(\omega t)$$

then

$$2M\ddot{q}_1 + 2K\ddot{q}_1 = 2P \sin(\omega t) \tag{8.9a}$$

$$2M\ddot{q}_2 + 6Kq_2 = 0 \tag{8.9b}$$

Thus the equations of motion for the two DOF mass–spring system have been transformed to two *decoupled* equations, each having the same form as the equation of motion for a one DOF system. It should be noted that the natural frequencies of the two equivalent one DOF systems represented by eqs (8.8a) and (8.8b) or eqs (8.9a) and (8.9b) are $\sqrt{(K/M)}$ and $\sqrt{(3K/M)}$ respectively, and therefore are the same as the first and second natural frequencies of the original two DOF system.

From eqs (4.15) and (4.12), the response of an undamped one DOF system to harmonic excitation, since $\xi = 0$, is

$$x = \frac{P_0}{K}\frac{1}{1-r^2}\sin(\omega t) \tag{8.10}$$

The solutions to eqs (8.9a) and (8.9b) are therefore

$$\begin{aligned} q_1 &= \frac{2P}{K - M\omega^2}\sin(\omega t) \\ q_2 &= 0 \end{aligned} \tag{8.11}$$

Substitution of the expressions for $q_1$ and $q_2$ into eq. (8.3) yields

$$\mathbf{X} = \begin{bmatrix} 1 & 1 \\ 1 & -1 \end{bmatrix} \begin{bmatrix} \dfrac{2P}{K - M\omega^2}\sin(\omega t) \\ 0 \end{bmatrix} = \begin{bmatrix} \dfrac{2P}{K - M\omega^2}\sin(\omega t) \\ \dfrac{2P}{K - M\omega^2}\sin(\omega t) \end{bmatrix} \tag{8.12}$$

It should be noted that the decoupling of the equations of motion in the way shown is achieved as a result of the orthogonality properties of the eigenvectors presented in chapter 7. When the eigenvectors are not normalized, this yields

$$\mathbf{X}^{\mathrm{T}}\mathbf{K}\mathbf{X} = \tilde{\mathbf{X}} \tag{8.13}$$

$$\mathbf{X}^{\mathrm{T}}\mathbf{M}\mathbf{X} = \tilde{\mathbf{M}} \tag{8.14}$$

where

$$\tilde{\mathbf{K}} = \mathrm{diag}\{K_{11}, K_{22}, \ldots, K_{ii}, \ldots, K_{NN}\} \tag{8.15}$$

$$\tilde{\mathbf{M}} = \mathrm{diag}\{M_{11}, M_{22}, \ldots, M_{ii}, \ldots, M_{NN}\} \tag{8.16}$$

The elements $K_{ii}$ and $M_{ii}$ are referred to as the *modal stiffness* and *modal mass* in the $i$th mode. It should also be noted that when the eigenvectors are normalized $K_{ii} = \omega_i^2$ and $M_{ii} = 1$.

An examination of eq. (8.3), which may be written as

$$\mathbf{X} = \begin{bmatrix} X_{11} & X_{12} \\ X_{21} & X_{22} \end{bmatrix} \begin{bmatrix} q_1 \\ q_2 \end{bmatrix} = \begin{bmatrix} X_{11}q_1 + X_{12}q_2 \\ X_{21}q_1 + X_{22}q_2 \end{bmatrix} \tag{8.17}$$

reveals that $q_1$ and $q_2$, which are scalars, when multiplied by the first and second eigenvector respectively yield the contribution by each mode to the total response.

## Forced vibration of damped two DOF systems

Consider the two DOF system shown in Fig. 8.2, where the damping mechanism is represented by two systems of equivalent viscous dampers. The first set ($C_1$) represents the damping caused by friction at the supports and any other forms of external damping forces, such as aerodynamic and hydrodynamic forces. The second set ($C_2$) represents the internal damping in the springs. In a real structure this would mainly be due to hysteresis losses and friction forces in member joints as well as in the cladding.

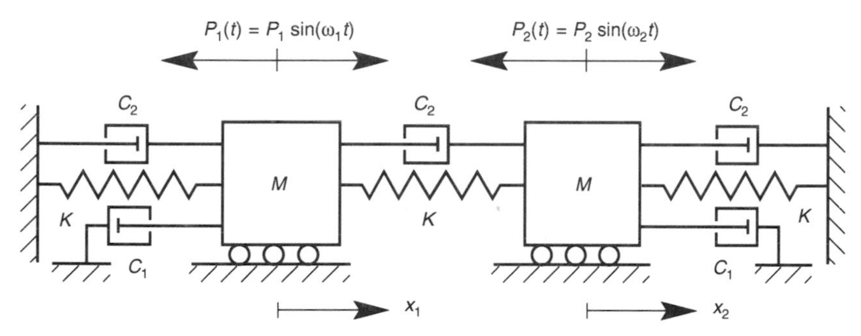


*Fig. 8.2. Two DOF damped lumped mass–spring systems acted on by harmonic forces*

The matrix formulation of the equations of motion for the system shown in Fig. 8.2 is

$$\begin{bmatrix} M & 0 \\ 0 & M \end{bmatrix} \begin{bmatrix} \ddot{x}_1 \\ \ddot{x}_2 \end{bmatrix} + \begin{bmatrix} (C_1 + 2C_2) & -C_2 \\ -C_2 & (C_1 + 2C_2) \end{bmatrix} \begin{bmatrix} \dot{x}_1 \\ \dot{x}_2 \end{bmatrix} + \begin{bmatrix} 2K & -K \\ -K & 2K \end{bmatrix} \begin{bmatrix} x_1 \\ x_2 \end{bmatrix}$$
$$= \begin{bmatrix} P_1(t) \\ P_2(t) \end{bmatrix} \tag{8.18}$$

or

$$\mathbf{M}\ddot{x} + \mathbf{C}\dot{x} + \mathbf{K}x = \mathbf{P}(t) \tag{8.19}$$

Now let as before

$$x = \mathbf{X}q$$

hence

$$\dot{x} = \mathbf{X}\dot{q} \qquad (8.20)$$
$$\ddot{x} = \mathbf{X}\ddot{q}$$

Hence

$$\mathbf{MX}\ddot{q} + \mathbf{CX}\dot{q} + \mathbf{KX}q = \mathbf{P}(t) \qquad (8.21)$$

Finally, premultiplication of each term in eq. (8.21) by $\mathbf{X}^{\mathrm{T}}$yields

$$\mathbf{X}^{\mathrm{T}}\mathbf{MX}\ddot{q} + \mathbf{X}^{\mathrm{T}}\mathbf{CX}\dot{q} + \mathbf{X}^{\mathrm{T}}\mathbf{KX}q = \mathbf{X}^{\mathrm{T}}\mathbf{P}(t) \qquad (8.22)$$

From examples 7.1 and 7.5

$$\mathbf{X} = \begin{bmatrix} 1 & 1 \\ 1 & -1 \end{bmatrix}$$

Substitution of this matrix for $\mathbf{X}$ into eq. (8.22) and the implied matrix multiplications yield

$$\begin{bmatrix} 2M & 0 \\ 0 & 2M \end{bmatrix}\begin{bmatrix} \ddot{q}_1 \\ \ddot{q}_2 \end{bmatrix} + \begin{bmatrix} 2(C_1 + C_2) & 0 \\ 0 & 2(C_1 + 3C_2) \end{bmatrix}\begin{bmatrix} \dot{q}_1 \\ \dot{q}_2 \end{bmatrix} + \begin{bmatrix} 2K & 0 \\ 0 & 6K \end{bmatrix}\begin{bmatrix} q_1 \\ q_2 \end{bmatrix}$$
$$= \begin{bmatrix} P_1(t) + P_2(t) \\ P_1(t) - P_2(t) \end{bmatrix} \qquad (8.23)$$

which may alternatively be written as

$$2M\ddot{q}_1 + 2(C_1 + C_2)\dot{q}_1 + 2Kq_1 = P_1(t) + P_2(t) \qquad (8.24a)$$
$$2M\ddot{q}_2 + 2(C_1 + 3C_2)\dot{q}_2 + 6Kq_2 = P_1(t) - P_2(t) \qquad (8.24b)$$

Thus the equations of motion have, as in the case of the equations for the undamped system shown in Fig. 8.1, been decoupled although damping has been included. Inspection of eq. (8.18) reveals that one part of the damping matrix is proportional to the mass matrix and one to the stiffness matrix. The damping matrix $\mathbf{C}$ may therefore be written as

$$\mathbf{C} = \begin{bmatrix} C_1 & 0 \\ 0 & C_1 \end{bmatrix} + \begin{bmatrix} 2C_2 & -C_2 \\ -C_2 & 2C_2 \end{bmatrix} = \alpha_1\begin{bmatrix} M & 0 \\ 0 & M \end{bmatrix} + \alpha_2\begin{bmatrix} 2K & -K \\ -K & 2K \end{bmatrix} \qquad (8.25)$$

where $\alpha_0$ and $\alpha_1$ are coefficients of proportionality. If the damping mechanism can be represented by a system of equivalent viscous dampers as shown in Fig. 8.2, it is therefore possible to model the damping mechanism as a function of the mass and stiffness of the system. In such cases, therefore, the damping may be expressed as

$$\mathbf{C} = \alpha_0\mathbf{M} + \alpha_1\mathbf{K} \qquad (8.26)$$

This is referred to as *Rayleigh damping*. Because the eigenvectors are orthogonal with respect to both the mass matrix and the stiffness matrix, it follows that for this form of damping they are also orthogonal with respect to the damping matrix. The orthogonality property of the eigenvectors with respect to the damping matrix is also the reason why the equations of motion for damped multi-DOF systems can be de-coupled. Inspection of eqs (8.24a) and (8.24b) shows that, as in the case of the undamped mass–spring system in Fig. 8.1, the natural angular frequencies are $\sqrt{(K/M)}$ and $\sqrt{(3K/M)}$. This, together with the fact that the values of $q$ when multiplied with the eigenvalue matrix yield the contribution from each mode to the total response, leads to the conclusion that the sum of the damping coefficients in eqs (8.24a) and (8.24b) is equal to the damping coefficients in the first and second modes respectively. It is therefore assumed, referring to eq. (3.22), that in eqs (8.24a) and (8.24b)

$$2(C_1 + C_2) = 2\xi_1\omega_1(2M) \tag{8.27a}$$

$$2(C_1 + 3C_2) = 2\xi_2\omega_2(2M) \tag{8.27b}$$

Thus the elements in the damping matrix in eq. (8.19) can be found by the matrix multiplication

$$\mathbf{C} = \mathbf{X}^{-\mathrm{T}}[2\xi\omega]\tilde{\mathbf{M}}\mathbf{X}^{-1} \tag{8.28}$$

where

$$\tilde{\mathbf{M}} = \mathbf{X}^{\mathrm{T}}\mathbf{M}\mathbf{X} \tag{8.29}$$

$$[2\xi\omega] = \begin{bmatrix} 2\xi_1\omega_1 & 0 \\ 0 & 2\xi_2\omega_2 \end{bmatrix} \tag{8.30}$$

With the expressions for the damping coefficients given by eqs (8.27a) and (8.27b), the uncoupled equations (8.24a) and (8.24b) may now be written as

$$M\ddot{q}_1 + 2\xi_1\omega_1 M\dot{q}_1 + Kq_1 = \tfrac{1}{2}\{P_1(t) + P_2(t)\} \tag{8.31a}$$

$$M\ddot{q}_2 + 2\xi_2\omega_2 M\dot{q}_2 + 3Kq_2 = \tfrac{1}{2}\{P_1(t) - P_2(t)\} \tag{8.31b}$$

If it is again assumed that

$$P_1 \sin(\omega_1 t) = P_2 \sin(\omega_2 t) = P_0 \sin(\omega t)$$

then

$$q_1 = \frac{P_0}{K}\frac{1}{\sqrt{\left[(1 - r_1{}^2)^2 + 4\xi^2 r_1{}^2\right]}}\sin(\omega t - \alpha_1) \tag{8.32a}$$

$$q_2 = 0 \tag{8.32b}$$

where

$$r_1 = \frac{\omega}{\omega_1} = \frac{\omega}{\sqrt{(K/M)}}$$

and, from eq. (4.12)

$$\alpha_1 = \tan^{-1}\frac{2\xi_1 r_1}{1 - r_1{}^2}$$

The maximum response occurs when

$$\sin(\omega t - \alpha) = 1$$

Thus the maximum response vector is given by

$$\begin{bmatrix} x_1 \\ x_2 \end{bmatrix} = \begin{bmatrix} 1 & 1 \\ 1 & -1 \end{bmatrix} \begin{bmatrix} \dfrac{P_0}{K} \dfrac{1}{\sqrt{\left[(1 - r_1{}^2)^2 + 4\xi^2 r_1{}^2\right]}} \\ 0 \end{bmatrix}$$

$$= \begin{bmatrix} \dfrac{P_0}{K} \dfrac{1}{\sqrt{\left[(1 - r_1{}^2)^2 + 4\xi^2 r_1{}^2\right]}} \\ \dfrac{P_0}{K} \dfrac{1}{\sqrt{\left[(1 - r_1{}^2)^2 + 4\xi^2 r_1{}^2\right]}} \end{bmatrix} \tag{8.33}$$

In general, when using the method of mode superposition it is more convenient to assume that

$$\tilde{\mathbf{C}} = \mathbf{X}^{\mathrm{T}}\mathbf{C}\mathbf{X} = [2\xi\omega]\tilde{\mathbf{M}} \tag{8.34}$$

where $\tilde{\mathbf{M}}$ is given by eq. (8.29), rather than assuming Rayleigh damping, which requires the calculation of the coefficients $\alpha_0$ and $\alpha_1$. Thus, for the DOF system considered

$$\tilde{\mathbf{C}} = \begin{bmatrix} 2\xi_1\omega_1\tilde{M}_1 & 0 \\ 0 & 2\xi_2\omega_2\tilde{M}_2 \end{bmatrix} \tag{8.35}$$

where

$$\tilde{\mathbf{M}}_1 = [1 \quad 1] \begin{bmatrix} M & 0 \\ 0 & M \end{bmatrix} \begin{bmatrix} 1 \\ 1 \end{bmatrix} = 2M$$

$$\tilde{\mathbf{M}}_2 = [1 \; -1] \begin{bmatrix} M & 0 \\ 0 & M \end{bmatrix} \begin{bmatrix} 1 \\ -1 \end{bmatrix} = 2M$$

The products $\mathbf{X}_i{}^{\mathrm{T}}\mathbf{M}\mathbf{X}_i = \tilde{\mathbf{M}}_i$ and $\mathbf{X}_i{}^{\mathrm{T}}\mathbf{C}\mathbf{X}_i = \tilde{\mathbf{C}}_i$ are referred to as the *modal mass* and *modal damping* in the $i$th mode; similarly, the product $\mathbf{X}_i{}^{\mathrm{T}}\mathbf{K}\mathbf{X}_i = \tilde{\mathbf{K}}_i$ is referred to as the *modal stiffness.*

## Forced vibration of multi-DOF systems with orthogonal damping matrices

In the previous section it is shown that the equations of motion for a two DOF system can be decoupled provided that the damping matrix can be diagonalized by premultiplication and postmultiplication of the mode-shape matrix. In the following the process of decoupling of the equations of motion is extended to a general $N$ DOF system. From the section on 'Orthogonality and normalization of eigenvectors' in chapter 7, it is known that premultiplication and postmultiplication of both the mass matrix and the stiffness matrix by the modeshape matrix will lead to diagonal matrices. Therefore the equations of motion can always be decoupled for systems with more DOF if the eigenvectors also are orthogonal with respect to the damping matrix. The general theory for decoupling and hence calculation of the response of multi-DOF systems to harmonic excitation can therefore easily be presented. Let the equations of motion for a general linear multi-DOF system be

$$\mathbf{M}\ddot{x} + \mathbf{C}\dot{x} + \mathbf{K}x = \mathbf{P}(t) \tag{8.36}$$

The corresponding eigenvalue equation for determination of the natural frequencies and modeshapes is

$$\mathbf{KX} - \omega^2\mathbf{MX} = 0 \tag{8.37}$$

which for an $N$ DOF system will yield the eigenvalues and eigenvectors

$$\omega^2 = \left[\omega_1{}^2, \omega_2{}^2, \ \ldots, \ \omega_N{}^2\right] \tag{8.38a}$$

$$\boldsymbol{X} = [\boldsymbol{X}_1, \boldsymbol{X}_2, \ \ldots, \ \boldsymbol{X}_N] \tag{8.38b}$$

In order to decouple the equations of motion, it is assumed that the damping matrix can be diagonalized, and that

$$\tilde{\mathbf{C}} = \boldsymbol{X}^{\mathrm{T}}\mathbf{C}\boldsymbol{X} = [2\xi\omega]\tilde{\mathbf{M}} \tag{8.39}$$

where

$$[2\xi\omega] = \mathrm{diag}\{2\xi_1\omega_1, \ 2\xi_2\omega_2, \ \ldots, \ 2\xi_N\omega_N\}$$

Now let

$$\boldsymbol{x} = \boldsymbol{X}\boldsymbol{q}$$
$$\dot{\boldsymbol{x}} = \boldsymbol{X}\dot{\boldsymbol{q}}$$
$$\ddot{\boldsymbol{x}} = \boldsymbol{X}\ddot{\boldsymbol{q}}$$

Substitution of the above expressions for $\boldsymbol{x}$, $\dot{\boldsymbol{x}}$ and $\ddot{\boldsymbol{x}}$ into eq. (8.39) and postmultiplication of each term by $\boldsymbol{X}^{\mathrm{T}}$ yields

$$\boldsymbol{X}^{\mathrm{T}}\mathbf{M}\boldsymbol{X}\ddot{\boldsymbol{q}} + \boldsymbol{X}^{\mathrm{T}}\mathbf{C}\boldsymbol{X}\dot{\boldsymbol{q}} + \boldsymbol{X}^{\mathrm{T}}\mathbf{K}\boldsymbol{X}\boldsymbol{q} = \boldsymbol{X}^{\mathrm{T}}\boldsymbol{P}(t) \tag{8.40}$$

From the orthogonality properties of eigenvectors

$$\boldsymbol{X}_i{}^{\mathrm{T}}\mathbf{M}\boldsymbol{X}_i = \tilde{\mathbf{M}}_i \quad \text{and} \quad \boldsymbol{X}_i{}^{\mathrm{T}}\mathbf{M}\boldsymbol{X}_j = 0 \quad \text{when } j \neq i$$
$$\boldsymbol{X}_i{}^{\mathrm{T}}\mathbf{K}\boldsymbol{X}_i = \tilde{\mathbf{K}}_i \quad \text{and} \quad \boldsymbol{X}_i{}^{\mathrm{T}}\mathbf{K}\boldsymbol{X}_j = 0 \quad \text{when } j \neq i$$

Hence eq. (8.40) reduces to

$$\tilde{\mathbf{M}}\ddot{\boldsymbol{q}} + \tilde{\mathbf{C}}\dot{\boldsymbol{q}} + \tilde{\mathbf{K}}\boldsymbol{q} = \boldsymbol{X}^{\mathrm{T}}\boldsymbol{P}(t) \tag{8.41}$$

Because $\tilde{\mathbf{M}}$, $\tilde{\mathbf{C}}$ and $\tilde{\mathbf{K}}$ are diagonal matrices, eq. (8.41) may also be written as

$$\tilde{M}_1\ddot{q}_1 + 2\xi_1\omega_1\tilde{M}_1\dot{q}_1 + \tilde{K}_1q_1 = \boldsymbol{X}_1{}^{\mathrm{T}}P(t)$$
$$\tilde{M}_2\ddot{q}_2 + 2\xi_2\omega_2\tilde{M}_2\dot{q}_2 + \tilde{K}_2q_2 = \boldsymbol{X}_2{}^{\mathrm{T}}P(t)$$
$$\cdots\cdots\cdots\cdots\cdots\cdots\cdots\cdots$$
$$\cdots\cdots\cdots\cdots\cdots\cdots\cdots\cdots$$
$$\tilde{M}_N\ddot{q}_N + 2\xi_N\omega_N\tilde{M}_N\dot{q}_N + \tilde{K}_Nq_N = \boldsymbol{X}_N{}^{\mathrm{T}}P(t)$$

The elements in the vector $\boldsymbol{q}$ may now be determined by solving the $N$ equivalent one DOF systems given by eq. (8.42), in which each equation represents a mass–spring system that will vibrate with the frequency and damping of the corresponding structural mode. Finally, the structural response vector for the $N$ DOF system is found by premultiplying the vector $\boldsymbol{q}$ by the modeshape matrix $\mathbf{X}$. Hence

$$\boldsymbol{x} = \mathbf{X}\boldsymbol{q} \tag{8.43}$$

The use of a normalized eigenvector to decouple the equations of motion will lead to a simplification of eqs (8.41) and (8.42). Let

$$\boldsymbol{x} = \boldsymbol{Z}\boldsymbol{q}$$
$$\dot{\boldsymbol{x}} = \boldsymbol{Z}\dot{\boldsymbol{q}}$$
$$\ddot{\boldsymbol{x}} = \boldsymbol{Z}\ddot{\boldsymbol{q}}$$

Substitution of the above expressions for $\boldsymbol{x}$, $\dot{\boldsymbol{x}}$ and $\ddot{\boldsymbol{x}}$ into eq. (8.36) and premultiplication of each term by $\mathbf{Z}^{\mathrm{T}}$ yields

$$\mathbf{Z}^{\mathrm{T}}\mathbf{M}\boldsymbol{Z}\ddot{\boldsymbol{q}} + \boldsymbol{Z}^{\mathrm{T}}\mathbf{C}\boldsymbol{Z}\dot{\boldsymbol{q}} + \boldsymbol{Z}^{\mathrm{T}}\mathbf{K}\boldsymbol{Z}\boldsymbol{q} = \boldsymbol{Z}^{\mathrm{T}}\boldsymbol{P}(\boldsymbol{t}) \tag{8.44}$$

From the properties of orthogonal normalized eigenvectors presented in chapter 7, and with the assumptions made above with respect to the damping matrix

$$\mathbf{Z}_i{}^{\mathrm{T}}\mathbf{M}\mathbf{Z}_i = 1 \quad \text{and} \quad \mathbf{Z}_i{}^{\mathrm{T}}\mathbf{M}\mathbf{Z}_j = 0 \quad \text{when } j \neq i$$
$$\mathbf{Z}_i{}^{\mathrm{T}}\mathbf{K}\mathbf{Z}_i = \omega_i{}^2 \quad \text{and} \quad \mathbf{Z}_i{}^{\mathrm{T}}\mathbf{K}\mathbf{Z}_j = 0 \quad \text{when } j \neq i$$
$$\mathbf{Z}_i{}^{\mathrm{T}}\mathbf{C}\mathbf{Z}_i = 2\xi_i\omega_i \quad \text{and} \quad \mathbf{Z}_i{}^{\mathrm{T}}\mathbf{C}\mathbf{Z}_j = 0 \quad \text{when } j \neq i$$

Hence eq. (8.44) can be written as

$$\mathbf{I}\ddot{\boldsymbol{q}} + 2\xi\omega\dot{\boldsymbol{q}} + \omega^2\boldsymbol{q} = \boldsymbol{Z}^{\mathrm{T}}\boldsymbol{P}(\boldsymbol{t}) \tag{8.45}$$

This represents, like eq. (8.41), a system of $N$ independent equations which can be written as

$$\ddot{q}_1 + 2\xi_1\omega_1\dot{q}_1 + \omega_1{}^2q_1 = \mathbf{Z}_1{}^{\mathrm{T}}\boldsymbol{P}(\boldsymbol{t})$$
$$\ddot{q}_2 + 2\xi_2\omega_2\dot{q}_2 + \omega_2{}^2q_2 = \mathbf{Z}_2{}^{\mathrm{T}}\boldsymbol{P}(\boldsymbol{t})$$
$$\cdots\cdots\cdots\cdots\cdots\cdots\cdots\cdots \tag{8.46}$$
$$\cdots\cdots\cdots\cdots\cdots\cdots\cdots\cdots$$
$$\ddot{q}_N + 2\xi_N\omega_N\dot{q}_N + \omega_N{}^2q_N = \mathbf{Z}_N{}^{\mathrm{T}}\boldsymbol{P}(\boldsymbol{t})$$

The use of normalized eigenvectors therefore results in decoupled equations of motion in which the mass is unity and the stiffnesses are equal to the eigenvalues. Finally, having determined the elements in $\boldsymbol{q}$ by solving the equations in eq. (8.46), the total response of the system is found from the transformation

$$\boldsymbol{x} = \mathbf{Z}\boldsymbol{q} \tag{8.47}$$

The choice, when analysing a structure between eq. (8.41) and eq. (8.45) is a matter of preference. They will lead to the same results.

The numerical effort becomes considerable, even for the most trivial of problems, if each of the elements in the forcing vector $\boldsymbol{P}(\boldsymbol{t})$ consists of a sum of harmonic functions such as

$$\boldsymbol{P}_i(\boldsymbol{t}) = \sum_{i=1}^{M} a_i \sin(\omega_i t) \tag{8.48}$$

and calculations will normally require the use of a computer.

**Example 8.1** Let the three-storey shear structure in example 2.5 (Fig. 2.14) be vibrated by a shaker positioned on the top floor. Calculate the response if the vibrator exerts a force $P(t) = 0{\cdot}6\sin(\omega_1 t)$ kN, where $\omega_1$ is the first natural frequency of the structure. The weight of each floor is 20·0 kN/m and the flexural rigidity $EI$ of each column is 89 100·0 kN m$^2$. The distance between the columns is 10·0 m and the height of the columns is 4·0 m. The damping in each mode is assumed to be 2·0% of critical.

The natural frequencies and modeshapes for the structure have been calculated in example 7.3, and the normalized modeshape matrix in example 7.9. Hence

$$\begin{bmatrix} \omega_1^2 \\ \omega_2^2 \\ \omega_3^2 \end{bmatrix} = \frac{89100{\cdot}0 \times 9{\cdot}81 \times 10^{-3}}{20{\cdot}0} \begin{bmatrix} 8{\cdot}1300 \\ 41{\cdot}1150 \\ 78{\cdot}8793 \end{bmatrix} = \begin{bmatrix} 355{\cdot}31 \\ 1796{\cdot}87 \\ 3447{\cdot}31 \end{bmatrix} \text{rad}^2/\text{s}^2$$

$$\begin{bmatrix} \omega_1 \\ \omega_2 \\ \omega_3 \end{bmatrix} = \begin{bmatrix} 18{\cdot}850 \\ 42{\cdot}390 \\ 58{\cdot}718 \end{bmatrix} \text{rad/s}$$

$$\mathbf{Z} = \begin{bmatrix} 1{\cdot}7454 & 3{\cdot}0818 & 1{\cdot}9498 \\ 3{\cdot}3157 & 0{\cdot}4330 & -3{\cdot}6531 \\ 4{\cdot}2336 & -4{\cdot}4926 & 3{\cdot}3105 \end{bmatrix} \times 10^{-3}$$

$$\begin{bmatrix} P_1(t) \\ P_2(t) \\ P_3(t) \end{bmatrix} = \begin{bmatrix} 0 \\ 0 \\ 600\sin(18{\cdot}85t) \end{bmatrix}$$

The decoupled equations of motion can therefore be written as

$$\ddot{q}_1+2\times0{\cdot}02\times18{\cdot}858\dot{q}_1+\ 355{\cdot}31q_2 = \ \ 4{\cdot}2336\times10^{-3}\times600\sin(18{\cdot}85t)$$

$$\ddot{q}_2+2\times0{\cdot}02\times42{\cdot}390\dot{q}_2+1796{\cdot}87q_2 = -4{\cdot}4926\times10^{-3}\times600\sin(18{\cdot}85t)$$

$$\ddot{q}_3+2\times0{\cdot}02\times58{\cdot}718\dot{q}_3+3447{\cdot}31q_3 = \ \ 3{\cdot}3105\times10^{-3}\times600\sin(18{\cdot}85t)$$

The general solution to these equations is

$$q = \frac{Z_{3i}\times600}{\omega_i^2}\mathrm{MF}_i\sin(18{\cdot}85t-\alpha_i)$$

where

$$\mathrm{MF}_1 = 1/2\xi_1 = 25{\cdot}0$$

$$\mathrm{MF}_2 = 1/\sqrt{\left[(1-r_2^2)^2+4\xi_2^2r_2^2\right]} = 1{\cdot}2462$$

$$\mathrm{MF}_3 = 1/\sqrt{\left[(1-r_3^2)^2+4\xi_3^2r_3^2\right]} = 1{\cdot}1148$$

$$\alpha_1 = \pi/2 = 1{\cdot}57080\,\mathrm{rad}$$

$$\alpha_2 = \tan^{-1}\left\{\frac{2\xi_2r_2}{1-r_2^2}\right\} = 0{\cdot}02217\ \mathrm{rad}$$

$$\alpha_3 = \tan^{-1}\left\{\frac{2\xi_3r_3}{1-r_3^2}\right\} = 0{\cdot}01432\,\mathrm{rad}$$

Because the contributions to the response from the second and third mode are obviously very small, the maximum response occurs when $\sin(18{\cdot}85t-\pi/2)\approx1$, i.e. when $t = 0{\cdot}1667\,\mathrm{s}$. Hence

$$q_1 = \frac{4{\cdot}2336\times10^{-3}\times600}{355{\cdot}31}\times25{\cdot}0\times1{\cdot}0 = 178{\cdot}7284\times10^{-3}\,\mathrm{m}$$

$$q_2 = \frac{-4{\cdot}4926\times10^{-3}\times600}{1796{\cdot}87}\times1{\cdot}55295\sin(18{\cdot}85\times0{\cdot}1667-0{\cdot}02217)$$

$$= -0{\cdot}0500\times10^{-3}\,\mathrm{m}$$

$$q_3 = \frac{3{\cdot}3105\times10^{-3}\times600}{3447{\cdot}31}\times1{\cdot}24274\sin(18{\cdot}85\times0{\cdot}1667-0{\cdot}01432)$$

$$= 0{\cdot}0098\times10^{-3}\,\mathrm{m}$$

The maximum response is therefore

$$\begin{bmatrix}x_1\\x_2\\x_3\end{bmatrix} = \begin{bmatrix}1{\cdot}7454 & 3{\cdot}0818 & 1{\cdot}9498\\3{\cdot}3157 & 0{\cdot}4330 & -3{\cdot}6531\\4{\cdot}2336 & -4{\cdot}4926 & 3{\cdot}3105\end{bmatrix}\begin{bmatrix}178{\cdot}7284\\-0{\cdot}0500\\0{\cdot}0098\end{bmatrix}\times10^{-6}$$

$$= \begin{bmatrix}0{\cdot}3118\\0{\cdot}5926\\0{\cdot}7569\end{bmatrix}\times10^{-3}\,\mathrm{m}$$

From the above calculations it can be seen that the contributions from the response in the second and third modes are negligible. This is to be expected, as the exciting frequency is equal to the first mode frequency. To eliminate response completely in the two higher modes would require a synchronized shaker system with a vibrator on each floor. Such systems are available but tend to be expensive.

**Problem 8.1** The three-storey shear structure in Fig. 7.4 has a first natural frequency of 2·0 Hz. The mass of each floor is 4000 kg. Calculate the response of the structure if it is vibrated by a harmonic force

$$P(t) = 1 \cdot 0 \sin(\omega_3 t)\,\mathrm{kN}$$

at the second floor level, where $\omega_3$ is the third natural angular frequency. Assume the damping in each mode to be 1·0% of critical.

**Problem 8.2** Calculate the response of the stepped antenna-mast in example 7.8 (Fig. 7.3) if it is excited by a harmonic force of $100 \sin(1 \cdot 1 \times \omega_1 t)$ N at the top of the mast, where $\omega_1$ is the first natural frequency of the mast. Assume the damping in the first and second modes to be 1·5% and 1·0% respectively of critical.

## References

Clough, R. W. *Dynamics of structures*. McGraw-Hill, London, 1975.

Coates, R. C., Coutie, M. G. & Kong, F. K. *Structural analysis*. Nelson, London, 1972.

Craig, R. R. Jr. *Structural dynamics*. Wiley, Chichester, 1981.

Harris, C. M. *Shock vibration*, 3rd edn. McGraw-Hill, London, 1988.

Paz, M. *Structural dynamics*. Van Nostrand Reinhold, New York, 1980.

# 9. Damping matrices for multi-degrees of freedom systems

## Incremental equations of motion for multi-DOF systems

The general incremental equations of motion for predicting the response of linear and nonlinear multi-DOF systems to load histories, assuming constant acceleration during a time step $\Delta t$, are given by eq. (6.61) as

$$\left[\mathbf{K}+\frac{2}{\Delta t}\mathbf{C}+\frac{4}{\Delta t^2}\mathbf{M}\right]\Delta\boldsymbol{x}=\Delta\mathbf{P}+2\mathbf{C}\dot{\boldsymbol{x}}+\mathbf{M}\left[\frac{4}{\Delta t}\dot{\boldsymbol{x}}+2\ddot{\boldsymbol{x}}\right] \tag{9.1}$$

where $\mathbf{K}$, $\mathbf{C}$ and $\mathbf{M}$ are the stiffness, damping and mass matrices for a multi-DOF structure, $\Delta\boldsymbol{x}$ is the incremental displacement vector, $\boldsymbol{x}$, $\dot{\boldsymbol{x}}$ and $\ddot{\boldsymbol{x}}$ are the displacement, velocity and acceleration vectors at time $t$, and $\Delta\boldsymbol{P}$ is the incremental load vector.

In the mode superposition method presented in chapter 8, the variables are separated by replacement of the displacement, velocity and acceleration vectors $\boldsymbol{x}$, $\dot{\boldsymbol{x}}$ and $\ddot{\boldsymbol{x}}$ in the equation of motion with a new set of generalized vectors $\boldsymbol{q}$, $\dot{\boldsymbol{q}}$ and $\ddot{\boldsymbol{q}}$, where

$$\begin{aligned}\boldsymbol{x}&=\mathbf{Z}\boldsymbol{q}\\ \dot{\boldsymbol{x}}&=\mathbf{Z}\dot{\boldsymbol{q}}\\ \ddot{\boldsymbol{x}}&=\mathbf{Z}\ddot{\boldsymbol{q}}\end{aligned} \tag{9.2}$$

and then postmultiplication of each term in the resulting equation by $\mathbf{Z}^{\mathrm{T}}$, the transpose of the normalized modeshape matrix $\mathbf{Z}$. This operation yielded $N$ independent equations representing $N$ one DOF systems, each with its own modal frequency and damping ratio. Thus there was no need to assemble a damping matrix with the same dimensions as the stiffness and mass matrices.

A similar transformation of eq. (9.1) yields

$$\left[\omega^2+\frac{2}{\Delta t}2\xi\omega+\frac{4}{\Delta t^2}\right]\Delta\boldsymbol{q}=\mathbf{Z}^{\mathrm{T}}\Delta\boldsymbol{P}=4\xi\omega\dot{\boldsymbol{q}}+\left[\frac{4}{\Delta t}\dot{\boldsymbol{q}}+2\ddot{\boldsymbol{q}}\right] \tag{9.3}$$

Thus for linear structures it is possible to limit the forward integration or step-by-step method to include only the response in significant modes. In the case of nonlinear structures this form of transformation is not permissible, because for such structures the natural frequencies and modeshapes vary with the amplitude of response. For nonlinear structures, therefore, it is necessary to assemble not only the stiffness and mass matrices, but also the structural damping matrices. This chapter presents two methods for modelling the structural damping in matrix form in terms of modal damping ratios, natural frequencies, stiffness and damping matrices. Theoretically such damping matrices ought to be updated at the end of each time step, as the stiffness, frequencies and damping ratios are functions of the amplitude of response. In practice, however, this is usually not necessary, because the damping ratios used will in most cases be only approximate values taken from codes of practice or the literature. However before methods of modelling structural damping by matrices are studied an outline is given of how damping ratios in higher modes are obtained.

## Measurement and evaluation of damping in higher modes

Damping in the first mode of multi-DOF systems can generally be evaluated as for one DOF systems: from decay functions, frequency sweeps, or by steady-state vibration at resonance. The use of only one vibrator will in most cases suffice to cause structures to vibrate with a modeshape that closely resembles the true one, and will therefore lead to reasonable values for the first damping ratio. In higher modes there are difficulties, however. Firstly, it is usually impossible to obtain decay functions for higher modes as most structures will, when excitation of a higher mode is stopped, revert to vibration in the first mode. Measurements of damping either by steady-state vibration at resonance or by frequency sweeps is generally also unsatisfactory, because the use of one vibrator will generally not be sufficient to cause a structure to vibrate in a pure mode. This is particularly noticeable when one is attempting to excite a structure in an antisymmetric mode, such as the second, fourth and sixth modes of a simply supported beam. That this is so can easily be seen or demonstrated by measuring the phase angle of response at different points on a structure. If the phase angle is 90° at the point of excitation it will usually be different at points away from the vibrator, with the difference increasing with increasing distance from the point of excitation. Thus, to obtain reasonably accurate values for damping in higher modes it is necessary to use more than one vibrator, whose force and frequency must be adjusted so that the structure at all points vibrates with phase angles of 90°. To achieve this, the vibrators must be controlled by a computer. As for one DOF systems, the damping can be measured by plotting the exciting force versus the amplitude of response for one cycle for each vibrator, as described in chapter 5, and then summing the work done by each vibrator. Thus the expression for the damping ratio for a structure vibrated by $N$ vibrators is given by

$$\xi = \frac{1}{2\pi\tilde{M}\omega_n^2}\sum_{i=1}^{N}\frac{A_{ni}}{x_{n0i}^2} \tag{9.4}$$

$$\xi = \frac{1}{2\pi\tilde{K}}\sum_{i=1}^{N}\frac{A_{ni}}{x_{n0i}^2} \tag{9.5}$$

where $\tilde{M}$ and $\tilde{K}$ are the modal mass and modal stiffness respectively and $A_{ni}$ is the area encompassed by the force–displacement curve for vibrator $n$ at the $i$th mode. Multi-point shaker systems are expensive, and are all mainly used by research institutions and industrial companies specializing in dynamic testing.

In general, values for damping ratios are obtained from codes of practice or the literature. The former usually give values only for damping ratios to be used in the dominant mode, with no guidance on values to be used in higher modes. It is therefore not uncommon to use the same damping ratio for all modes.

## Damping matrices

In chapter 8 it is shown that the dynamic response of linear multi-DOF structures can be determined by decoupling the equations of motion, and summing the responses in each mode. Thus it is necessary only to assign values to the damping ratios for the modes contributing to the total response, without having to set up a damping matrix. The implied orthogonality of the modeshapes with respect to the damping matrix enables realistic numerical modelling of structural damping, provided it can be assumed that the damping does not couple the modes. This assumption is usually correct, provided aerodynamic and hydrodynamic damping, when significant, are modelled separately.

In general it is necessary only to model the *structural* damping by a damping matrix when undertaking the form of dynamic response analysis indicated by eq. (9.1), in which case damping due to external forces such as those caused by air and water can be taken into account separately. When this is the case the construction of orthogonal damping matrices is a convenient method of modelling the structural damping. In the case of nonlinear structures the principle of orthogonality no longer applies, as the modeshapes as well as the frequencies are functions of the amplitude of response. For weakly nonlinear structures the nonlinearity is not significant, as assumed damping ratios will at best be only approximately correct. Also, a considerable amount of experimental evidence indicates that modal damping ratios vary with the amplitude of response.

## Modelling of structural damping by orthogonal damping matrices

### *First method*

In chapter 8 it is shown that the equations of motion for a multi-DOF structure can be written as

$$\mathbf{M}\ddot{x} + \mathbf{C}\dot{x} + \mathbf{K}x = \boldsymbol{P}(t) \tag{9.6}$$

and the equations can be uncoupled provided the damping matrix has the same orthogonal properties with respect to the modeshape vectors as have the mass and stiffness matrices. When this is the case

$$\mathbf{Z}_i^{\mathrm{T}}\mathbf{C}\mathbf{Z}_i = 2\xi_i\omega_i \tag{9.7}$$

$$\mathbf{Z}^{\mathrm{T}}\mathbf{C}\mathbf{Z} = [2\xi\omega] \tag{9.8}$$

Thus

$$\mathbf{C} = \mathbf{Z}^{-1}[2\xi\omega]\mathbf{Z}^{-1} \tag{9.9}$$

The inversion of the modeshape matrix $\mathbf{Z}$ can be avoided. From eq. (7.57)

$$\mathbf{Z}^{\mathrm{T}}\mathbf{M}\mathbf{Z} = \mathbf{I} \tag{9.10}$$

Hence

$$\mathbf{Z}^{-\mathrm{T}} = \mathbf{M}\mathbf{Z} \tag{9.11}$$

$$\mathbf{Z}^{-1} = \mathbf{Z}^{\mathrm{T}}\mathbf{M} \tag{9.12}$$

Substitution of the expressions for $\mathbf{Z}^{-\mathrm{T}}$ and $\mathbf{Z}^{-1}$ into eq. (9.9) yields the following expression for the damping matrix

$$\mathbf{C} = \mathbf{M}\mathbf{Z}[2\xi\omega]\mathbf{Z}^{\mathrm{T}}\mathbf{M} \tag{9.13}$$

*Second method*

Another way to construct a damping matrix with orthogonal properties from modal damping ratios is to assume that the damping is a function of both the mass and the stiffness, and to make use of the general orthogonal relationship

$$\mathbf{Z}_i^{\mathrm{T}}\mathbf{M}(\mathbf{M}^{-1}\mathbf{K})^q\mathbf{Z}_j = 0 \tag{9.14}$$

which is satisfied when $i \neq j$, and $q = \ldots, -2, -1, 0, 1, 2, \ldots$

When $q = 0$ and $q = 1$, eq. (9.12) yields the previously obtained orthogonality conditions of the modeshape vectors with respect to the mass matrix and the stiffness matrix. Inspection of eq. (9.14) indicates that it is possible to formulate an orthogonal damping matrix of the form

$$\mathbf{C} = \sum_{q=0}^{N-1} \alpha_q\mathbf{M}(\mathbf{M}^{-1}\mathbf{K})^q \tag{9.15}$$

which will contain the correct damping in $N$ modes provided the corresponding values for $\alpha$ can be determined. An expression for calculating values of $\alpha$ can be developed by first premultiplying and postmultiplying both sides of eq. (9.15) by $\mathbf{Z}_i^{\mathrm{T}}$ and $\mathbf{Z}_i$ respectively. This yields the $i$ mode contribution to the total damping as

$$\mathbf{C}_i = 2\xi_i\omega_i = \sum_{q=0}^{N-1} \mathbf{Z}_i^{\mathrm{T}}\alpha_q\mathbf{M}(\mathbf{M}^{-1}\mathbf{K})^q\mathbf{Z}_i \tag{9.16}$$

In order to simplify the right-hand side of eq. (9.16), the two sides of the frequency equation for the $i$th mode

$$\mathbf{KZ}_i = \omega_i^2 \mathbf{MZ}_i \tag{9.17}$$

are multiplied by $\alpha_q$, transposed, postmultiplied by $(\mathbf{M}^{-1}\mathbf{K})^q \mathbf{Z}_i$, and then written in reverse order. This yields the relationship

$$\omega_i^2 \mathbf{Z}_i^{\mathrm{T}} \alpha_q \mathbf{M}(\mathbf{M}^{-1}\mathbf{K})^q \mathbf{Z}_i = \mathbf{Z}_i^{\mathrm{T}} \alpha_q \mathbf{K}(\mathbf{M}^{-1}\mathbf{K})^q \mathbf{Z}_i \tag{9.18}$$

Substitution of $\omega_i^2 \mathbf{M}$ for $\mathbf{K}$ in the right-hand side of eq. (9.18) gives

$$\mathbf{Z}_i^{\mathrm{T}} \alpha_q \mathbf{M}(\mathbf{M}^{-1}\mathbf{K})^q \mathbf{Z}_i = \alpha_q \omega_i^{2q} \tag{9.19}$$

which finally, on substitution of the left-hand side of eq. (9.19) into eq. (9.16), yields

$$2\xi_i \omega_i = \sum_{q=0}^{N-1} \alpha_q \omega_i^{2q} \tag{9.20}$$

from which as many values of $\alpha$ can be found as there are known damping ratios. Thus, given the damping ratios for, say, the first four modes of an $N$ DOF structure, only four values for $\alpha$, namely $\alpha_0$, $\alpha_1$, $\alpha_2$ and $\alpha_3$ can be determined by using eq. (9.20) that the modal damping in the $i$th mode, given four values for $\xi$, is given by the polynomial

$$2\xi_i \omega_i = \alpha_0 + \alpha_1 \omega_i^2 + \alpha_2 \omega_i^4 + \alpha_3 \omega_i^6 \tag{9.21}$$

where the four values for $\alpha$ may be calculated from the matrix equation

$$\begin{bmatrix} 2\xi_1\omega_1 \\ 2\xi_2\omega_2 \\ 2\xi_3\omega_3 \\ 2\xi_4\omega_4 \end{bmatrix} \begin{bmatrix} 1 & \omega_1^2 & \omega_1^4 & \omega_1^6 \\ 1 & \omega_2^2 & \omega_2^4 & \omega_2^6 \\ 1 & \omega_3^2 & \omega_3^4 & \omega_3^6 \\ 1 & \omega_4^2 & \omega_4^4 & \omega_4^6 \end{bmatrix} \begin{bmatrix} \alpha_0 \\ \alpha_1 \\ \alpha_2 \\ \alpha_3 \end{bmatrix} \tag{9.22}$$

Assigning values to lower mode damping ratios only may result in values for damping ratios in higher modes that are very different from the real ones. This however is not important provided the damping ratios for the modes in which a structure mainly responds are correct. In practice it is often assumed that only one or two values for $\alpha$ are different from zero. When this is the case eq. (9.15) is reduced to one of the following

$$\mathbf{C} = \alpha_0 \mathbf{M} \tag{9.23a}$$

$$\mathbf{C} = \alpha_1 \mathbf{K} \tag{9.23b}$$

$$\mathbf{C} = \alpha_0 \mathbf{M} + \alpha_1 \mathbf{K} \tag{9.23c}$$

This may result in adequate modelling of the damping for a large number of civil engineering structures that vibrate only in a few of the lower modes, but will not suffice for such structures as guyed masts, cable-stayed bridges, and cable and membrane roofs that respond significantly in a large number of modes. The expression for damping given by eq. (9.23) is referred to as Rayleigh damping, and is mentioned in chapter 8. When this form of damping is assumed, the damping in any mode can be calculated from

$$2\xi_i\omega_i = \alpha_0 + \alpha_1\omega_i^2 \tag{9.24}$$

Rayleigh damping is a convenient form for modelling the damping of weakly nonlinear structures, as it leads to damping matrices with the same banding as the stiffness matrix.

From eq. (9.28) it can be seen that when the damping is assumed to be proportional to the mass only, i.e. when $q = 0$, the damping ratios decrease with increasing mode frequencies; and that when the damping is assumed proportional to the stiffness only, i.e. when $q = 1$, and damping ratios increase with increasing mode frequencies. Equation (9.24) indicates that eq. (9.23c) will model the structural damping mechanism correctly if values of $2\xi\omega$, when plotted against $\omega^2$, yield a straight line.

**Example 9.1** Construct the damping matrix for the three-storey shear structure shown in example 2.5 (Fig. 2.14), by using eq. (9.13), if the weight of the floors is 20·0 kN/m and the damping in each mode is assumed to be 1·0% of critical. The natural angular frequencies and the normalized modeshape matrix for the structure are given in example 8.1 as

$$\begin{bmatrix}\omega_1\\ \omega_2\\ \omega_3\end{bmatrix} = \begin{bmatrix}18{\cdot}850\\ 42{\cdot}390\\ 58{\cdot}718\end{bmatrix}\text{rad/s}\quad \mathbf{Z} = \begin{bmatrix}1{\cdot}7454 & 3{\cdot}0818 & 1{\cdot}9498\\ 3{\cdot}3157 & 0{\cdot}4330 & -3{\cdot}6531\\ 4{\cdot}2336 & -4{\cdot}4926 & 3{\cdot}3105\end{bmatrix}\times 10^{-3}$$

The mass matrix for the structure is

$$\mathbf{M} = 20\,387{\cdot}36\begin{bmatrix}3 & 0 & 0\\ 0 & 2 & 0\\ 0 & 0 & 1\end{bmatrix}\text{kg}$$

$$[2\xi\omega] = \begin{bmatrix}2\times 0{\cdot}01\times 18{\cdot}850 & 0 & 0\\ 0 & 2\times 0{\cdot}01\times 42{\cdot}390 & 0\\ 0 & 0 & 2\times 0{\cdot}01\times 58{\cdot}714\end{bmatrix}$$

$$= \begin{bmatrix}0{\cdot}37700 & 0 & 0\\ 0 & 0{\cdot}84780 & 0\\ 0 & 0 & 1{\cdot}17428\end{bmatrix}$$

Hence

$$\mathbf{C} = 415{\cdot}64444 \times \begin{bmatrix} 3 & 0 & 0 \\ 0 & 2 & 0 \\ 0 & 0 & 1 \end{bmatrix} \begin{bmatrix} 1{\cdot}7454 & 3{\cdot}0818 & 1{\cdot}9498 \\ 3{\cdot}3157 & 0{\cdot}4330 & -3{\cdot}6531 \\ 4{\cdot}2336 & -4{\cdot}4926 & 3{\cdot}3105 \end{bmatrix}$$

$$\times \begin{bmatrix} 0{\cdot}37700 & 0 & 0 \\ 0 & 0{\cdot}84780 & 0 \\ 0 & 0 & 1{\cdot}17428 \end{bmatrix} \begin{bmatrix} 1{\cdot}7454 & 3{\cdot}3157 & 4{\cdot}2336 \\ 3{\cdot}0818 & 0{\cdot}4330 & -4{\cdot}4926 \\ 1{\cdot}9498 & -3{\cdot}6531 & 3{\cdot}3105 \end{bmatrix}$$

$$\times \begin{bmatrix} 3 & 0 & 0 \\ 0 & 2 & 0 \\ 0 & 0 & 1 \end{bmatrix} \text{N s/m}$$

i.e.

$$\mathbf{C} = \begin{bmatrix} 51{\cdot}1165 & -12{\cdot}5967 & -1{\cdot}7114 \\ -12{\cdot}5967 & 33{\cdot}2093 & -8{\cdot}7771 \\ -1{\cdot}7114 & -8{\cdot}7771 & 15{\cdot}2700 \end{bmatrix} \times 10^3 \text{ N s/m}$$

Note that the matrix is not only full, but also symmetric.

**Example 9.2** Use eqs (9.15) and (9.20) to construct the damping matrix for the shear structure in example 2.5 (Fig. 2.14). The *EI* value for the columns and the weight of the floor are as the previous examples: 89 100·00 kN m$^2$ and 20·0 kN/m respectively. The natural angular frequencies for the structure are given in examples 8.1 and 9.1.

From eqs (9.15) and (9.20)

$$\mathbf{C} = \sum_{q=0}^{N-1} \alpha_q \mathbf{M}(\mathbf{M}^{-1}\mathbf{K})^q$$

$$2\xi_i\omega_i = \sum_{q=0}^{N-1} \alpha_q \omega_i^{2q}$$

$$\mathbf{M} = 20\,387{\cdot}36 \begin{bmatrix} 3 & 0 & 0 \\ 0 & 2 & 0 \\ 0 & 0 & 1 \end{bmatrix} \text{kg}$$

$$\mathbf{K} = 16\,706{\cdot}25 \begin{bmatrix} 7 & -3 & 0 \\ -3 & 5 & -2 \\ 0 & -2 & 2 \end{bmatrix} \text{kN/m}$$

$$\omega = \begin{bmatrix} 18{\cdot}850 \\ 42{\cdot}390 \\ 58{\cdot}718 \end{bmatrix} \text{rad/s}$$

Given the damping ratios in three modes, eq. (9.20) may be written in matrix form as

$$\begin{bmatrix} 2\xi_1\omega_1 \\ 2\xi_2\omega_2 \\ 2\xi_3\omega_3 \end{bmatrix} = \begin{bmatrix} 1 & \omega_1{}^2 & \omega_1{}^4 \\ 1 & \omega_2{}^2 & \omega_2{}^4 \\ 1 & \omega_3{}^2 & \omega_3{}^4 \end{bmatrix} \begin{bmatrix} \alpha_0 \\ \alpha_1 \\ \alpha_2 \end{bmatrix}$$

$$\begin{bmatrix} \alpha_0 \\ \alpha_1 \\ \alpha_2 \end{bmatrix} = \begin{bmatrix} 1 & 18{\cdot}850^2 & 18{\cdot}850^4 \\ 1 & 42{\cdot}390^2 & 42{\cdot}390^4 \\ 1 & 58{\cdot}718^4 & 58{\cdot}718^4 \end{bmatrix}^{-1} \begin{bmatrix} 0{\cdot}37700 \\ 0{\cdot}84780 \\ 1{\cdot}07436 \end{bmatrix}$$

$$= \begin{bmatrix} 0{\cdot}2218644 \\ 4{\cdot}58361 \times 10^{-4} \\ -6{\cdot}12287 \times 10^{-8} \end{bmatrix}$$

For three values of $\alpha$, eq. (9.15) may be written as

$$\mathbf{C} = \alpha_0\mathbf{M} + \alpha_1\mathbf{K} + \alpha_2\mathbf{M}^{-1}\mathbf{K}^2$$

$$\alpha_0\mathbf{M} = 0{\cdot}2218644 \times 20\,387{\cdot}36 \begin{bmatrix} 3 & 0 & 0 \\ 0 & 2 & 0 \\ 0 & 0 & 1 \end{bmatrix}$$

$$= \begin{bmatrix} 13\,569{\cdot}688 & 0 & 0 \\ 0 & 9406{\cdot}459 & 0 \\ 0 & 0 & 4523{\cdot}229 \end{bmatrix}$$

$$\alpha_1\mathbf{K} = 4{\cdot}58361 \times 10^{-4} \times 16\,706{\cdot}25 \begin{bmatrix} 7 & -3 & 0 \\ -3 & 5 & -2 \\ 0 & -2 & 2 \end{bmatrix}$$

$$= \begin{bmatrix} 53\,602{\cdot}454 & -22\,972{\cdot}480 & 0 \\ -22\,972{\cdot}480 & 38\,287{\cdot}467 & -15\,314{\cdot}987 \\ 0 & -15\,314{\cdot}987 & 15\,314{\cdot}987 \end{bmatrix}$$

$$\alpha_2\mathbf{M}^{-1}\mathbf{K}^2 = -\,6{\cdot}12287 \times 10^{-8} \times \frac{16\,706{\cdot}25^2}{20\,387{\cdot}36} \times$$

$$\begin{bmatrix} 1/3 & 0 & 0 \\ 0 & 1/2 & 0 \\ 0 & 0 & 1 \end{bmatrix} \begin{bmatrix} 7 & -3 & 0 \\ -3 & 5 & -2 \\ 0 & -2 & 2 \end{bmatrix}^2$$

$$\alpha_2\mathbf{M}^{-1}\mathbf{K}^2 = \begin{bmatrix} -16\,205{\cdot}362 & 10\,058{\cdot}501 & -1676{\cdot}417 \\ 15\,087{\cdot}571 & -15\,528{\cdot}959 & 5867{\cdot}459 \\ -5029{\cdot}250 & 11\,734{\cdot}917 & -6705{\cdot}667 \end{bmatrix}$$

Substitution of the matrices for $\alpha_2\mathbf{M}$, $\alpha_1\mathbf{K}$ and $\alpha_2\mathbf{M}^{-1}\mathbf{K}^2$ into the expression for $\mathbf{C}$ and their addition yields

$$\mathbf{C} = \begin{bmatrix} 50\,966{\cdot}780 & -12\,913{\cdot}979 & -1676{\cdot}417 \\ -7884{\cdot}909 & 32\,164{\cdot}968 & -9447{\cdot}528 \\ -5029{\cdot}250 & -3580{\cdot}070 & 13\,132{\cdot}549 \end{bmatrix} \mathrm{N\,s/m}$$

Inspection of the above damping matrix reveals that it is not symmetric. The reason for this is that the term $\mathbf{M}^{-1}\mathbf{K}$ is not symmetric unless all the elements in $\mathbf{M}$ are equal.

**Problem 9.1** For the structure in examples 9.1 and 9.2, assume the damping in the first two modes to be 1·0% of critical. Construct the damping matrix assuming Rayleigh damping and hence calculate the implied damping ratio in the third mode.

**Problem 9.2** Use the damping matrix calculated in problem 9.1 to set up the dynamic matrix for the structure in examples 9.1 and 9.2. Assume the time step to be approximately equal to one-tenth of the period of the highest frequency.

## References

Clough, R. W. & Penzien, J. *Dynamics of structures*. McGraw-Hill, London, 1975.
Craig, R. R. Jr. *Structural dynamics*. Wiley, Chichester, 1981.
Harris, C. M. *Shock vibration*, 3rd edn. McGraw-Hill, London, 1988.
Paz, M. *Structural dynamics*. Van Nostrand Reinhold, New York, 1980.

# 10. The nature and statistical properties of wind

## Introduction

Wind is unsteady and exhibits random fluctuations in both time and space domains. Because wind can be considered to possess stationary characteristics, it is possible to describe its functions in statistical terms. Advances in computational techniques have made it possible to carry out statistical analysis of wind records and to determine their statistical characteristics, such as those described by the variance of fluctuations, auto-correlation and spectral density functions, the last of which are also commonly referred to as power spectra. Further advances in computational techniques have made it possible to generate wind histories and wind fields with the same statistical characteristics as real wind.

For linear structures, reasonable estimates of the response to wind can be made through a stochastic approach, in which the statistical characteristics of the response are determined in terms of the statistical properties of wind. This form of analysis is carried out in the frequency domain, and is the method most used by practising engineers.

However, for nonlinear structures such as membrane, cable and cable-stayed structures, whose structural characteristics vary with the amplitude of response and hence with time, reliable estimates of response to wind can only be made using a determined approach, in which the structural properties are updated at the end of each time step. In deterministic analysis single wind histories and wind fields simulating real time are generated from spectral density functions for fluctuating wind speeds. Basically, there are two distinct methods for generating wind histories. The first is one of superposition of harmonic waves; the second is based on filtering sequences of white noise.

## The nature of wind

Wind is a phenomenon caused by the movement of air particles in the earth's atmosphere. The movement of air in the atmospheric boundary layer, which extends to about 1 km above the earth's surface, is referred to as surface wind. The wind derives its energy primarily from the sun. Solar radiation accompanied by radiation away from earth produces temperature differences and consequently pressure gradients that cause acceleration of the air. Away from the ground, the pressure system is relatively stationary, because the pressure gradients are balanced by the centripetal and Coriolis accelerations. The centripetal acceleration is due to the curvature of the isobars; and Coriolis acceleration is due to the earth's rotation. This balance of forces results in a steady-state condition that causes the air to flow in a direction parallel to the isobars.

Near the ground, the balance of the pressure system is disturbed by drag forces caused by the earth's surface roughness. Ground surface roughness, whether occurring naturally — mountains, hills and forests — or as man-made obstructions such as buildings, bridges and dams, causes so much mechanical stirring of the air movement that

- the wind speed near the surface is retarded
- the wind direction changes and is no longer parallel to the isobars
- the flow conditions become unsteady and the wind exhibits instantaneous random variations in magnitude and direction

The rougher the surface, the more prominent these effects are. The effects decrease with increasing height above the ground. The height at which the effects have virtually vanished is referred to as the gradient height, and ranges from 300 m to 600 m depending on the degree of surface roughness.

Examination of wind records shows that the velocity of wind fluctuates and that the fluctuations vary both with the wind speed and with the roughness of the ground. It has therefore been found convenient to express the wind velocity as the sum of the mean velocity $U(z, x)$ in the along-wind direction at height $z$ and the fluctuating time-dependent velocity components $u(z, x, t)$, $u(z, y, t)$ and $u(z, z, t)$, where $x$ represents the along-wind, $y$ the horizontal across-wind and $z$ the vertical across-wind directions at height $Z$. Hence

$$\begin{bmatrix} U(z,x,t) \\ U(z,y,t) \\ U(z,z,t) \end{bmatrix} = \begin{bmatrix} U(z;x) \\ 0 \\ 0 \end{bmatrix} + \begin{bmatrix} u(z,x,t) \\ u(z,y,t) \\ u(z,z,t) \end{bmatrix} \tag{10.1}$$

or

$$\mathbf{U}(\boldsymbol{z}, \boldsymbol{t}) = \mathbf{U}(\boldsymbol{z}) + \mathbf{u}(\boldsymbol{z}, \boldsymbol{t}) \tag{10.2}$$

In cases where the horizontal and vertical across-wind fluctuations are of secondary importance, the instantaneous wind velocity can be treated as a scalar quantity, in which case, omitting the direction indicator, $x$, the instantaneous velocity at height $z$ is given by

$$U(z, t) = U(z) + u(z, t) \tag{10.3}$$

Research has revealed that the long-term statistical properties of wind are general and independent of type of terrain, wind strength and site location. This significant conclusion emerged from power spectral analysis of wind recorded over several years and at different locations. The resulting spectrum, in which the square of the amplitudes of each frequency was plotted against the frequency, provides a measure of the distribution of the energy of the random fluctuations of the wind velocity

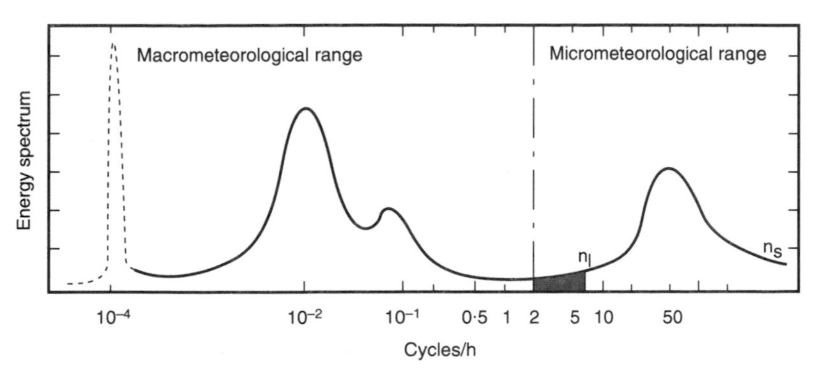


*Fig. 10.1. Spectrum of longitudinal wind fluctuations: the full line of the spectrum is after van der Hoven*

in the frequency domain. A typical spectrum, whose full line is known as the van der Hoven power spectrum, is shown in Fig. 10.1. An examination of Fig. 10.1 yields the following information.

- The energy is distributed in two main humps separated by a gap, the so-called *spectral gap*, which exists for periods between 10 min and 2 h. This implies that the fluctuations in the mean velocity of wind can be measured by calculating the mean velocities of wind speed signals recorded over periods ranging from only 10 min to 2 h. In this way, fluctuations due to the high-frequency components are eliminated so that only those due to the long-term fluctuations can be observed. Thus, as mentioned above, the wind velocity can be divided into two parts: an average steady-state velocity that varies with the long-term fluctuations due to macrometeorological causes, and a fluctuating velocity with high-frequency components due to turbulence.
- The first peak is linked to the annual variation. The second peak is linked to the *four day* period. This is the time of passage of a complete macrometeorological system, i.e. the duration of an average storm. The third peak is due to *day and night* thermal fluctuations. The fourth peak, which is in the micrometeorological range, is centred around a frequency of nearly one cycle/min and is caused by ground roughness.

As a result of the properties of wind outlined above, the response calculations of structures can be divided into two parts: (*a*) the calculation of the quasi-static response caused by the mean velocity component of wind and due to the very low-frequency fluctuations in the macrometeorological system; (*b*) the calculation of the response due to the high-frequency components, which are the source of dynamic excitation.

## Mean wind speed and variation of mean velocity with height

It has been established that recording periods between 10 min and 2 h provide reasonably stable values for the mean component of the wind speed. A period of 1 h lies nearly in the middle of this range and is the recording period adopted in the UK where meteorological stations in different parts of the country record and summarize the maximum daily wind speeds. The hourly wind speeds are recorded at a height of 10 m, but wind speed increases with increasing altitude above the ground until it reaches the velocity $V_g$ at the gradient height. Several laws have been used to describe the way in which the mean velocity varies with height. Today the most generally adopted one is the *logarithmic law*, which gives the mean speed $U(z)$ at height $z$ above the ground as

$$U(z) = 2{\cdot}5u_* \ln(z/z_0) \tag{10.4}$$

where

$$u_* = \frac{U(10)}{2{\cdot}5 \ln(10/z_0)} \tag{10.5}$$

or

$$u_* = U(10)\sqrt{k} \tag{10.6}$$

and $u_*$ is the shear velocity or friction velocity, $z_0$ is the roughness length (for values see Table 10.1), $k$ is the surface drag coefficient (for values see Table 10.1), and $U(10)$ is the reference mean velocity 10 m above ground level.

If the surface drag coefficient $k$ is known, then the corresponding value for $z_0$ can be found by using eq. (10.4). Thus

$$z_0 = z\mathrm{e}^{-U(z)/2{\cdot}5u_*} \tag{10.7}$$

The logarithmic law is applicable to heights in excess of 10 m. Below this height the velocity is assumed to be constant and equal to $U(10)$. In some of the more recent codes the logarithmic law has been modified, and the mean velocity at height $z$ is given by

$$U(z) = 2{\cdot}5u_* \Big[\ln(z/z_0) + 5{\cdot}75(z/H) - 1{\cdot}87(z/H)^2 - 1{\cdot}33(z/H)^3 + 0{\cdot}25(z/H)^4\Big] \tag{10.8}$$

*Table 10.1. Roughness lengths and surface drag coefficients for various types of terrain*

| Type of terrain | $z_0$ (m) | $k \times 10^3$ |
|---|---|---|
| Sand | 0·0001–0·001 | 1·2–1·9 |
| Sea surface | 0·005 | 0·7–2·6 |
| Low grass | 0·01–0·04 | 3·4–5·2 |
| High grass | 0·04–0·10 | 5·2–7·6 |
| Pine forest | 0·90–1·00 | 28·0–30·0 |
| Suburban areas | 0·20–0·40 | 10·5–15·4 |
| Centres of cities | 0·35–0·45 | 14·2–16·6 |
| Centres of large cities | 0·60–0·80 | 20·2–25·1 |

where $H$, the gradient height, can be determined from

$$H = u_*/2\beta\omega \sin \phi \tag{10.9}$$

and $\omega$ is the angular rotation of the earth ($= 7{\cdot}2722 \times 10^{-5}$ rad/s), $\phi$ is the local angle of latitude, and $\beta$ is a constant $= 6$.

For the lowest 200 m of the atmosphere, the contributions from the square, cubic and fourth order terms can be omitted, in which case eq. (10.8) reduces to

$$U(z) = 2{\cdot}5u_*[\ln(z/z_0) + 5{\cdot}75(z/H)] \tag{10.10}$$

When the expression for $H$ given by eq. (10.9) is substituted into eq. (10.10), and $z = 10$ m, the following relationship between $u_*$ and $z_0$ is obtained when using the above values for $\omega$ and $\beta$

$$u_* = \frac{U(10) - 0{\cdot}1254454 \sin \phi}{2{\cdot}5 \ln(10/z_0)} \tag{10.11}$$

As $-1{\cdot}0 < \sin \phi < 1{\cdot}0$, eq. (10.11) can for most applications be simplified to

$$u_* = \frac{U(10)}{2{\cdot}5 \ln(10/z_0)} \tag{10.12}$$

**Example 10.1** A guyed mast is instrumented with anemometers at 10 m and 100 m above the ground. From the analysis of the records of one set of hourly readings, it was found that for wind from north-east the mean velocity at 10 m was 19·6 m/s, while the corresponding mean velocity at 100 m was 33·6 m/s. Assuming the variation of the mean wind speed with height as expressed by eq. (10.4) to be correct, calculate the roughness length $z_0$ and surface drag coefficient $k$ for the site for the given wind direction. Calculate also the gradient height if the latitude of the site is 53°.

From eq. (10.4)

$$U(10) = 19{\cdot}60 = 2{\cdot}5u_* \ln(10/z_0)$$
$$U(100) = 33{\cdot}64 = 2{\cdot}5u_* \ln(100/z_0)$$

form which

$$\ln(100/z_0) = 1{\cdot}7155 \ln(10/z_0)$$

Hence

$$(100/z_0) = (10/z_0)^{1{\cdot}7155}$$

Thus

$$z_0^{1{\cdot}7155} = 0{\cdot}5193976$$
$$\underline{z_0 = 0{\cdot}400\,\text{m}}$$

The surface drag coefficient is determined using eq. (10.6). This requires that the shear velocity $u_*$ be determined first by substitution of the values for $U(10)$ and $z_0$ into eq. (10.5), yielding

$$u_* = \frac{19{\cdot}61}{2{\cdot}5\ln(10/0{\cdot}4)} = 2{\cdot}437\,\mathrm{m/s}$$

Hence the value of surface drag coefficient is

$$k = u_*^2/U^2(10) = 2{\cdot}437^2/19{\cdot}61^2 = \underline{0{\cdot}0154}$$

which agrees with the value corresponding to $z_0 = 0{\cdot}4\,\mathrm{m}$ given in Table 10.1.

The gradient height for the site is found by using eq. (10.5). Thus

$$H = 2{\cdot}437/12 \times 7{\cdot}2722 \times 10^{-5} \sin 57^\circ = \underline{3329{\cdot}79\,\mathrm{m}}$$

This calculated value is considerably greater than the gradient height of 900 m assumed in the wind map issued by the Meteorological Office for the UK.

**Example 10.2** For a site at longitude 57° the surface drag coefficient $k = 0{\cdot}01$. The estimated maximum wind speed occurring during a 50 year period at a height of 10 m is 25·0 m/s. Determine and compare the corresponding mean wind profiles obtained from the ground and up to a height of 100 m, using equations (10.4) and (10.10).

From eq. (10.6)

$$u_* = 25{\cdot}0\sqrt{0{\cdot}01} = 2{\cdot}50\,\mathrm{m/s}$$

The corresponding value for $z_0$ is found by substituting the values $U(10) = 25{\cdot}0\,\mathrm{m/s}$, $z = 10$ and $u_* = 2{\cdot}5\,\mathrm{m/s}$ into eq. (10.7). This yields

$$z_0 = 10{\cdot}0 \times \mathrm{e}^{-25{\cdot}0/2{\cdot}5\times 2{\cdot}50} = 0{\cdot}183\,\mathrm{m}$$

From eq. (10.4)

$$U(z) = 2{\cdot}5u_* \ln(z/z_0)$$

Hence

$$U(z) = 2{\cdot}5 \times 2{\cdot}50\ln(z/0{\cdot}183)$$

From eq. (10.10)

$$U(z) = 2{\cdot}5u_*[\ln(z/z_0) + 5{\cdot}75(z/H)]$$

where, from eq. (10.9)

$$H = u_*/2\beta\omega\sin\phi$$

Hence

$$H = 2{\cdot}50/2 \times 6 \times 7{\cdot}2722 \times 10^{-5} \sin 57^\circ = 3415{\cdot}872\,\text{m}$$

and thus

$$U(z) = 2{\cdot}5 \times 2{\cdot}50[\ln(z/0{\cdot}183) + 5{\cdot}75(z/3415{\cdot}872)]$$

Substitution in turn of the values 20 m, 40 m, 60 m, 80 m, 100 m, 150 m, 200 m and 250 m for $z$ into the above two expressions for $U(z)$ yields the values given in Table 10.2. Thus the use of eq. (10.10) becomes more significant in the design of very tall structures such as towers and guyed masts.

*Table 10.2. Example 10.2 data*

| $U(z)$ | Eq. (10.4) | Eq. (10.10) | Difference |
|---|---|---|---|
| $U(20)$ | 29·34 m/s | 29·55 m/s | 0·716% |
| $U(40)$ | 33·67 m/s | 34·09 m/s | 1·247% |
| $U(60)$ | 36·21 m/s | 36·84 m/s | 1·740% |
| $U(80)$ | 38·00 m/s | 38·84 m/s | 2·211% |
| $U(100)$ | 39·40 m/s | 40·45 m/s | 2·665% |
| $U(150)$ | 41·93 m/s | 43·51 m/s | 3·768% |
| $U(200)$ | 43·73 m/s | 45·83 m/s | 4·802% |
| $U(250)$ | 45·12 m/s | 47·75 m/s | 5·829% |

## Statistical properties of the fluctuating velocity component of wind

In the description of the nature of wind above, it is explained that the velocity of wind could be considered to consist of a constant or mean wind speed component and a fluctuating velocity component due to the turbulence or gusting caused by the ground roughness. Recordings of wind have shown that the velocity of wind can be considered as a stationary random process. Thus

$$\frac{1}{T}\int_0^T U(t)\mathrm{d}t = U \qquad (10.13)$$

$$\frac{1}{T}\int_0^T u(t)\mathrm{d}t = 0 \qquad (10.14)$$

Because of this the characteristics of the fluctuating component of wind can be quantified by statistical functions. The most important of these for the dynamic analyst are

- the variance $\sigma^2$ and the standard deviation $\sigma$
- the auto-covariance function $C_u(\tau)$ for the fluctuating velocity component $u(t)$
- the spectral density function or power spectrum $S_u(n)$
- the cross-covariance function $C_{uv}(\tau)$ of the fluctuating velocity components $u(t)$ and $v(t)$

- the cross-spectral density function or cross-power spectrum $S_{uv}(n)$
- the coherence function $coh_{uv}(n)$
- the probability density function $p(u)$ and peak factor $\kappa$ for $u(t)$
- the cumulative distribution function $P(U)$ of $U(t)$

where $n$ is the frequency of a constituent harmonic wind component, as opposed to $f$ which, in this book, is used to denote a structural modeshape frequency. The definitions and mathematical formulations of the above functions are given in the following sections.

### *Variance and standard deviation*

The variance of the fluctuating or gust velocity component is defined as

$$\sigma^2(u) = \frac{1}{T}\int_0^T \mathbf{u(t)}^{\mathrm{T}}\mathbf{u(t)}\mathrm{d}t = \sigma^2(u_x) + \sigma^2(u_y) + \sigma^2(u_z) \tag{10.15}$$

where

$$\mathbf{u(t)} = \begin{bmatrix} u_x(t) \\ u_y(t) \\ u_z(t) \end{bmatrix} \tag{10.16}$$

The variances along the $x$-axis, $y$-axis and $z$-axis are therefore equal to the mean square value of the fluctuations in these directions. From recorded data it has been observed that the greatest part of the variance is associated with the fluctuations of the velocity in the direction of the mean flow. If the direction along the flow parallel to the ground is the $x$-direction, the direction perpendicular to the flow and parallel to the ground is the $y$-direction, and the direction perpendicular to the flow is the $z$-direction, then it can be stated that usually

$$\begin{aligned} \sigma^2(u_x) &\approx 10\sigma^2(u_y) \\ \sigma^2(u_y) &> \sigma^2(u_z) \end{aligned} \tag{10.17}$$

In general it is therefore assumed that

$$\sigma_u{}^2 \approx \sigma^2(u_x) = \frac{1}{T}\int_0^T u_x{}^2(t)\mathrm{d}t \tag{10.18}$$

The variance $\sigma^2(u)$ is obviously a function of the ground roughness and may be expressed in terms of the shear velocity $u_*$ as

$$\sigma_u{}^2 = \beta u_*{}^2 \tag{10.19}$$

Previously it was generally assumed that $\sigma_u$ was independent of height, and that for engineering purposes the constant $\beta \approx 6{\cdot}0$ when the averaging time was 1 h. The reader should, however, be aware that, particularly over rough ground, values as low as $\beta \approx 4{\cdot}0$ have been reported in the literature. Nowadays it is generally accepted that the variance varies with height and not only with ground roughness and mean wind speed. An expression that takes this dependence on height into account is

$$\sigma_u(z) = 2{\cdot}63u_*\eta(0{\cdot}538 + z/z_0)^{\eta 16} \tag{10.20}$$

where $\eta = 1 - z/H$ and the gradient height $H$ is given by eq. (10.9).

The standard deviation at $\sigma(z)$ at height $z$ provides a measure of the dispersion of the wind speed around its mean value $U(z)$ and is used as a measure of the turbulence intensity $I(z)$, which is given by

$$I_u(z) = \sigma_u(z)/U_u(z) \tag{10.21}$$

### *Auto-correlation and auto-covariance functions*

Two other important statistical concepts are as follows: the so-called *auto-correlation function* $R(\tau)$, where

$$R_U(\tau)_{T\to\infty} = \frac{1}{T}\int_{-\infty}^{\infty} [U + u(t)][U + u(t+\tau)]\mathrm{d}t \tag{10.22}$$

and the *auto-covariance function* $C_u(\tau)$, where

$$C_u(\tau)_{T\to\infty} = \frac{1}{T}\int_{-\infty}^{\infty} [u(t)u(t+\tau)]\mathrm{d}t \tag{10.23}$$

The function $C_u(\theta)$ provides a measure of the interdependence of the fluctuating velocity component $u$ of the wind at times $t$ and $t + \tau$. From eq. (10.18) it follows that when $\tau = 0$

$$C_u(\tau) = C_u(0) = \sigma^2(u) \tag{10.24}$$

Because wind histories are considered as stationary random processes, with statistical properties independent of time, it follows that $R_U(\tau) = R_U(-\tau)$ and $C_u(\theta = C_u(0)$.

It has also been found convenient to define an *auto-covariance coefficient* which is defined as the ratio of $C(\tau)$ to $C(0)$. Hence the expression for the auto-covariance coefficient is given by

$$c_u(\tau) = C_u(\tau)/C_u(0) = C_u(\tau)/{\sigma_u}^2 \tag{10.25}$$

Thus when $\tau = 0$, $c_u(\tau) = 1{\cdot}0$. In the limit when $\tau \to \infty$, $c_u(\tau) \to 0$. The auto-covariance coefficient can therefore be regarded as a measure of the extent to which the fluctuation of the wind at time $t$ is a function of the fluctuation at time $t + \tau$. If the value of $c_u(\tau)$ is small, then the two quantities are almost independent, while if $c_u(\tau) = 1{\cdot}0$ they are completely dependent on each other. For wind the auto-covariance coefficient decreases with increasing values of $\tau$, as shown in Fig. 10.2, where $c_u(\tau)$ is plotted against the time lag $\tau$ for a recorded along-wind and across-wind trajectory.

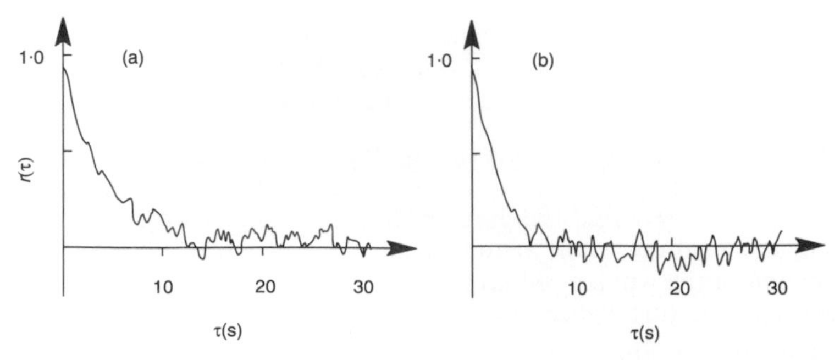


*Fig. 10.2. Auto-correlation functions for (a) along-wind and (b) across-wind components of recorded wind*

## Spectral density functions of longitudinal velocity fluctuations

*Spectral density functions*, also referred to as *power spectra*, are another important function that defines the random nature of wind. A spectral density function is denoted by $S_u(n)$, where the variable $n$ is the frequency of the sinusoidal velocity components of the fluctuating part of the wind velocity. Spectral density functions give a measure of the energy distribution of the harmonic velocity components, and form the basis for dynamic response analysis of linear structures in the *frequency domain.* It can be shown that they can be expressed as Fourier transforms of the auto-covariance function $C_u(\tau)$. Thus

$$S_u(n) = 4\int_0^\infty C_u(\tau)\cos(2\pi n\tau)\mathrm{d}\tau \tag{10.26}$$

$$c_u(\tau) = \int_0^\infty S_u(n)\cos(2\pi n\tau)\mathrm{d}n \tag{10.27}$$

When the time lag $\tau = 0$, $C_u(\tau) = C_u(0) = \sigma_u{}^2$. Hence eq. (10.27) yields

$$\int_0^\infty S_u(n)\,\mathrm{d}n = \sigma_u{}^2 \tag{10.28}$$

Davenport (see ref. 10.2) suggested the following formulation for the spectral density function

$$S_u(n) = \frac{4u_*{}^2 f^2}{n(1+f^2)^{4/3}} \tag{10.29}$$

where

$$f = \frac{1200n}{U(10)}$$

Harris (see ref. 10.1, vol. 2) modified the formulation by Davenport and suggested the formulation

$$S_u(n) = \frac{4u_*{}^2 f^2}{n(2+f^2)^{5/6}} \tag{10.30}$$

where

$$f = \frac{1800n}{U(10)}$$

Both the above expressions for the spectral density functions depend only on the mean wind speed $U(10)$ and the ground roughness $z_0$; they are independent of the height $z$. This is contrary to experimental evidence. Thus the use of the constant *length scales* $L = 1200\,\text{m}$ and $L = 1800\,\text{m}$ was doubted. As a result, Deaves and Harris (see ref. 10.1, vol. 2) introduced a length scale that varied with height, and developed the following expression for the spectral density function

$$S_u(n) = \frac{0{\cdot}115\sigma_u{}^2(z)T_u(z)}{\{0{\cdot}0141 + n^2T_u{}^2(z)\}^{5/6}} \tag{10.31}$$

where the *time scale* $T_u(z)$ is determined by integration of the auto-covariance coefficient $c_u(z)$. Thus

$$T_u(z) = \int_0^\infty c_u(z, \tau)\mathrm{d}\tau \tag{10.32}$$

and $\sigma_u{}^2(z)$ can be calculated by use of eq. (10.20). The time scale $T_u(z)$ is related to the length scale $L_u(z)$ through

$$L_u(z) = T_u(z)U(z) \tag{10.33}$$

Thus the dependence of the length scale on height is implied in the expression for the spectral density function given by eq. (10.31). The evaluation of $T_u(z)$ by integration of $c_u(z)$ is not a practical proposition for design purposes, and ref. 10.1, vol. 2, gives a method for calculating $L_u(z)$ from which $T_u(z)$ can be calculated using eq. (10.33). The method is lengthy and is considered to be outside the scope of this book.

A more convenient formulation of a spectral density function that varies with height is that suggested by Kaimal (see ref. 10.2) and given by

$$S_u(z,n) = \frac{200u_*{}^2f(z,n)}{n[1 + 50f(z,n)]^{5/3}} \tag{10.34}$$

where

$$f(z,n) = \frac{zn}{U(z)}$$

In the higher frequency range in which structures are likely to respond, this function approximates very closely to spectra of recorded wind histories. However, it is suspect in the lower frequency range. Another spectral density function, that also varies with height, is based on the current ESDU (European Statistical Data Unit) model, which is given by

$$S_u(z,n) = \frac{U(10)^{4{\cdot}4}(1 + S_{\text{top}}\phi)^{2{\cdot}66}}{2500z_0{}^{1/25}n^{1/3}U(z)^2} \tag{10.35}$$

where $S_{top}$ is a topographic factor and $\phi$ is the hill slope. Values for the spectral density functions given by eqs (10.29), (10.30), (10.34) and (10.35) are compared in Table 10.3 for a mean velocity $U(10) = 25{\cdot}0$ m/s and a roughness length $z_0 = 0{\cdot}3$ m.

As can be seen from Table 10.3, the values of $S_u(z, n)$ obtained using eqs (2.34) and (2.35) decrease with increasing height. It can be observed that for the lower frequencies the spectrum based on the ESDU model yields much lower values for the power spectral density function in the lower frequency range than do the other three spectra.

In Fig. 10.3, eqs (10.29), (10.30) and (10.34) are plotted in nondimensional form for turbulent wind with $U(10) = 30$ m/s and $z_0 = 0{\cdot}08$ m.

*Table 10.3. Variation in spectral density function values for $U(10) = 25{\cdot}0\,m/s$, $z_0 = 0{\cdot}3\,m$ and $S_{top}\phi = 0$*

| Equation | Height | 0·1 Hz | 0·5 Hz | 1·0 Hz | 2·0 Hz | 3·0 Hz | 4·0 Hz |
|---|---|---|---|---|---|---|---|
| 10.29 | | 108·0275 | 7·8016 | 2·4616 | 0·7757 | 0·3947 | 0·2244 |
| 10.30 | | 84·5362 | 5·9598 | 1·8790 | 0·5920 | 0·3012 | 0·1865 |
| 10.34 | 100 m | 54·1373 | 4·1139 | 1·3136 | 0·4166 | 0·2124 | 0·1317 |
| 10.34 | 200 m | 38·9267 | 2·8276 | 0·8975 | 0·2838 | 0·1446 | 0·0896 |
| 10.32 | 300 m | 31·6070 | 2·2567 | 0·7147 | 0·2257 | 0·1149 | 0·0712 |
| 10.35 | 100 m | 1·5322 | 0·5499 | 0·3464 | 0·2182 | 0·1663 | 0·1375 |
| 10.35 | 200 m | 1·2833 | 0·4389 | 0·2765 | 0·1742 | 0·1329 | 0·1097 |
| 10.35 | 300 m | 1·1371 | 0·3889 | 0·2450 | 0·1543 | 0·1178 | 0·0972 |

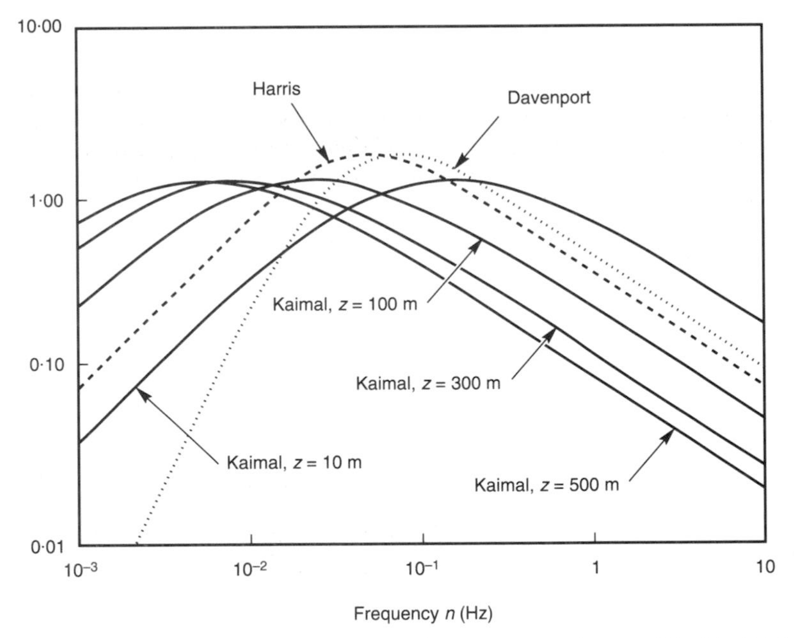


*Fig. 10.3. Comparison of spectral density functions given by eqs (10.29), (10.30) and (10.34) for $U(10) = 30\,m/s$ and $z_0 = 0{\cdot}08\,m$*

### Cross-correlation and cross-covariance functions

The cross-correlation and cross-covariance of two continuous records $[U + u(t)]_j$ and $[V + v(t)]_k$, recorded at two different stations $j$ and $k$ in space, are measures of the degree to which the two records are correlated in the amplitude domain. The cross-correlation function is given by

$$R^{jk}_{UV}(\tau)_{T\to\infty} = \frac{1}{T}\int_{-\infty}^{\infty} [U + u(t)]_j[V + v(t+\tau)]_k \mathrm{d}t \tag{10.36}$$

and the cross-covariance function by

$$C^{jk}_{uv}(\tau)_{t\to\infty} = \frac{1}{T}\int_{-\infty}^{\infty} u_j(t)v_k(t+\tau)\mathrm{d}t \tag{10.37}$$

When $\tau = 0$

$$C^{jk}_{uv}(\tau) = C^{jk}_{uv}(0) = \sigma^2(u_j, v_k) = {\sigma_{uv}}^2 \tag{10.38}$$

where ${\sigma_{uv}}^2$ is the cross-variance.

### Cross-spectral density and coherence functions for longitudinal velocity fluctuations

The cross-covariance function between the fluctuating velocity components of wind at stations $j$ and $k$ at zero time lag having been defined, it can be shown that

$$C^{jk}_{uv}(0) = \int_0^{\infty} S^{\mathrm{cr}}_{u_jv_k}(n)\mathrm{d}n = \int_0^{\infty} S^C_{u_j}v_k(n)\,\mathrm{d}n + \iota\int_0^{\infty} S^Q_{u_jv_k}(n)\,\mathrm{d}n \tag{10.39}$$

where $\iota = \sqrt{-1}$ and $S^{\mathrm{cr}}_{u_jv_k}(n)$ is the *cross-spectral density function*, which is a measure of the degree to which two histories $u(t)$ and $v(t)$, recorded at stations $j$ and $k$ respectively, are correlated in the frequency domain. The terms $S^C_{u_jv_k}(n)$ and $S^Q_{u_jv_k}(n)$ are known as the *co-spectrum* and *quadrature spectrum* respectively. In wind engineering the quadrature spectrum is usually assumed to be negligible compared to the co-spectrum. Thus eq. (10.39) may be reduced to

$$C^{jk}_{uv}(0) = \int_0^{\infty} S^{\mathrm{cr}}_{u_jv_k}(n)\mathrm{d}n = \int_0^{\infty} S^C_{u_jv_k}(n)\mathrm{d}n \tag{10.40}$$

On the basis of wind tunnel measurements, it has been suggestd that it is reasonable to assume in engineering calculations that

$$S^{\mathrm{cr}}_{u_jv_k}(n) = \sqrt{[S_{u_j}(n) \times S_{v_k}(n)]}\,\mathrm{e}^{-\phi} \tag{10.41}$$

where $\mathrm{e}^{-\phi}$, known as the *narrow-band cross-correlation*, is the square root of the *coherence function* $\mathrm{e}^{-2\phi} = coh^2u_{jk}(n)$, and

$$\phi = \frac{2n\sqrt{\left[{C_x}^2(x_j - x_k)^2 + {C_y}^2(y_j - y_k)^2 + {C_z}^2(z_j - z_k)^2\right]}}{U(z_j) + U(z_k)} \tag{10.42}$$

where the exponential decay coefficients $C_z = 10$ and $C_y = 16$. Full-scale measurements, however, indicate that $C_y$ and $C_z$ decrease with increasing height, and increase with increasing wind speed and increasing ground roughness. Different wind codes may therefore recommend other values for $C_y$ and $C_z$ than those given above. In chapter 11 it is shown that the response of multi-DOF systems is a function of both spectral and cross-spectral density functions. It is therefore of interest to see how the value of $e^{-\phi}$ varies with the distance between two points. In general it can be observed that the value of $e^{-\phi}$ decreases with (*a*) increasing distance between two points, (*b*) increasing frequencies and (*c*) decreasing wind speeds. Substitution of different values for $n$, $(y_j - y_k)$, $(z_j - z_k)$, $U(z_j)$ and $U(z_k)$ into eq. (10.42) indicates that for values of $U(10) \approx 25{\cdot}0\,\text{m/s}$ and frequencies greater than approximately 1·5 Hz, the correlation between two histories is negligible when the distance between two stations is greater than, say, 5·0 m. For many civil engineering structures, therefore, this seems to imply that the effect of cross-correlation can frequently be ignored when undertaking dynamic analysis in the frequency domain.

**Example 10.3** A 45 m tall mast, whose first natural frequency is 1·0 Hz and second natural frequency is 2·0 Hz, is subjected to a mean wind speed of 25 m/s 10 m above ground level. Calculate the values of the spectral density function corresponding to the two first natural frequencies of the mast at points $P_1 = 25{\cdot}0\,\text{m}$, $P_2 = 35{\cdot}0\,\text{m}$, $P_3 = 40{\cdot}0\,\text{m}$, and $P_4 = 45{\cdot}0\,\text{m}$ along the length of the mast. Hence calculate the values of the coherence and hence the cross-spectral density functions for points $P_4$ and $P_3$, points $P_4$ and $P_2$, and points $P_4$ and $P_1$. Assume the ground roughness length to be 0·3 m and use eq. (10.34) when calculating the spectral density values.

Before calculating the spectral density, coherence and cross-spectral density functions, one must first calculate the shear velocity $u_*$ and the velocities $U(25)$, $U(35)$, $U(40)$ and $U(45)$. From eq. (10.5)

$$u_* = \frac{U(10)}{2{\cdot}5\ln(10/z_0)} = \frac{25{\cdot}0}{2{\cdot}5\ln(10/0{\cdot}3)} = 2{\cdot}8518\,\text{m/s}$$

Hence

$$U(25) = 2{\cdot}5 \times 2{\cdot}8518 \times \ln(25/0{\cdot}3) = 31{\cdot}5327\,\text{m/s}$$
$$U(35) = 2{\cdot}5 \times 2{\cdot}8518 \times \ln(35/0{\cdot}3) = 33{\cdot}9316\,\text{m/s}$$
$$U(40) = 2{\cdot}5 \times 2{\cdot}8518 \times \ln(40/0{\cdot}3) = 34{\cdot}8836\,\text{m/s}$$
$$U(45) = 2{\cdot}5 \times 2{\cdot}8518 \times \ln(45/0{\cdot}3) = 35{\cdot}7233\,\text{m/s}$$

The spectral density function given by eq. (10.34) is

$$S_u(z,n) = \frac{200u_*^2 f(z,n)}{n[1 + 50f(z,n)]^{5/3}}$$

where

$$f(z,n) = \frac{zn}{U(z)}$$

Hence

$$f(25{\cdot}0, 1{\cdot}0) = \frac{25{\cdot}0 \times 1{\cdot}0}{31{\cdot}5327} = 0{\cdot}79283$$

$$S_u(25{\cdot}0, 1{\cdot}0) = \frac{200 \times 2{\cdot}8518^2 \times 0{\cdot}79283}{1{\cdot}0[1 + 50 \times 0{\cdot}79283]^{5/3}} = 2{\cdot}6843\,\mathrm{m^2/s}$$

$$f(35{\cdot}0, 1{\cdot}0) = \frac{35{\cdot}0 \times 1{\cdot}0}{33{\cdot}9316} = 1{\cdot}03149$$

$$S_u(35{\cdot}0, 1{\cdot}0) = \frac{200 \times 2{\cdot}8518^2 \times 1{\cdot}03149}{1{\cdot}0[1 + 50 \times 1{\cdot}03149]^{5/3}} = 2{\cdot}2739\,\mathrm{m^2/s}$$

$$f(40{\cdot}0, 1{\cdot}0) = \frac{50{\cdot}00 \times 1{\cdot}0}{34{\cdot}8836} = 1{\cdot}14667$$

$$S_u(40{\cdot}0, 1{\cdot}0) = \frac{200 \times 2{\cdot}8518^2 \times 1{\cdot}14667}{1{\cdot}0[1 + 50 \times 1{\cdot}14667]^{5/3}} = 2{\cdot}1257\,\mathrm{m^2/s}$$

$$f(45{\cdot}0, 1{\cdot}0) = \frac{45{\cdot}0 \times 1{\cdot}0}{35{\cdot}7233} = 1{\cdot}25968$$

$$S_u(45{\cdot}0, 1{\cdot}0) = \frac{200 \times 2{\cdot}8518^2 \times 1{\cdot}25968}{1{\cdot}0[1 + 50 \times 1{\cdot}25968]^{5/3}} = 2{\cdot}0017\,\mathrm{m^2/s}$$

The cross-spectral density function and the square root of the coherence function $e^{-\phi}$ are given by eqs (10.41) and (10.42). Because the points $P_1$, $P_2$, $P_3$, and $P_4$ lie on a vertical line, the expression for $e^{-\phi}$ reduces to

$$e^{-\phi} = \exp\{-2nC_z(z_j - z_k)/[U(Z_j) - U(Z_k)]\}$$

If it is assumed that $C_z = 10$, then

$$coh_{uv}(45{\cdot}0, 40{\cdot}0, 1{\cdot}0) = \exp\left\{\frac{-2 \times 1{\cdot}0 \times 10(45{\cdot}0 - 40{\cdot}0)}{(35{\cdot}7233 + 34{\cdot}8836)}\right\} = 0{\cdot}24261$$

$$coh_{uv}(45{\cdot}0, 35{\cdot}0, 1{\cdot}0) = \exp\left\{\frac{-2 \times 1{\cdot}0 \times 10(45{\cdot}0 - 35{\cdot}0)}{(35{\cdot}7233 + 33{\cdot}9316)}\right\} = 0{\cdot}05663$$

$$coh_{uv}(45{\cdot}0, 25{\cdot}0, 1{\cdot}0) = \exp\left\{\frac{-2 \times 1{\cdot}0 \times 10(45{\cdot}0 - 25{\cdot}0)}{(35{\cdot}7233 + 31{\cdot}5327)}\right\} = 0{\cdot}00261$$

The values for the different cross-spectral density functions at $n = 1{\cdot}0$ Hz can now be calculated by use of eq. (10.42), which for two points along a vertical line may be written as

$$S^C_{uv}(z_1, z_2, n) = e^{-\phi}[S_u(z_1, n)S_v(z_2, n)]^{1/2}$$

thus

$$S^C_{uv}(45{\cdot}0, 40{\cdot}0, 1{\cdot}0) = 0{\cdot}24261 \times [2{\cdot}0017 \times 2{\cdot}1257]^{1/2} = 0{\cdot}5005\,\mathrm{m^2/s}$$

$$S^C_{uv}(45{\cdot}0, 35{\cdot}0, 1{\cdot}0) = 0{\cdot}05663 \times [2{\cdot}0017 \times 2{\cdot}2739]^{1/2} = 0{\cdot}1181\,\mathrm{m^2/s}$$

$$S^C_{uv}(45{\cdot}0, 25{\cdot}0, 1{\cdot}0) = 0{\cdot}00261 \times [2{\cdot}0017 \times 2{\cdot}6843]^{1/2} = 0{\cdot}0061\,\mathrm{m^2/s}$$

The above calculations repeated for $n = 2{\cdot}0$ Hz yield

$$S_u(25{\cdot}0, 2{\cdot}0) = 0{\cdot}8631\,\mathrm{m^2/s}$$

$$S_u(35{\cdot}0, 2{\cdot}0) = 0{\cdot}7277\,\mathrm{m^2/s}$$

$$S_u(40{\cdot}0, 2{\cdot}0) = 0{\cdot}6792\,\mathrm{m^2/s}$$

$$S_u(45{\cdot}0, 2{\cdot}0) = 0{\cdot}6388\,\mathrm{m^2/s}$$

$$coh_{uv}(45{\cdot}0, 40{\cdot}0, 2{\cdot}0) = 0{\cdot}058861$$

$$coh_{uv}(45{\cdot}0, 35{\cdot}0, 2{\cdot}0) = 0{\cdot}003206$$

$$coh_{uv}(45{\cdot}0, 25{\cdot}0, 2{\cdot}0) = 0{\cdot}000007$$

$$S^C_{uv}(45{\cdot}0, 40{\cdot}0, 2{\cdot}0) = 0{\cdot}038771\,\mathrm{m^2/s}$$

$$S^C_{uv}(45{\cdot}0, 35{\cdot}0, 2{\cdot}0) = 0{\cdot}002185\,\mathrm{m^2/s}$$

$$S^C_{uv}(45{\cdot}0, 25{\cdot}0, 2{\cdot}0) = 0{\cdot}000005\,\mathrm{m^2/s}$$

The above calculations reveal that the values of the cross-spectral density functions, as well as the ratios of the same functions to the spectral density functions, decrease with increasing distance between two histories and with increasing frequency.

## Probability density function and peak factor for fluctuating component of wind

Let the range of the amplitudes of the fluctuating velocity component of wind $u(t)$ associated with a given record be divided into equal intervals $\Delta u(t)$ and let the number of times the amplitude of $u(t)$ lies within the interval $u_i(t)$ to $u_{i+1}(t)$ be $n_i$. A graph in which the numbers of $n_i$ are plotted against the interval $u_i(t)$ and $u_{i+1}(t)$ as shown in Fig. 10.4(a) is called a *histogram*. If $n_i$ is divided by the total number or readings $n$, and the interval $\Delta u(t)$ is made so small that it may be written as $\mathrm{d}u(t)$, the histogram becomes a smooth curve as shown in Fig. 10.4. The curve is referred to as the *probability density function* or *probability distribution function* and is denoted $p(u)$. Because of its derivation it follows that

$$\int_{-\infty}^{\infty} p(u)\mathrm{d}u = 1 \tag{10.43}$$

$$\int_{-\infty}^{\infty} u(t)p(u)\mathrm{d}u = \sigma_u{}^2 \tag{10.44}$$

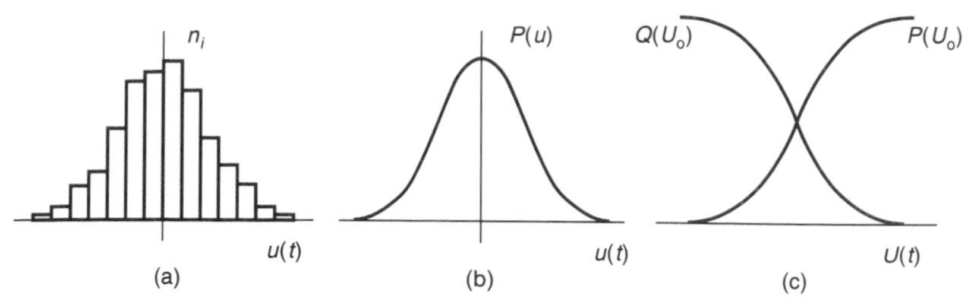


*Fig. 10.4. (a) Histogram, (b) probability density function, (c) cumulative distribution functions*

In wind engineering the fluctuating component of wind is considered as a normally distributed stationary random signal with zero mean and standard deviation $\sigma_u$. The probability density function can therefore be assumed to be Gaussian, in which case it can be shown that

$$p(z) = \frac{1}{\surd(2\pi)\exp[-z^2/2]} \tag{10.45}$$

where

$$z = u(t)/\sigma_u{}^2 \tag{10.46}$$

The magnitude of the amplitude of the maximum fluctuation that may occur within a given time interval $T$ of such a process is expressed at

$$u(t)_{\max} = \kappa\sigma_u \tag{10.47}$$

where

$$\kappa = \surd(2\ln\nu T) + 0{\cdot}577\surd(2\ln\nu T) \tag{10.48}$$

$$\nu = \left\{\int_0^\infty n^2 S_u(n)\mathrm{d}n \Big/ \int_0^\infty S_u(n)\mathrm{d}n\right\}^{1/2} \tag{10.49}$$

For weakly damped structures $\nu$ may be assumed to be equal to $f = 2\pi\omega_\mathrm{n}$. When this is the case

$$u(t)_{\max} = \{\surd[2\ln(2\pi\omega_\mathrm{n}T)] + 0{\cdot}577\surd[2\ln(2\pi\omega_\mathrm{n}T)]\sigma_u \tag{10.50}$$

## Cumulative distribution function

In many codes a design value for wind is defined as a value with a stated probability of being exceeded. A useful tool for this purpose is the *cumulative distribution function*, which is usually denoted $P(U_0)$. If $p(U)$ is the probability distribution function for the total wind velocity $U(T)$, then

$$P(U_0) = \mathrm{Prob}[U < U_0] = \int_{-\infty}^{U_0} p(U)\mathrm{d}U \tag{10.51}$$

Alternatively, a design value may be defined as a value that has a stated probability of not being exceeded. If this is denoted $Q(U_0)$, the cumulative distribution function in this case is given by

$$Q(U_0) = \text{Prob}[U > U_0] = \int_{U_0}^{-\infty} p(U)\mathrm{d}U \qquad (10.52)$$

Thus $P(U_0)$ yields the probability that the wind speed $U(t)$ is less than $U_0$, and $Q(U_0)$ the probability that $U(t)$ is greater than $U_0$. Diagrams of both types of cumulative distribution function are shown in Fig. 10.4(c).

## Pressure coefficients

The fluctuating pressure caused by wind is given by

$$p(t) = \tfrac{1}{2}\rho C_\mathrm{p}(U(t) - \dot{x})^2 \qquad (10.53)$$

where $\rho$ is the density of air, $C_\mathrm{p}$ is the pressure coefficient for a given point, and $\dot{x}$ is the velocity of the structure at the same point and in the direction of the wind. The integration of the pressure $p(t)$ over the surface of a structure or structural element will yield the resultant force exerted by the wind.

The force components parallel and perpendicular to the along-wind direction are given respectively as

$$F_\mathrm{d} = \tfrac{1}{2}\rho C_\mathrm{d} A_\mathrm{d}(U(t) - \dot{x})^2 \qquad (10.54a)$$

$$F_\mathrm{l} = \tfrac{1}{2}\rho C_\mathrm{l} A_\mathrm{l}(U(t) - \dot{x})^2 \qquad (10.54b)$$

where $C_\mathrm{d}$ is the drag coefficient, $C_\mathrm{l}$ is the lift coefficient, $A_\mathrm{d}$ is the area projected on to a plane perpendicular to the direction of the wind, and $A_\mathrm{l}$ is the area projected on to a plan in the along-wind direction by a unit length of structure or structural element.

**Problem 10.1** Given that the wind speed 10 m above the ground is 25·0 m/s and the surface drag coefficient $k = 28{\cdot}0 \times 10^{-3}$, calculate the shear velocity $u_*$ and the roughness length $z_0$.

**Problem 10.2** A 100 m tall transmission tower is situated in an area with pine forest for which the roughness length may be assumed to be 1·0 m. The design wind speed 10 m above the ground is 28 m/s. Calculate the corresponding shear velocity, and the wind velocities at 50 m, 90 m and 100 m heights.

**Problem 10.3** Let the first natural frequency of the tower in problem 10.2 be 1·0 Hz. Calculate the values of Davenport and Harris power spectra for this frequency and for the design wind speed given in problem 10.2. Compare the values obtained with those calculated for heights of 90 m and 100 m using Kaimal's spectrum.

**Problem 10.4** Use the data obtained in problems 10.2 and 10.3 to calculate the values of the square root of the coherence function and of the cross-spectral density function at the first natural frequency of the tower for heights of 90 m and 100 m. Assume the value of the exponential decay coefficient $C_z$ in eq. (10.32) to be 8.

## References

Lawson, T. V. *Wind effects on buildings*, vols 1 and 2. Applied Science, London, 1980.

Simue, E. & Scanlan, R. H. *Wind effects on structures*. Wiley, Chichester, 1978.

# 11. Dynamic response to turbulent wind: frequency domain analysis

## Aeroelasticity and dynamic response

Wind acting on structures induces stresses and deflections. If the deformations caused alter the boundary conditions of the incident wind to such an extent that they alter the flow pattern, and this gives rise to succeeding deflections of an oscillating nature, a phenomenon referred to as *aeroelastic instability* is said to occur. All aeroelastic instabilities result from aerodynamic forces that are influenced by the motion of the structure. The main types of aeroelastic instability are *cross-wind galloping*, *torsional divergence* and *flutter*. *Buffeting*, which is defined as the random loading of structures due to velocity fluctuations in the oncoming wind, may be aeroelastically stable or unstable.

Cross-galloping is mainly associated with slender sections having special cross-sections such as rectangular or D sections, or the effective sections of greased or ice-coated cables. Such structures can exhibit amplitudes of vibration many times their cross-sectional dimensions, and at frequencies much less than those of vortex shedding from the same sections.

Torsional divergence is divergence in which structures subjected to the lift, drag and torsional forces due to wind will tend to twist the structures such as to increase the angle of attack. As the wind velocity increases the structure will twist further until the structure may be twisted to destruction. The wind speed at which structural collapse occurs is referred to as the *critical divergence velocity* of wind. In most cases of interest to the structural engineer the critical divergence velocities are extremely high, and much greater than the wind velocities considered in design.

The term 'flutter' covers a whole class of aeroelastic oscillations such as *classical* flutter, *single-DOF* flutter and *panel* flutter. Classical flutter implies an aeroelastic phenomenon in which wind causes and couples together oscillations of a structure in one vertical and one rotational DOF. Single-DOF flutter is associated with bluff, unstreamlined bodies causing flow separation. Notable examples are the decks of cable-suspended span bridges, which can exhibit single-degree torsional instability. Panel flutter is sustained vibration of panels caused by the passing of wind. Most prominent has been the high-speed passage of air in supersonic flows. In civil engineering, panel flutter has mainly been associated with membrane structures and prestressed cable net roofs with insufficient local or global anticlastic curvature.

Galloping, flutter, and also vortex-induced vibration, when the motion of the structure controls the vortex shedding, are referred to as *self-excited vibration*. If the flow of air results in an initial disturbance the oscillations will either diverge or decay, according to whether the energy of motion extracted from the flow is greater or less than the energy dissipated by the level of structural damping. In the case of flutter, wind speeds that cause neither decaying nor diverging oscillations are referred to as *critical* flutter velocities.

Buffeting, as mentioned above, is defined as the random loading due to the velocity fluctuations in turbulent wind. Aeroelastic instability in the case of buffeting is mainly associated with nonlinear structures such as slender towers, the decks of cable-suspended span bridges and insufficiently tensioned cables in cable beam and cable net roof structures.

A great deal of research has been undertaken in order to develop and improve methods for predicting the response for the different types of aeroelastic instability. As yet many of the problems are only partially understood, and the solution of a particular problem usually requires the use of wind tunnels in order to generate numerical models with properties similar to the prototype. Aeroelastic unstable problems are inherently nonlinear and, although important they are outside the scope of this book. For further information the interested reader is referred to books on wind engineering such as Lawson (1990) and Simue & Scalan (1978). A unified view point is that in general the solution of aeroelastic instability problems can be tackled only by forward integration in time, which enables wind speeds, structural deformations and structural stiffness, aerodynamic coefficients and aerodynamic damping to be updated at the end of successive time increments. In most textbooks the theoretical solution of aeroelastic instability problems is confined to two-dimensional problems.

## Dynamic response analysis of aeroelastically stable structures

The dynamic response to wind of aeroelastically stable structures may be predicted by either a time domain or a frequency domain approach. The former requires the generation of spatially correlated wind histories. The latter is based on the use of spectral density or power spectra for wind. Of the two the frequency domain is more generally used, although time domain methods are more powerful. The reason why the latter approach so far is not much used is that insufficient research has been undertaken (*a*) to generate statistically correct spatially correlated wind fields, and (*b*) to develop efficient commercially available computer programs.

## Frequency domain analysis of single-DOF systems

The aim of this approach, originally proposed by Davenport (1961), is to predict the statistical properties of the structural response starting from the knowledge of the statistical properties of the forces due to wind.

Assuming that the fluctuating nature of the wind velocity is stationary, forces due to wind are fully defined by their mean values, their probability distributions and their spectrum of fluctuations. The method is applicable only to structures whose response can be assumed to be linear. When it is applied to nonlinear structures it is assumed that the dynamic response is small compared to the static one, the nonlinearities being taken into account only when calculating the latter. The total response is calculated by superimposing the dynamic response on the static one.

The frequency domain method is based on the following hypotheses

- the dynamic response of the structure is linear
- the mean aerodynamic force due to turbulent wind is the same as that in a steady flow with the same mean velocity
- the relationships between the velocity and force fluctuations are linear
- the probability distributions of the wind speed fluctuations are Gaussian.

The second hypothesis implies that the effect of the acceleration of the wind is negligible. If required, this effect can be accounted for by an additional pressure term $\rho C_{\mathrm{m}}(A/B)\mathrm{d}u(t)/\mathrm{d}t$, where $C_{\mathrm{m}}$ is an additional mass coefficient, $A$ is a reference area, and $B$ is a reference dimension. The existence of this term follows from consideration of the dynamic equilibrium condition of the wind. It represents the force that the wind flowing around a building exerts on the structure as a consequence of the change in wind velocity. The third hypothesis requires that the velocity fluctuations $u$ should be negligible compared with the mean velocity $U$.

The prediction of statistical response requires knowledge of the mean response, the response spectrum, and the probability distribution of the response. The mean response is determined by considering the load due to the mean wind speed $U$ as a static load, while the response due to the fluctuating component $u(t)$ of wind is determined by first calculating the variance of response. The reason for this is that the relationships between velocity, force and displacement fluctuations are assumed to be linear, and the distribution of the velocity fluctuations is assumed to be Gaussian. Thus the distribution of the amplitudes of the fluctuating wind force must also be Gaussian, as must be the distribution of the amplitudes of the fluctuating component of the response. From eq. (10.25), the variance of the fluctuating component of wind is given by

$$\sigma_u^{\,2} = \int_0^{\infty} S_u(n)\mathrm{d}n \tag{11.1}$$

Similarly, the variances of a drag force $f_{\mathrm{d}}(t)$ and response $x(t)$ are found from integration of the force and response spectra respectively. Thus

$$\sigma_f^{\,2} = \int_0^{\infty} S_f(n)\mathrm{d}n \tag{11.2}$$

$$\sigma_x^{\,2} = \int_0^{\infty} S_x(n)\mathrm{d}n \tag{11.3}$$

## Relationships between response, drag force and velocity spectra for one DOF systems

The fluctuating along-wind drag force acting on the area $A$ of a one DOF system vibrating with a velocity $\dot{x}(t)$ is given by

$$f_d(t) = \tfrac{1}{2}\rho C_d A[U(t) - \dot{x}(t)]^2 - \tfrac{1}{2}\rho C_d A U^2 \tag{11.4}$$

or

$$f_d(t) = \tfrac{1}{2}\rho C_d A\left[U^2 + u^2(t) + \dot{x}^2(t) + 2Uu(t) - 2U\dot{x}(t) - 2u(t)\dot{x}(t) - U^2\right] \tag{11.5}$$

When it can be assumed that $u(t)$ and $\dot{x}(t)$ are small compared to $U$, the terms $u^2(t)$, $\dot{x}^2(t)$ and $2u(t)\dot{x}(t)$ are ignored and the expression for $f_d(t)$ is written as

$$f_d(t) = \tfrac{1}{2}\rho C_d A[2Uu(t) - 2U\dot{x}(t)] \tag{11.6}$$

Thus the equation of motion for a one DOF system subjected to a fluctuating drag force may be written as

$$M\ddot{x} + 2\xi_s\omega_n M\dot{x} + Kx = \tfrac{1}{2}\rho C_d[2Uu(t) - 2U\dot{x}(t)] \tag{11.7}$$

Since $\dot{x}(t) = \dot{x}$, the terms in eq. (11.7) may be rearranged and the equation written as

$$M\ddot{x} + (2\xi_s\omega_n M\dot{x} + \rho C_d A U)\dot{x} + Kx = \tfrac{1}{2}\rho C_d A[2Uu(t)] \tag{11.8}$$

or

$$M\ddot{x} + 2\omega_n M(\xi_s + \xi_a)\dot{x} + Kx = \tfrac{1}{2}\rho C_d A[2Uu(t)] \tag{11.9}$$

where

$$\xi_a = \frac{\rho C_d A U}{2\omega_n M} \tag{11.10}$$

and $\xi_a$ is the *equivalent viscous aerodynamic damping ratio*, which for light flexible structures can contribute considerably to the total damping.

Inspection of eq. (11.9) shows that the resulting dynamic force acting on the structure, when the term $\rho C_d A U\dot{x}(t)$ is considered as part of the total damping mechanism, is

$$f_d(t) = \rho C_d A U u(t) \tag{11.11}$$

or

$$f_d(t) = 2\frac{F_d}{U}u(t) \tag{11.12}$$

where

$$F_d = \tfrac{1}{2}\rho C_d A U^2 \tag{11.13}$$

In order to obtain a relationship between the spectrum of the fluctuating component of the drag force and the spectrum of the fluctuating velocity component, the frequency spans of the fluctuating wind and force components are divided into unit frequency intervals, with each interval centred at the frequency $n$. If only one frequency interval is considered, then

$$u(t) = u\sin(2\pi nt) \tag{11.14}$$

$$f_d(t) = f_d\sin(2\pi nt) \tag{11.15}$$

since $f_d$ varies linearly with $u(t)$. Substitution of the expressions for $u(t)$ and $f_d$ into eq. (11.12) yields

$$f_d = 2F_d(u/U) \tag{11.16}$$

Thus the relationship between the amplitudes of force and velocity is

$$\frac{f_d}{F_d} = 2\frac{u}{U} \tag{11.17}$$

or

$$\frac{{f_d}^2}{{F_d}^2} = 4\frac{u^2}{U^2} \tag{11.18}$$

As the coordinates of spectral density functions are proportional to the square of the amplitudes and inversely proportional to the frequency of each of the constituent harmonics, it follows that

$$\frac{S_{f_d}(n)}{{F_d}^2} = 4\frac{S_u(n)}{U^2} \tag{11.19}$$

which may be written in nondimensional form by multiplying each term by the frequency $n$.

The effects of the spatial variation in the wind velocity and the frequency dependence of the drag coefficient, both of which are important for structures with large surfaces, may be taken into account by introducing and *aerodynamic admittance function* $A(n)$. Thus eq. (11.19) may be rewritten as

$$\frac{nS_{f_d}(n)}{{F_d}^2} = 4A(n)\frac{nS_u(n)}{U^2} \tag{11.20}$$

The literature gives little information on the proper values to be used for the aerodynamic admittance function. It appears that more research is required in this field. Experimental values proposed by Davenport (1961) and Vickery (1965) are given in Fig. 11.1.

Having developed an expression for the load spectrum in terms of the velocity spectrum, it remains to express the response spectrum in terms of the load spectrum. From the theory of forced vibrations of damped linear one DOF systems (see eq. (4.15)), the response $x(t)$ to a force

$$f_d(t) = f_d\sin(2\pi nt) \tag{11.21}$$

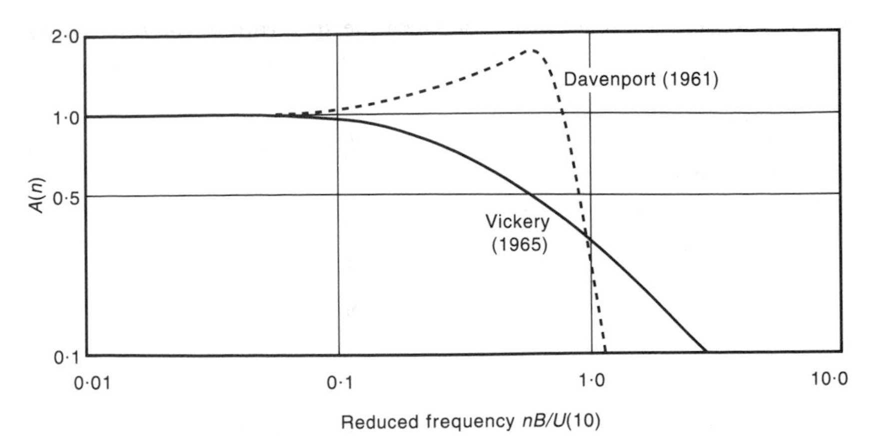


*Fig. 11.1. Variation of the aerodynamic admittance factor $A(n)$ with the reduced frequency $nB/U(10)$: the value $B$ is a structural reference dimension, $n$ corresponds to a structural mode frequency, and $U(10)$ is the reference wind velocity*

is

$$x(t) = \frac{f_d}{K} \frac{1}{\sqrt{\left[(1 - r^2)^2 + (2\xi r)^2\right]}} \sin(2\pi nt - \alpha) \tag{11.22}$$

or

$$x(t) = \frac{f_d}{K} MF(n) \sin(2\pi nt - \alpha) \tag{11.23}$$

Thus the maximum value of $x(t)$, which occurs when $\sin(2\pi nt) = 1$, is

$$x = \frac{f_d}{K} MF(n) \tag{11.24}$$

where $x$ and $f_d$ are the amplitudes of the harmonic response and force components associated with the unit frequency interval centred at the frequency $n$. Since $K = F_d/x_s$, eq. (11.24) may also be written in the form

$$\frac{x}{x_s} = \frac{f_d}{F_d} MF(n) \tag{11.25}$$

Squaring of each term in eq. (11.25) yields

$$\frac{x^2}{{x_s}^2} = \frac{{f_d}^2}{{F_d}^2} MF^2(n) \tag{11.26}$$

where $MF^2(n) = M(n)$ is referred to as the *mechanical admittance factor*. Because, as mentioned above, the coordinates of power spectra are proportional to the square of the amplitudes of the constituent harmonics, it follows that

$$\frac{S_x(n)}{{x_s}^2} = \frac{S_{f_d}(n)}{{F_d}^2} M(n) \tag{11.27}$$

or, if each term in eq. (11.27) is multiplied by $n$, in nondimensional form as

$$\frac{nS_x(n)}{x_s^2} = 4M(n)A(n)\frac{S_u(n)}{U^2} \tag{11.29}$$

The variance of the fluctuating component of the response is now determined by integration of both sides of eq. (11.29) with respect to $n$. Thus

$$\sigma_x^2 = \int_0^\infty S_x(n)\mathrm{d}n = 4\frac{x_s^2}{U^2}\int_0^\infty M(n)A(n)S_u(n)\mathrm{d}n \tag{11.30}$$

For weakly damped structures the expression for $\sigma_x^2$ can be approximated to

$$\sigma_x^2 = \int_0^\infty S_x(n)\mathrm{d}n \approx 4\frac{x_s^2}{U^2}M(n)A(n)S_u(n)\Delta n \tag{11.31}$$

where

$$\Delta n = \xi\pi n$$
$$M(n) = 1/4\xi^2$$

in which case

$$\sigma_x^2 = \frac{x_s^2}{U^2}\frac{\pi n}{\xi}A(n)S_u(n) \tag{11.32}$$

Hence the maximum probable displacement is given by

$$x_{\max} = \kappa\sigma_x \tag{11.33}$$

where $\kappa$ is a peak factor for weakly damped structures (see eq. (10.50)).

**Example 11.1** A motorway sign of dimensions shown in Fig. 11.2 may be assumed to vibrate as a one DOF system, in the along-road direction. The supporting structure is designed as a portal frame with a horizontal beam, which can be considered to be rigid. The $EI$ value for each column is 228 799·08 kN m$^2$, and the equivalent lumped mass, 9·0 m above the ground, is 5·3 t. At the point where the sign is positioned the motorway runs through woodland, so the roughness length $z_0$ may be taken as 0·9 m. If the design wind speed $U(10) = 30{\cdot}0$ m/s, determine (i) the maximum dynamic and hence maximum total response; (ii) the maximum shear force and bending moment occurring at the foot of each column. Use the power spectrum proposed by Davenport (1961) (eq. (10.26)) and the curve for the aerodynamic admittance factor proposed by Vickery (1965) (Fig. 11.1) when calculating the variance of response. The drag coefficient for the 20·0 × 2·0 m motorway sign $C_d = 2{\cdot}03$. The specific density of air is 1·226 kg/m$^3$.

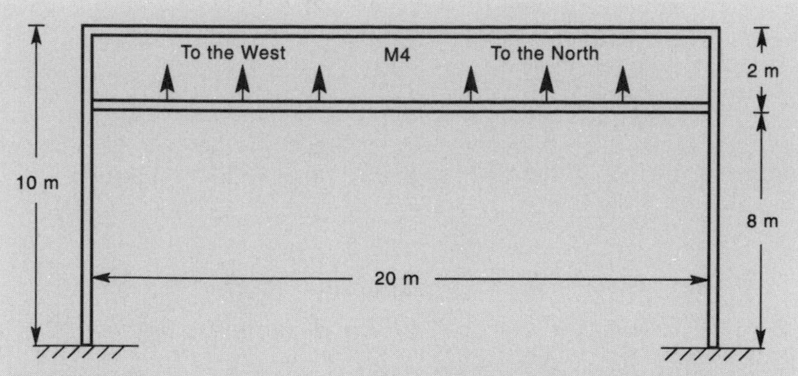


*Fig. 11.2. Motorway sign*

The prediction of the along-wind dynamic response to wind tends to be lengthy. At this stage, therefore, the reader may find it helpful to have a listing of the expressions and equations needed, as they form the framework for the required calculations.

The total response $x$ is given by

$$x = x_s + x_d = F_d/K + \kappa\sigma_x$$

where

$$F_d = \tfrac{1}{2}\rho C_d A U^2$$

$$K = 2(3EI/L^3) = 6EI/L^3$$

$$\sigma_x{}^2 = \frac{x_s{}^2}{U^2}\frac{\pi n}{\xi}A(n)S_u(n)$$

$$n = \frac{1}{2\pi}\sqrt{\left(\frac{K}{M}\right)}$$

$$\xi = \xi_{st} + \xi_a = \xi_{st} + \frac{\rho C_d A U}{2\omega_n M} = \xi_{st} + \frac{F_d}{2\pi n U M}$$

$$S_u(n) = \frac{4u_*{}^2 f^2}{n(1+f^2)^{4/3}}$$

$$u_* = \frac{U(10)}{2{\cdot}5\ln(10/z_0)}$$

$$f = 1200n/U(10)$$

$$\kappa = \sqrt{[2\ln(nT)]} + 0{\cdot}577/\sqrt{[2\ln(nT)]}$$

where $T = 3600$ s, if the value of $U(10)$ is based on records of 1 h duration.

*Calculation of $F_d$, $K$ and $x_s$*

$$F_{\mathrm{d}} = \tfrac{1}{2} \times 1{\cdot}226 \times 2{\cdot}03 \times 40{\cdot}0 \times 30{\cdot}0^2 = 44\,798{\cdot}04\,\mathrm{N}$$

$$K = 6 \times 228\,799{\cdot}08/9{\cdot}0^3 = 1883{\cdot}12\,\mathrm{kN/m}$$

$$x_{\mathrm{s}} = 44\,789{\cdot}04/1883{\cdot}12 \times 1000 = 0{\cdot}238\,\mathrm{m}$$

*Calculation of the along-wind natural frequency $f_n$*

$$f_{\mathrm{n}} = n = \frac{1}{2\pi}\sqrt{\left(\frac{1883{\cdot}12 \times 1000}{5300{\cdot}0}\right)} = 3{\cdot}0\,\mathrm{Hz}$$

*Determination of the aerodynamic ratio $\xi_a$ and total damping ratio $\xi$*

$$\xi_{\mathrm{d}} = \frac{44\,798{\cdot}04}{2\pi \times 3{\cdot}0 \times 30{\cdot}0 \times 5300{\cdot}0} = 0{\cdot}0149$$

$$\xi = 0{\cdot}01 + 0{\cdot}0149 = 0{\cdot}0249$$

*Calculation of the shear velocity $u_*$*

$$u_* = \frac{30{\cdot}0}{2{\cdot}5\ln(10/0{\cdot}9)} = 4{\cdot}984\,\mathrm{m/s}$$

*Determination of the value of Davenport's spectrum at $n = f_n$*

$$f = 1200n/U(10) = 1200 \times 3{\cdot}0/30{\cdot}0 = 120{\cdot}0$$

$$S_u(n) = \frac{4 \times 4{\cdot}984^2 \times 120{\cdot}0^2}{3{\cdot}0(1 + 120{\cdot}0^2)^{4/3}} = 1{\cdot}3612\,\mathrm{m^2/s}$$

*Calculation of the aerodynamic admittance factor $A(n)$*

$$\text{reduced frequency } nB/U(10) = 3{\cdot}0 \times 20{\cdot}0/30{\cdot}0 = 2{\cdot}0\,\mathrm{Hz}$$

Hence from Fig. 11.1

$$A(u) = A(3{\cdot}0) = 0{\cdot}139$$

*Evaluation of the variance ${\sigma_x}^2$ and standard variation of response $\sigma_x$*

$${\sigma_x}^2 = \frac{0{\cdot}0238^2}{30{\cdot}0^2} \times \frac{\pi \times 3{\cdot}0}{0{\cdot}0259} \times 0{\cdot}139 \times 1{\cdot}3612 = 4{\cdot}3333 \times 10^{-5}\,\mathrm{m}^2$$

$$\sigma_x = 0{\cdot}006583\,\mathrm{m}$$

*Determination of the peak factor $\kappa$*

$$\kappa = \sqrt{[2\ln(3{\cdot}0 \times 3600)]} + \frac{0{\cdot}577}{\sqrt{[2\ln(3{\cdot}0 \times 3600)]}} = 4{\cdot}441$$

*Calculation of the maximum response $x$, the maximum bending moment $M_{max}$, and the maximum shear force $SF_{max}$ in each column*

$$x = x_{\mathrm{s}} + \kappa\sigma_x = 0{\cdot}0238 + 4{\cdot}441 \times 0{\cdot}006583 = \underline{0{\cdot}0530351\,\mathrm{m}}$$

$$SF_{\max} = \tfrac{1}{2}Kx = \tfrac{1}{2} \times 1883{\cdot}12 \times 0{\cdot}0530351 = \underline{49{\cdot}936\,\mathrm{kN}}$$

$$M_{\max} = \tfrac{1}{2}(Kx)H = \tfrac{1}{2} \times 1883{\cdot}12 \times 0{\cdot}0530351 \times 9{\cdot}0 = \underline{449{\cdot}422\,\mathrm{kN\,m}}$$

It is worth noting that in the above example the air contributes significantly to the total damping, and that the dynamic response is 1·23 times the response due to the mean wind, although the natural frequency of the structure lies within the part of the frequency spectrum of the wind where the energy of the wind fluctuations is considerably reduced (see Fig. 10.3). The level of energy may be more fully appreciated by the following example, in which the eccentricity and eccentric mass of a variable speed motor, which will produce the same maximum amplitude of vibration as that caused by the wind, are calculated.

**Example 11.2** Calculate the product value of the eccentric mass times the eccentricity of a variable speed vibrator that will vibrate the motorway sign in example 11.1 at resonance with the same maximum amplitude as that caused by the wind, and hence calculate the value of the eccentric mass at an eccentricity of 0·25 m.

The maximum dynamic response of a one DOF system to harmonic excitation caused by an eccentric mass vibrator is

$$x_{\max} = \frac{me\omega^2}{K}\frac{1}{2\xi}$$

Ignoring the aerodynamic damping, which is a function of the mean wind velocity

$$me = \frac{2\xi K x_{\max}}{\omega_n{}^2} = \frac{2 \times 0{\cdot}01 \times 1883{\cdot}12 \times 1000 \times (4{\cdot}441 \times 0{\cdot}006583)}{(2\pi \times 3{\cdot}0)^2}$$

$$= \underline{3{\cdot}099\ \text{kg m}}$$

Assuming an eccentricity of 0·25 m, the size of the eccentric mass would be

$$m = 3{\cdot}099/0{\cdot}25 = \underline{12{\cdot}396\ \text{kg}}$$

Thus a very large vibrator would be needed to produce the same maximum amplitudes of vibration as those caused by the wind.

### Extension of the frequency domain method to multi-DOF systems

The response of multi-DOF systems can now be calculated in a similar manner to that of one DOF systems by first decoupling the equations of motion (see chapter 8), and then considering each modal equation as the equation of motion of a single-DOF system.

The equations of motion for a multi-DOF system subjected to the drag forces caused by the fluctuating component wind can be written in matrix form as

$$\mathbf{M}\ddot{x} = \mathbf{C}\dot{x} + \mathbf{K}x = f_{\mathrm{d}}(t) \tag{11.34}$$

In order to decouple the equations of motion, let

$$\boldsymbol{x} = \mathbf{Z}\boldsymbol{q} \tag{11.35}$$

where $\mathbf{Z}$ is the normalized modeshape matrix and $\boldsymbol{q}$ is the principal coordinate vector of the system. Substitution of the expression for $\boldsymbol{x}$ into eq. (11.34) and postmultiplication of each term in the same equation by $\mathbf{Z}^{\mathrm{T}}$ yields the following system of decoupled equations which govern the response

$$\begin{aligned}
&\ddot{q}_1 + 2\xi_1\omega_1\dot{q}_1 + \omega_1{}^2 q_1 = f_{q1}(t)\\
&\ddot{q}_2 + 2\xi_2\omega_2\dot{q}_2 + \omega_2{}^2 q_2 = f_{q2}(t)\\
&\dots\dots\dots\dots\dots\dots\dots\dots\\
&\dots\dots\dots\dots\dots\dots\dots\dots\\
&\ddot{q}_i + 2\xi_i\omega_i\dot{q}_i + \omega_i{}^2 q_i = f_{qi}(t)\\
&\dots\dots\dots\dots\dots\dots\dots\dots\\
&\dots\dots\dots\dots\dots\dots\dots\dots\\
&\ddot{q}_N + 2\xi_N\omega_N\dot{q}_N + \omega_N{}^2 q_N = f_{qN}(t)
\end{aligned} \tag{11.36}$$

where

$$f_{q_i}(t) = \mathbf{Z}_i{}^{\mathrm{T}}\boldsymbol{f}_{\mathrm{d}}(\boldsymbol{t}) \tag{11.37}$$

The relationship between the global and principal coordinates is given by eq. (11.35). Thus

$$x_j = \sum_{i=1}^{N} Z_{ji} q_i \tag{11.38}$$

If the terms in eq. (11.38) are squared and the cross-coupling terms between the modes are neglected

$$x_j{}^2 = \sum_{i=1}^{N} Z_{ji}{}^2 q_i{}^2 \tag{11.39}$$

The spectrum $S_{x_j}(n)$ can therefore be computed as the superposition of the spectra $S_{q_i}(n)$ as follows

$$S_{x_j}(n) = \sum_{i=1}^{N} Z_{ji}{}^2 S_{q_i}(n) \tag{11.40}$$

The spectrum associated with each principal coordinate $q_i$ is dependent on the spectrum $S_{f_{qi}}(n)$ of the corresponding force component $f_{qi}(t)$ in the modal force vector. For a one DOF system the response spectrum is given in terms of the force spectrum by eq. (11.27), which may be rewritten as

$$S_x(n) = \frac{1}{K^2} M(n) S_f(n) \tag{11.41}$$

Thus, similarly for the $i$th principal coordinate $q_i$

$$S_{q_i}(n) = \frac{1}{(\omega_i^2)^2} M_i(n) S_{f_{qi}}(n) \qquad (11.42)$$

Having obtained an expression for the spectrum of the generalized coordinate $q_i$, it remains to determine an expression for the spectra for the modal force component $f_{q_i}$. The integral of the spectrum of the $i$th component $f_{q_i}(t)$ in the modal force vector is

$$\int_0^\infty S_{f_{qi}}(n)\mathrm{d}n = \sigma_{f_{di}}{}^2 = \frac{1}{T}\int_0^T f_{q_i}(t) f_{q_i}(t)\mathrm{d}t \qquad (11.43)$$

From eq. (11.37)

$$f_{q_i}(t) = \sum_{j=1}^{N} Z_{ji} f_{dj}(t) \qquad (11.44)$$

Substitution of this expression for $f_{q_i}(t)$ into eq. (11.43) yields

$$\int_0^\infty S_{f_{qi}}(n)\mathrm{d}n = \frac{1}{T}\int_0^T \sum_{j=1}^{N} Z_{ji} f_j(t) \times \sum_{k=1}^{N} Z_{ki} f_k(t)\mathrm{d}t \qquad (11.45)$$

Since in this equation only the global forces $f_j(t)$ and $f_k(t)$ vary with time, it may be written as

$$\begin{aligned}\int_0^\infty S_{f_{qi}}(n)\mathrm{d}n &= \sum_{j=1}^{N}\sum_{k=1}^{N} Z_{ji} Z_{ki} \times \frac{1}{T_0}\int_0^T f_j(t) \times f_k(t)\mathrm{d}t \\ &= \sum_{i=1}^{N}\sum_{k=1}^{N} Z_{ji} Z_{ki} R_{f_\mathrm{d}}^{ik}(0)\end{aligned} \qquad (11.46)$$

where $R_{f_\mathrm{d}}^{jk}(0)$ is the *cross-covariance* between the fluctuating global loads at stations $j$ and $k$ at zero time lag. Lawson (1990) shows that

$$R_{f_\mathrm{d}}^{jk}(0) = \int_0^\infty S_{f_{jk}}^{\mathrm{cr}}(n)\mathrm{d}n = \int_0^\infty S_{f_{jk}}^{\mathrm{c}}(n)\mathrm{d}n + \iota\int_0^\infty S_{f_{jk}}^{q}(n)\mathrm{d}n \qquad (11.47)$$

where $\iota = \sqrt{-1}$ and $S_{f_{jk}}^{\mathrm{cr}}(n)$ is the *cross-spectral density function* or *cross-power spectrum*, $S_{f_{jk}}^{\mathrm{c}}(n)$ is the *co-spectrum*, and $S_{f_{jk}}^{q}(n)$ is the *quadrature spectrum* for the wind forces at stations $j$ and $k$. In wind engineering the quadrature spectrum of the load is generally assumed to be negligible compared with the co-spectrum. Hence

$$S_{f_{jk}}^{\mathrm{cr}}(n)\mathrm{d}n = S_{f_{jk}}^{\mathrm{c}}(n) \qquad (11.48)$$

Equation (11.46) can therefore be written as

$$\int_0^\infty S_{f_{jk}}(n)\mathrm{d}n = \sum_{j=1}^{N}\sum_{k=1}^{N} Z_{ji} Z_{ki}\int_0^\infty S_{f_{jk}}^{\mathrm{c}}(n)\mathrm{d}n \qquad (11.49)$$

Differentiation of eq. (11.49) with respect to $n$ yields the spectrum of the modal force $f_{q_i}$. Thus

$$S_{f_{qi}}(n) = \sum_{j=1}^{N} \sum_{k=1}^{N} Z_{ji} Z_{ki} S_{f_{jk}}^{c}(n) \tag{11.50}$$

Wind tunnel testing and full-scale measurements indicate that for civil engineering purposes it is sufficient to use the following formulation for the co-spectrum

$$S_{f_{jk}}^{c}(n) = \sqrt{[S_{f_j}(n) \times S_{f_k}(n)]} \times coh f_{jk}(n) \tag{11.51}$$

From eq. (11.20)

$$S_{f_j}(n) = 4 \frac{F^2(z_j)}{U^2(z_j)} A(n) S_{u_j}(n) \tag{11.52a}$$

and

$$S_{f_k}(n) = 4 \frac{F^2(z_k)}{U^2(z_k)} A(n) S_{u_k}(n) \tag{11.52b}$$

and from eq. (10.40)

$$coh f_{jk}(n) = \mathrm{e}^{-\phi_{jk}(n)} \tag{11.53}$$

where $\phi_{jk}(n)$, which is a function of the frequency $n$, the station coordinates $x_j$, $y_j$, $z_j$ and $x_k$, $y_k$, $z_k$, the mean wind speeds at stations $j$ and $k$, and the coefficients $C_x$, $C_y$ and $C_z$, is given by eq. (10.42) as

$$\phi_{jk}(n) = \frac{2n\sqrt{\left[C_x{}^2(x_j - x_k)^2 + C_y{}^2(y_j - y_k)^2 + C_z{}^2(z_j - z_k)^2\right]}}{U(z_j) + U(z_K)} \tag{11.54}$$

## Summary of expressions used in the frequency domain method for multi-DOF systems

Below is listed the sequence of equations needed to predict the response of a structure, with a natural angular frequency vector $\boldsymbol{\omega} = [\omega_1, \omega_2, \ldots \omega_i, \ldots \omega_N]^{\mathrm{T}}$ and a normalized modeshape matrix $\mathbf{Z} = [\mathbf{Z}_1, \mathbf{Z}_2, \ldots \mathbf{Z}_i, \ldots \mathbf{Z}_N]^{\mathrm{T}}$, to the buffeting of turbulent wind.

$$\boldsymbol{x} = \mathbf{Z}\boldsymbol{q} \tag{11.55}$$

$$q_i = \kappa_{q_i} \sigma_{q_i} \tag{11.56}$$

$$\sigma_{q_i}{}^2 = \int_0^\infty S_{q_i}(n) \mathrm{d}n \tag{11.57}$$

$$S_{q_i}(n) = \frac{1}{(\omega_i{}^2)^2} M_i(n) S_{f_{qi}}(n) \tag{11.58}$$

$$S_{f_{qi}}(n) = \sum_{j=1}^{N} \sum_{k=1}^{N} Z_{ji} Z_{ki} S_{f_{jk}}^{c}(n) \tag{11.59}$$

$$S_{f_{jk}}^{c}(n) = \sqrt{[S_{f_j}(n) \times S_{f_k}(n)]} \times \mathrm{e}^{-\phi_{jk}(n)} \tag{11.60}$$

$$S_{f_j}(n) = 4\frac{F^2(z_j)}{U^2(z_j)}A_j(n)S_{u_j}(n) \tag{11.61a}$$

$$S_{f_k}(n) = 4\frac{F^2(z_k)}{U^2(z_k)}A_k(n)S_{u_k}(n) \tag{11.61b}$$

$$\phi_{jk}(n) = \frac{2n\sqrt{\left[C_x{}^2(x_j - x_k)^2 + C_y{}^2(y_j - y_k)^2 + C_z{}^2(z_j - z_k)^2\right]}}{U(z_j) + U(z_k)} \tag{11.62}$$

For structures that are weakly damped, as is normally the case, it is usually sufficient to assume that

$$\sigma_{q_i}{}^2 = \int_0^\infty S_{q_i}(n)\mathrm{d}n = S_{q_i}(n_i)\Delta n = S_{q_i}(\omega_i)\Delta\omega \tag{11.63}$$

where

$$\Delta\omega = \tfrac{1}{2}\xi_i\omega_i \tag{11.64}$$

in which case eqs (11.58)–(11.62) may be written as

$$S_{q_i}(\omega_i) = \frac{1}{(\omega_i{}^2)^2}M_i(\omega_i)S_{f_{qi}}(\omega_i) \tag{11.65}$$

$$S_{f_{qi}}(\omega_i) = \sum_{j=1}^{N}\sum_{k=1}^{N} Z_{ji}Z_{ki}S^{\mathrm{c}}_{f_{jk}}(\omega_i) \tag{11.66}$$

$$S^{\mathrm{c}}_{f_{jk}}(\omega_i) = \sqrt{\left[S_{f_j}(\omega_i) \times S_{f_k}(\omega_i)\right]}\mathrm{e}^{-\phi_{jk}(\omega_i)} \tag{11.67}$$

$$S_{f_j}(\omega_i) = 4\frac{F^2(z_j)}{U^2(z_j)}\Delta_j(\omega_i)S_{u_j}(\omega_i) \tag{11.68a}$$

$$S_{f_k}(\omega_i) = 4\frac{F^2(z_k)}{U^2(z_k)}A_k(\omega_i)S_{u_k}(\omega_i) \tag{11.68b}$$

$$\phi_{jk}(\omega_i) = \frac{\omega_i\sqrt{\left[C_x{}^2(x_j - x_k)^2 + C_y{}^2(y_j - y_k)^2 + C_z{}^2(z_j - z_k)^2\right]}}{\pi\left[U(z_j) + U(z_k)\right]} \tag{11.69}$$

## Modal force spectra for two DOF systems

From eq. (11.59), the expression for the modal force spectrum in the $i$th mode is given by

$$S_{f_{qi}}(n) = \sum_{j=1}^{N}\sum_{k=1}^{N} Z_{ji}Z_{ki}S^{\mathrm{c}}_{f_{jk}}(n) \tag{11.70}$$

Thus the first modal force spectrum for a two DOF system having the modeshape matrix

$$\mathbf{Z} = \begin{bmatrix} Z_{11} & Z_{12} \\ Z_{21} & Z_{22} \end{bmatrix} \tag{11.71}$$

is

$$Sf_{q1}(n) = \sum_{j=1}^{2}\sum_{k=1}^{2} Z_{j1}Z_{k1}S^{c}_{f_{jk}}(n) \tag{11.72}$$

or

$$Sf_{q1}(n) = Z_{11}{}^{2}S^{c}_{f_{11}} + Z_{11}Z_{21}S^{c}_{f_{12}}(n) + Z_{21}Z_{11}S^{c}_{f_{21}}(n) + Z_{21}{}^{2}S^{c}_{f_{22}}(n) \tag{11.73}$$

From eq. (11.60) it follows that

$$S^{c}_{f_{11}}(n) = \sqrt{[S_{f_1}(n) \times S_{f_1}(n)]} \times e^{-0} = S_{f_1}(n) \tag{11.74a}$$

$$S^{c}_{f_{22}}(n) = \sqrt{[S_{f_2}(n) \times S_{f_2}(n)]} \times e^{-0} = S_{f_2}(n) \tag{11.74b}$$

$$S^{c}_{f_{12}}(n) = S^{c}_{f_{21}}(n) = \sqrt{[S_{f_1}(n) \times S_{f_2}(n)]} \times e^{-\phi_{12}(n)} \tag{11.75}$$

Hence

$$S_{f_{q1}}(n) = Z_{11}{}^{2}S_{f_1}(n) + 2Z_{11}Z_{21}\sqrt{[S_{f_1}(n) \times S_{f_2}(n)]} \times e^{-\phi_{12}(n)} + Z_{21}{}^{2}S_{f_2}(n) \tag{11.76a}$$

Similarly, the second modal force spectrum is given by

$$S_{f_{q2}}(n) = Z_{12}{}^{2}S_{f_1}(n) + 2Z_{12}Z_{22}\sqrt{[S_{f_1}(n) \times S_{f_2}(n)]} \times e^{-\phi_{12}(n)} + Z_{22}{}^{2}S_{f_2}(n) \tag{11.76b}$$

### Modal force spectra for three DOF systems

The expressions for the first, second and third modal force spectra for a three DOF system can be developed similarly. Thus

$$\begin{aligned} S_{f_{q1}}(n) = {} & Z_{11}{}^{2}S_{f_1}(n) + Z_{21}{}^{2}S_{f_2}(n) + Z_{31}{}^{2}S_{f_3}(n) \\ & + 2Z_{11}Z_{21}\sqrt{[S_{f_1}(n) \times S_{f_2}(n)]} \times e^{-\phi_{12}(n)} \\ & + 2Z_{11}Z_{32}\sqrt{[S_{f_1}(n) \times S_{f_3}(n)]} \times e^{-\phi_{13}(n)} \\ & + 2Z_{21}Z_{31}\sqrt{[S_{f_2}(n) \times S_{f_3}(n)]} \times e^{-\phi_{23}(n)} \end{aligned} \tag{11.77a}$$

$$\begin{aligned} S_{f_{q2}}(n) = {} & Z_{12}{}^{2}S_{f_1}(n) + Z_{22}{}^{2}S_{f_2}(n) + Z_{32}{}^{2}S_{f_3}(n) \\ & + 2Z_{12}Z_{22}\sqrt{[S_{f_1}(n) \times S_{f_2}(n)]} \times e^{-\phi_{12}(n)} \\ & + 2Z_{12}Z_{32}\sqrt{[S_{f_1}(n) \times S_{f_3}(n)]} \times e^{-\phi_{13}(n)} \\ & + 2Z_{22}Z_{32}\sqrt{[S_{f_2}(n) \times S_{f_3}(n)]} \times e^{-\phi_{23}(n)} \end{aligned} \tag{11.77b}$$

$$\begin{aligned} S_{f_{q3}}(n) = {} & Z_{13}{}^{2}S_{f_1}(n) + Z_{23}{}^{2}S_{f_2}(n) + Z_{33}{}^{2}S_{f_3}(n) \\ & + 2Z_{13}Z_{23}\sqrt{[S_{f_1}(n) \times S_{f_2}(n)]} \times e^{-\phi_{12}(n)} \\ & + 2Z_{13}Z_{32}\sqrt{[S_{f_1}(n) \times S_{f_3}(n)]} \times e^{-\phi_{13}(n)} \\ & + 2Z_{23}Z_{33}\sqrt{[S_{f_2}(n) \times S_{f_3}(n)]} \times e^{-\phi_{23}(n)} \end{aligned} \tag{11.77c}$$

Thus the expressions for the force spectra for a multi-DOF system may be quite lengthy. Fortunately, in the case of practical engineering problems most of the cross-spectral density terms are negligible and may be ignored because of the distance between the load stations (see example 10.3).

## Aerodynamic damping of multi-DOF systems

Equation (11.10) gives an expression for the aerodynamic damping of one DOF systems, and in example 11.1 it is shown that the level of aerodynamic damping compared with that of structural damping can be considerable. This is also the case with some multi-DOF structures, such as guyed masts, where the damping caused by the relative velocity of the structure to that of the air flow is of greater importance than the structural one. In chapter 9 it is shown how to construct damping matrices that permit the equations of motion to be decoupled. The difficulty in including the damping due to air is that the aerodynamic damping terms couple the equations of motion. This leads, as will be shown, to an iterative solution method of the modal equations, unless assumptions are made that may not be justified.

The matrix equation of motion for a multi-DOF structure subjected to turbulent wind is given by

$$\mathbf{M}\ddot{\boldsymbol{x}} + \mathbf{C}\dot{\boldsymbol{x}} + \mathbf{K}\boldsymbol{x} = \left[\tfrac{1}{2}\rho\mathbf{C}_{\mathrm{d}}\mathbf{A}\left\{(\mathbf{U} + \boldsymbol{u}(t) - \dot{\boldsymbol{x}}(t))^2 - \mathbf{U}^2\right\}\right] \tag{11.78}$$

Let

$$\mathbf{F}_{\mathrm{d}} = \tfrac{1}{2}\rho\mathbf{C}_{\mathrm{d}}\mathbf{A}\mathbf{U}^2 \tag{11.79}$$

Substitution of eq. (11.79) into eq. (11.78), ignoring the terms with $u^2(t)$, $\dot{x}^2(t)$ and $u(t)x(t)$, yields, since $\dot{x}(t) = \dot{x}$

$$\mathbf{M}\ddot{\boldsymbol{x}} + \mathbf{C}\dot{\boldsymbol{x}} + [2\mathbf{F}_{\mathrm{d}}/\mathbf{U}]\dot{\boldsymbol{x}} + \mathbf{K}\boldsymbol{x} = [2\mathbf{F}_{\mathrm{d}}/\mathbf{U}]\boldsymbol{u}(t) \tag{11.80}$$

Postmultiplication of each term in eq. (11.80) by $\mathbf{Z}^{\mathrm{T}}$, where $\mathbf{Z}$ is the normalized modeshape matrix, and substitution of the following expressions for $\boldsymbol{x}$, $\dot{\boldsymbol{x}}$ and $\ddot{\boldsymbol{x}}$

$$\boldsymbol{x} = \mathbf{Z}\boldsymbol{q}$$
$$\dot{\boldsymbol{x}} = \mathbf{Z}\dot{\boldsymbol{q}}$$
$$\ddot{\boldsymbol{x}} = \mathbf{Z}\ddot{\boldsymbol{q}}$$

into the resulting matrix equation yields

$$\begin{aligned}
&\ddot{q}_1 + 2\xi_{s1}\omega_1\dot{q}_1 + \mathbf{Z}_1{}^{\mathrm{T}}[2\mathbf{F}_{\mathrm{d}}/\mathbf{U}]\mathbf{Z}\dot{\boldsymbol{q}} + \omega_1{}^2 q_1 = \sum_{i=1}^{N} Z_{i1}(2F_{\mathrm{d}i}/U_i)u_i(t)\\
&\ddot{q}_2 + 2\xi_{s2}\omega_2\dot{q}_2 + \mathbf{Z}_2{}^{\mathrm{T}}[2\mathbf{F}_{\mathrm{d}}/\mathbf{U}]\mathbf{Z}\dot{\boldsymbol{q}} + \omega_2{}^2 q_2 = \sum_{i=1}^{N} Z_{i2}(2F_{\mathrm{d}i}/U_i)u_i(t)\\
&\cdots\cdots\cdots\cdots\cdots\cdots\cdots\cdots\cdots\cdots\cdots\cdots\\
&\ddot{q}_n + 2\xi_{sn}\omega_n\dot{q}_n + \mathbf{Z}_n{}^{\mathrm{T}}[2\mathbf{F}_{\mathrm{d}}/\mathbf{U}]\mathbf{Z}\dot{\boldsymbol{q}} + \omega_n{}^2 q_n = \sum_{i=1}^{N} Z_{in}(2F_{\mathrm{d}i}/U_i)u_i(t)
\end{aligned} \tag{11.81}$$

Thus in the $r$th modal equation the aerodynamic damping term is given by

$$\mathbf{Z}_r^{\mathrm{T}}[2\mathbf{F}_{\mathrm{d}}/\mathbf{U}]\mathbf{Z}\dot{\boldsymbol{q}} = \sum_{i=1}^{N}\sum_{j=1}^{N} Z_{jr}[2F_{\mathrm{d}j}/U_j]Z_{ji}\dot{q}_i \quad (11.82)$$

or

$$\begin{aligned}\mathbf{Z}_r^{\mathrm{T}}[2\mathbf{F}_{\mathrm{d}}/\mathbf{U}]\mathbf{Z}\dot{\boldsymbol{q}} &= \alpha_{1r}\dot{q}_1 + \alpha_{2r}\dot{q}_2 + \ldots + \alpha_{rr}\dot{q}_r + \ldots + \alpha_{nr}\dot{q}_n \\ &= \sum_{i=1}^{N} \alpha_{ir}q_i\end{aligned} \quad (11.83)$$

where

$$\alpha_{ir} = \sum_{j=1}^{N} Z_{jr}[2F_{\mathrm{d}j}/U_j]Z_{ji} \quad (11.84)$$

Equation (11.83) may also be written as

$$\mathbf{Z}_r^{\mathrm{T}}[2\mathbf{F}_{\mathrm{d}}/\mathbf{U}]\mathbf{Z}\dot{\boldsymbol{q}} = \left(\alpha_{1r}\frac{\dot{q}_1}{\dot{q}_r} + \alpha_{2r}\frac{\dot{q}_2}{\dot{q}_r} + \ldots + \alpha_{rr} + \ldots + \alpha_{nr}\frac{\dot{q}_n}{\dot{q}_r}\right)\dot{q}_r \quad (11.85)$$

which shows that damping due to air couples the modal equations. These can therefore be solved only by making certain assumptions. If the motion in each mode is assumed to be simple harmonic or sinusoidal, then any of the terms in eq. (11.85), say term $i$, may be written as

$$\alpha_{ir}\frac{\dot{q}_i}{\dot{q}_r} = \alpha_{ir}\frac{q_i\omega_i\cos(\omega_i t - \phi_i)}{q_r\omega_r\cos(\omega_r t - \phi_r)} \quad (11.86)$$

where $\phi$ is a random phase angle. Since the value of $\cos(\omega t - \phi)$ may vary between $-1$ and $+1$, it follows that the value of the ratio $\dot{q}_i/\dot{q}_r$ may vary between $-\infty$ and $+\infty$. To decouple the modal equations it is therefore necessary to assume that the average values of the terms $\alpha_{ir}(\dot{q}_i/\dot{q}_r)$ are zero. As the fluctuations in wind velocities are random, the wind velocity itself is assumed to be stationary, and since in the frequency domain in the first instant only the variance of response is calculated, this assumption does not seem to be unreasonable.

Thus the modal eq. (11.80) may be written as

$$\begin{aligned}\ddot{q}_1 + 2(\xi_{\mathrm{s}1} + \xi_{\mathrm{a}1})\omega_1\dot{q}_1 + \omega_1^2 q_1 &= \sum_{i=1}^{N} Z_{i1}(2F_{\mathrm{d}i}/U_i)u_i(t) \\ \ddot{q}_2 + 2(\xi_{\mathrm{s}2} + \xi_{\mathrm{a}2})\omega_2\dot{q}_2 + \omega_2^2 q_2 &= \sum_{i=1}^{N} Z_{i2}(2F_{\mathrm{d}i}/U_i)u_i(t) \\ &\ldots\ldots\ldots \\ &\ldots\ldots\ldots \\ \ddot{q}_n + 2(\xi_{\mathrm{s}n} + \xi_{\mathrm{a}n})\omega_n\dot{q}_n + \omega_{\mathrm{n}}^2 q_{\mathrm{n}} &= \sum_{i=1}^{N} Z_{in}(2F_{\mathrm{d}i}/U_i)u_i(t)\end{aligned} \quad (11.87)$$

where the simplified modal aerodynamic damping ratio in the $r$ the mode given by either

$$\xi_{ar} = \frac{\alpha_{rr}}{2\omega_r} = \frac{1}{2\omega_r}\mathbf{Z}_r^{\mathrm{T}}[2F/U]\mathbf{Z}_r \qquad (11.88a)$$

or

$$\xi_{ar} = \frac{\alpha_{rr}}{2\omega_r} = \frac{1}{2\omega_r}\sum_{j=1}^{N} Z_{jr}[2F_{\mathrm{d}j}/U_j]Z_{jr} \qquad (11.88b)$$

**Example 11.2** The mast in Fig. 11.3 supports two discs, one at 10 m and one at 20 m above the ground. The diameter of each disc is 4·0 m, and the drag coefficient $C_{\mathrm{d}} = 2{\cdot}0$. The mast is situated in an area where the roughness length is assumed to be 1·0 m. Calculate the lateral response of each disc when the mast is subjected to a mean wind of 30 m/s at a height of 10 m above ground level. Assume the exponential decay coefficient for the wind speed and ground roughness to be $C_z = 8$. Use the logarithmic law (eq. (10.11)) to calculate the mean wind profile, and Kaimal's power spectrum in order to take account of the variation of the spectral density function with height. The condensed stiffness matrix $\tilde{\mathbf{K}}$, the normalized modeshape matrix $\mathbf{Z}$ and the angular frequency vector $\omega$ for the tower are given below.

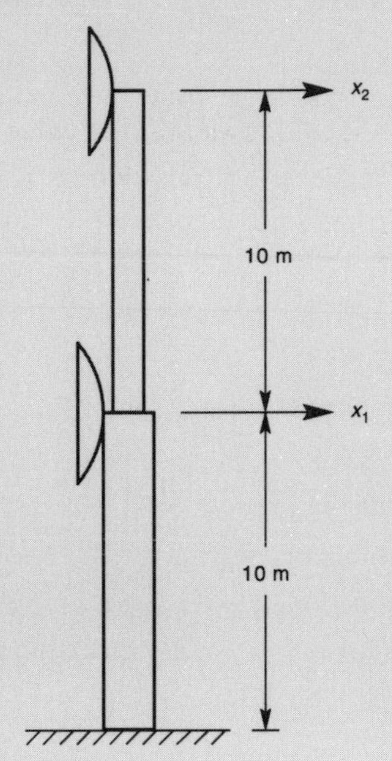


*Fig. 11.3. Tower supporting two discs*

Assume the damping in the first and second modes to be 1·0% of critical, and the aerodynamic admittance factor to be 0·5 in the first mode and 0·25 in the second. The wind load on the mast itself may be ignored.

$$\tilde{\mathbf{K}} = 765{\cdot}79891 \times \begin{bmatrix} 36 & -10 \\ -10 & 4 \end{bmatrix} \text{kN/m}$$

$$\mathbf{Z} = \begin{bmatrix} 3{\cdot}443 & 6{\cdot}521 \\ 10{\cdot}109 & -10{\cdot}753 \end{bmatrix} \times 10^{-3}$$

$$\omega = \begin{bmatrix} 25{\cdot}133 \\ 119{\cdot}098 \end{bmatrix} \text{rad/s}$$

*Determination of the shear velocity* $u_*$. From eq. (10.5)

$$u_* = \frac{30{\cdot}0}{2{\cdot}0 \ln(10{\cdot}0/1{\cdot}0)} = 5{\cdot}212\,\text{m/s}$$

The mean wind velocity at a height of 20 m above ground level is found using eq. (10.4). Thus

$$\text{U}(20) = 2{\cdot}5 \times 5{\cdot}212 \ln(20{\cdot}0/1{\cdot}0) = 39{\cdot}034\,\text{m/s}$$

The force vector due to the mean wind velocity is therefore

$$\begin{bmatrix} F_{d1} \\ F_{d2} \end{bmatrix} = \begin{bmatrix} \frac{1}{2} \times 1{\cdot}226 \times 2{\cdot}0 \times \pi \times 2{\cdot}0^2 \times 30{\cdot}000^2 \\ \frac{1}{2} \times 1{\cdot}226 \times 2{\cdot}0 \times \pi \times 2{\cdot}0^2 \times 39{\cdot}034^2 \end{bmatrix} \times 10^{-3}$$

$$= \begin{bmatrix} 13{\cdot}866 \\ 23{\cdot}474 \end{bmatrix} \text{kN}$$

The inverse of the condensed stiffness matrix $\tilde{\mathbf{K}}$ is

$$\mathbf{K}^{-1} = \begin{bmatrix} 0{\cdot}118711 & 0{\cdot}296778 \\ 0{\cdot}296778 & 1{\cdot}068400 \end{bmatrix} \times 10^{-3}\,\text{m/kN}$$

Hence the displacements due to the mean wind velocity are given by

$$\begin{bmatrix} x_{s1} \\ x_{s2} \end{bmatrix} = \begin{bmatrix} 0{\cdot}118711 & 0{\cdot}296778 \\ 0{\cdot}296778 & 1{\cdot}068400 \end{bmatrix} \begin{bmatrix} 13{\cdot}866 \\ 23{\cdot}474 \end{bmatrix} \times 10^{-3}$$

$$= \begin{bmatrix} 8{\cdot}613 \\ 29{\cdot}195 \end{bmatrix} \times 10^{-3}\,\text{m}$$

The decoupled equations of motion for the mast are given by

$$\ddot{q}_1 + 2\xi_{s1}\omega_1\dot{q}_1 + \alpha_{11}q_1 + \omega_1{}^2 q_1 = Z_{11}(2F_{d1}/U_1)u_1(t) + Z_{21}(2F_{d2}/U_2)u_2(t)$$

$$\ddot{q}_2 + 2\xi_{s2}\omega_2\dot{q}_2 + \alpha_{22}q_2 + \omega_2{}^2 q_2 = Z_{12}(2F_{d1}/U_1)u_1(t) + Z_{22}(2F_{d2}/U_2)u_2(t)$$

where from eq. (11.87b)

$$\alpha_{rr} = \sum_{j=1}^{N} Z_{jr}[2F_{dj}/U_j]Z_{jr}$$

Hence

$$\alpha_{11} = Z_{11}[2F_{d1}/U_1]Z_{11} + Z_{21}[2F_{d2}/U_2]Z_{21}$$
$$\alpha_{22} = Z_{12}[2F_{d1}/U_1]Z_{12} + Z_{22}[2F_{d2}/U_2]Z_{22}$$

$$\alpha_{11} = \{3{\cdot}443[2 \times 13\,866{\cdot}0/30{\cdot}000]3{\cdot}443 + 10{\cdot}109[2 \times 23\,474{\cdot}0/39{\cdot}034]10{\cdot}109\} \times 10^{-6} = 0{\cdot}1339$$
$$\alpha_{22} = \{6{\cdot}521[2 \times 13\,866{\cdot}0/30{\cdot}000]6{\cdot}521 + 10{\cdot}753[2 \times 23\,474{\cdot}0/39{\cdot}034]10{\cdot}753\} \times 10^{-6} = 0{\cdot}1784$$

Thus the damping force in the first mode is

$$2\xi_1\omega_1\dot{q}_1 + \alpha_{11}\dot{q}_1 = 2\left\{\xi_1 + \frac{\alpha_{11}}{2\omega_1}\right\}\omega_1\dot{q}_1 = 2\left\{0{\cdot}01 + \frac{0{\cdot}1339}{2 \times 25{\cdot}133}\right\}\omega_1\dot{q}_1$$

Hence

$$\xi_1 = \xi_{s1} + \xi_{a1} = 0{\cdot}01226$$

The damping force in the second mode is

$$2\xi_2\omega_2\dot{q}_2 + \alpha_{22}\dot{q}_2 = 2\left\{\xi_2 + \frac{\alpha_{22}}{2\omega_2}\right\}\omega_2\dot{q}_2 = 2\left\{0{\cdot}01 + \frac{0{\cdot}1784}{2 \times 119{\cdot}098}\right\}\omega_2\dot{q}_2$$

Hence

$$\xi_2 = \xi_{s2} + \xi_{a2} = 0{\cdot}01075$$

The calculation of the principal coordinates $q_1$ and $q_2$ requires first the calculation of the values of the power spectrum for wind velocities at heights 10 m and 20 m for $\omega_1 = 25{\cdot}133$ rad/s and $\omega_2 = 119{\cdot}098$ rad/s. The spectrum proposed by Kaimal given by eq. (10.34), is

$$S_u(z,n) = \frac{200{u_*}^2 f}{n(1+50f)^{5/3}}$$

where

$$f = \frac{zn}{U(z)}$$

For $H = 10m$ and $\omega_1 = 25{\cdot}133$ rad/s

$$f = 10 \times 25{\cdot}133/2\pi \times 30{\cdot}0 = 1{\cdot}3333471$$

$$S_u(10, 25{\cdot}133) = \frac{200 \times 5{\cdot}212^2 \times 1{\cdot}3333471}{4{\cdot}000(1 + 50 \times 1{\cdot}3333471)^{5/3}} = 1{\cdot}6117216\,\text{m}^2/\text{s}$$

For $H = 10m$ and $\omega_2 = 119{\cdot}098$ rad/s

$$f = 10 \times 119{\cdot}098/2\pi \times 30{\cdot}0 = 6{\cdot}3183451$$

$$S_u(10, 119{\cdot}098) = \frac{200 \times 5{\cdot}212^2 \times 6{\cdot}3183451}{18{\cdot}955(1 + 50 \times 6{\cdot}3183451)^{5/3}} = 0{\cdot}1229349\,\mathrm{m^2/s}$$

For $H = 20m$ and $\omega_1 = 25{\cdot}133$ rad/s

$$f = 20 \times 25{\cdot}133/2\pi \times 39{\cdot}034 = 2{\cdot}0495164$$

$$S_u(20, 25{\cdot}133) = \frac{200 \times 5{\cdot}212^2 \times 2{\cdot}0495164}{4{\cdot}000(1 + 50 \times 2{\cdot}0495164)^{5/3}} = 1{\cdot}2205777\,\mathrm{m^2/s}$$

For $H = 20m$ and $\omega_2 = 119{\cdot}098$ rad/s

$$f = 20 \times 119{\cdot}098/2\pi \times 39{\cdot}034 = 9{\cdot}7120641$$

$$S_u(20, 119{\cdot}098) = \frac{200 \times 5{\cdot}212^2 \times 9{\cdot}7120641}{18{\cdot}955(1 + 50 \times 9{\cdot}7120641)^{5/3}} = 0{\cdot}0924701\,\mathrm{m^2/s}$$

Having calculated the values of the velocity spectrum, the next step is to calculate the values of the force spectra at stations $H = 10$ m and $H = 20$ m, at frequencies $\omega_1 = 25{\cdot}133$ rad/s and $\omega_2 = 119{\cdot}098$ rad/s. The expression for the force spectrum given by eqs (11.65) and (11.66), omitting the subscript d, is

$$S_{f_d}(z, n) = 4\frac{F^2(z)}{U^2(z)}A(n)S_u(z, n)$$

Thus at $H = 10$ m and $\omega_1 = 25{\cdot}133$ rad/s

$$S_{f_1}(10, 25{\cdot}133) = 4 \times \frac{13{\cdot}866^2 \times 10^6}{30{\cdot}000^2} \times 0{\cdot}50 \times 1{\cdot}6117216$$
$$= 0{\cdot}6886204 \times 10^6\,\mathrm{N\,s}$$

At $H = 10$ m and $\omega_2 = 119{\cdot}098$ rad/s

$$S_{f_1}(10, 119{\cdot}098) = 4 \times \frac{13{\cdot}866^2 \times 10^6}{30{\cdot}000^2} \times 0{\cdot}25 \times 0{\cdot}1229349$$
$$= 0{\cdot}0262624 \times 10^6\,\mathrm{N\,s}$$

At $H = 20$ m and $\omega_1 = 25{\cdot}133$ rad/s

$$S_{f_2}(20, 25{\cdot}133) = 4 \times \frac{23{\cdot}474^2 \times 10^6}{39{\cdot}034^2} \times 0{\cdot}50 \times 1{\cdot}2205777$$
$$= 0{\cdot}8828430 \times 10^6\,\mathrm{N\,s}$$

At $H = 20$ m and $\omega_2 = 119{\cdot}098$ rad/s

$$S_{f_2}(20, 119{\cdot}098) = 4 \times \frac{23{\cdot}474^2 \times 10^6}{39{\cdot}034^2} \times 0{\cdot}25 \times 0{\cdot}0924701$$
$$= 0{\cdot}0334419 \times 10^6\,\mathrm{N\,s}$$

The square root of the coherence function is given by eq. (11.69). For the wind forces at $H = 10\,\text{m}$ and $H = 20\,\text{m}$, the function $\phi(z_1, z_2, n)$ reduces to

$$\phi(z_1, z_2, n) = \frac{\omega_i C_z(z_2 - z_1)}{\pi[U(z_2) + U(z_1)]}$$

Thus for $\omega_1 = 25{\cdot}133\,\text{rad/s}$

$$\phi(10, 20, 25{\cdot}133) = \frac{25{\cdot}133 \times 8 \times (20 - 10)}{\pi[39{\cdot}034 + 30{\cdot}000]} = 9{\cdot}2708896$$

$$e^{-\phi(10, 20, 25{\cdot}133)} = e^{-9{\cdot}2708896} = 9{\cdot}41247 \times 10^{-5}$$

For $\omega_2 = 119{\cdot}098\,\text{rad/s}$

$$\phi(10, 20, 119{\cdot}098) = \frac{119{\cdot}098 \times 8 \times (20 - 10)}{\pi[39{\cdot}034 + 30{\cdot}000]} = 43{\cdot}932058$$

$$e^{-\phi(10, 20, 119{\cdot}098)} = e^{-43{\cdot}932058} = 8{\cdot}32817 \times 10^{-20}$$

One is now in a position to calculate the force spectra in the principal modes at frequencies $\omega_1 = 25{\cdot}133$ rad/s and $\omega_2 = 63{\cdot}369$ rad/s.

For a two DOF system the expression for the force spectrum in the first mode is given by eq. (11.76a) as

$$S_{f_{qi}}(n_1) = Z_{11}{}^2 S_{f_1}(n_1) + 2Z_{11}Z_{21}\sqrt{[S_{f_1}(n_1) \times S_{f_2}(n_1)]} \times e^{-\phi_{12}(n_1)} + Z_{21}{}^2 S_{f_2}(n_1)$$

$$\begin{aligned} S_{f_{q1}}(25{\cdot}133) = \{&4{\cdot}515^2 \times 0{\cdot}6886204 + 2 \times 4{\cdot}515 \\ &\times 8{\cdot}873\sqrt{(0{\cdot}6886204 \times 0{\cdot}8828430)} \times 9{\cdot}41247 \times 10^{-5} \\ &+ 8{\cdot}873^2 \times 0{\cdot}8828430\} \times 10^{-3} \times 10^{-3} \times 10^6 \\ = {}&\ldots\ldots\ldots = \underline{83{\cdot}549905\,\text{N s}} \end{aligned}$$

Similarly, the expression for the force spectrum in the second mode is given by eq. (11.76b) as

$$S_{f_{q2}}(n_2) = Z_{12}{}^2 S_{f_1}(n_2) + 2Z_{12}Z_{22}\sqrt{[S_{f_1}(n_2) \times S_{f_2}(n_2)]} \times e^{-\phi_{12}(n_2)} + Z_{22}{}^2 S_{f_2}(n_2)$$

$$\begin{aligned} S_{f_{q2}}(119{\cdot}098) = \{&5{\cdot}838^2 \times 0{\cdot}0262624 - 2 \times 5{\cdot}838 \times \\ &12{\cdot}453\sqrt{(0{\cdot}0262624 \times 0{\cdot}0334419)} \times 8{\cdot}32817 \times 10^{-20} \\ &+ 12{\cdot}453^2 \times 0{\cdot}0334419\} \times 10^{-3} \times 10^{-3} \times 10^6 \\ = {}&\ldots\ldots\ldots = \underline{6{\cdot}081158\,\text{N s}} \end{aligned}$$

The expression for the response spectra for the principal coordinates $q_1$ and $q_2$ is given by eq. (11.62)

$$S_{q_i}(\omega_i) = \frac{1}{(\omega_i{}^2)^2} M_i(\omega_i) S_{f_{qi}}(\omega_i)$$

$$S_{q1}(\omega_1) = \frac{1}{\omega_1{}^4} M_1(\omega_1) S_{f_{q1}}(\omega_1) = \frac{1}{25{\cdot}133^4} \times \frac{1}{0{\cdot}01226^2} \times 83{\cdot}549905$$
$$= 348{\cdot}27936 \times 10^{-3}\,\mathrm{m^2\,s}$$

$$S_{q2}(\omega_2) = \frac{1}{\omega_2{}^4} M_2(\omega_2) S_{f_{q2}}(\omega_2) = \frac{1}{119{\cdot}098^4} \times \frac{1}{0{\cdot}01075^2} \times 6{\cdot}081158$$
$$= 6{\cdot}538700 \times 10^{-5}\,\mathrm{m^2\,s}$$

For lightly damped structures the variance of $q_1$ is given by

$$\sigma_{q_i}{}^2 = \int_0^\infty S_{q_i}(\omega_i)\mathrm{d}\omega = S_{q_i}(\omega_i)\Delta\omega$$

where

$$\Delta\omega = \tfrac{1}{2}\xi\omega_i$$

Hence

$$\sigma_{q1}{}^2 = \tfrac{1}{2} \times 348{\cdot}27936 \times 10^{-3} \times 0{\cdot}01226 \times 25{\cdot}133 = 53{\cdot}657761 \times 10^{-3}\,\mathrm{m^2}$$
$$\sigma_{q2}{}^2 = \tfrac{1}{2} \times 6{\cdot}53870 \times 10^{-5} \times 0{\cdot}01075 \times 119{\cdot}098 = 0{\cdot}041858 \times 10^{-3}\,\mathrm{m^2}$$

$$\sigma_{q1} = 0{\cdot}2316414$$
$$\sigma_{q2} = 0{\cdot}0064697\,\mathrm{m}$$

Hence

$$q_1 = \{\surd[2\ln(4{\cdot}0000 \times 3600)] + 0{\cdot}577/\surd[2\ln(4{\cdot}0000 \times 3600)]\} \times 0{\cdot}2316414 = 1{\cdot}0442\,\mathrm{m}$$
$$q_2 = \{\surd[2\ln(18{\cdot}9550 \times 3600] + 0{\cdot}577\surd[2\ln(18{\cdot}9550 \times 3600)]\} \times 0{\cdot}0064697 = 0{\cdot}0313\,\mathrm{m}$$

$$\begin{bmatrix} x_{d1} \\ x_{d2} \end{bmatrix} = \begin{bmatrix} 4{\cdot}515 & 5{\cdot}838 \\ 8{\cdot}873 & -12{\cdot}543 \end{bmatrix} \begin{bmatrix} 1{\cdot}0442 \\ 0{\cdot}0313 \end{bmatrix} \times 10^{-3}$$
$$= \begin{bmatrix} 4{\cdot}897 \\ 8{\cdot}873 \end{bmatrix} \times 10^{-3}\,\mathrm{m}$$

Hence the total displacements at $H = 10\,\mathrm{m}$ and $H = 20\,\mathrm{m}$ are

$$\begin{bmatrix} x_1 \\ x_2 \end{bmatrix} = \begin{bmatrix} x_{d1} \\ x_{d2} \end{bmatrix} + \begin{bmatrix} x_{d1} \\ x_{d2} \end{bmatrix} = \begin{bmatrix} 8{\cdot}613 \\ 29{\cdot}915 \end{bmatrix} \times 10^{-3} + \begin{bmatrix} 4{\cdot}897 \\ 8{\cdot}873 \end{bmatrix} \times 10^{-3}$$
$$= \begin{bmatrix} 13{\cdot}510 \\ 38{\cdot}068 \end{bmatrix} \times 10^{-3}\,\mathrm{m}$$

### Simplified wind response analysis of linear multi-DOF structures in the frequency domain

In examples 11.2 and 10.3 it can be seen that if two stations are as much as 10 m apart, the values of the cross-spectral density function for the wind forces at the two stations are negligible compared to the direct spectral density functions. Also, when the structures are heavy, as in the case of the stepped mast, the aerodynamic damping is small compared to the structural one. It can further be noticed that in higher modes the reduced frequencies result in aerodynamic admittance factors which, together with the fact that the energy of the wind at higher frequencies is very much reduced, cause structures to respond mainly in the first mode. A simplified and less time-consuming explorative response analysis can therefore be undertaken.

From eq. (11.16), for weakly damped structures

$$\sigma_{q_i}{}^2 = \int_0^\infty S_{q_i}(\omega)\mathrm{d}\omega \approx \tfrac{1}{2}\xi_i\omega_i S_{q_i}(\omega) \tag{11.89}$$

where, from eq. (11.42)

$$S_{q_i}(\omega_i) = \frac{1}{\tilde{K}_i^2} M(\omega_i) S_{f_{qi}}(\omega_i) \tag{11.90}$$

and $\tilde{K}_i = \omega_i{}^2$ if the modeshape vectors are normalized.

In cases where the load stations are so far apart that the cross-spectral density functions for the wind forces can be ignored, eq. (11.50) may be written as

$$S_{f_{qi}}(\omega_i) = \sum_{j=1}^{N} Z_{ji}{}^2 \left(\frac{2F_{\mathrm{d}j}}{U_j}\right)^2 A_j(\omega_i) S_{u_j}(\omega_i) \tag{11.91}$$

Hence

$$S_{q_i}(\omega_i) = \frac{1}{(\omega_i{}^2)^2} M(\omega_i) \sum_{j=1}^{N} Z_{ji}{}^2 \left(\frac{2F_{\mathrm{d}j}}{U_j}\right)^2 A_j(\omega_i) S_{u_j}(\omega_i) \tag{11.92}$$

$$\sigma_{q_i}{}^2 = \frac{1}{\omega_i{}^3}\frac{1}{2\xi_i} \sum_{j=1}^{N} \left(\frac{2F_{\mathrm{d}j}}{U_j}\right)^2 A_j(\omega_i) S_{u_j}(\omega_i) \tag{11.93}$$

**Example 11.3** Let the three-storey shear structure shown in Fig. 11.4 represent a condensed numerical model of a 30 m tall tower block, with width equal to depth equal to 10 m, with the floors in the building lumped together in the numerical model as three floors 10 m apart. The mass of each of equivalent floor is 120 000 kg, and the corresponding total shear stiffness of the columns between each floor is $12{\cdot}0 \times 10^6$ h/m. Calculate the response to turbulent wind having a mean velocity of 25 m/s at a height of 10 m above the ground, if the surface drag coefficient for the area is 0·015. Assume the structural damping in the first, second and third modes to be respectively 1·5%, 1·0% and 1·0% of critical. The drag coefficient at all levels of the building may be taken as $C_d = 1{\cdot}3$. The density of air is 1·226 kg/m$^3$. Aerodynamic damping and the cross-correlation of wind may be ignored. Use the power spectral density function proposed by Kaimal to take account of the variation of the power spectrum of wind with height. The natural angular frequencies and normalized modeshape matrix for the model structures are

$$\omega = \begin{bmatrix} 4{\cdot}439 \\ 12{\cdot}466 \\ 18{\cdot}025 \end{bmatrix} \text{rad/s}$$

$$\omega^2 = \begin{bmatrix} 19{\cdot}70 \\ 155{\cdot}40 \\ 324{\cdot}90 \end{bmatrix} \text{rad}^2/\text{s}^2$$

$$\mathbf{Z} = \begin{bmatrix} 0{\cdot}947 & 2{\cdot}128 & 1{\cdot}703 \\ 1{\cdot}706 & 0{\cdot}950 & -2{\cdot}128 \\ 2{\cdot}128 & -1{\cdot}703 & 0{\cdot}953 \end{bmatrix} \times 10^{-3}$$

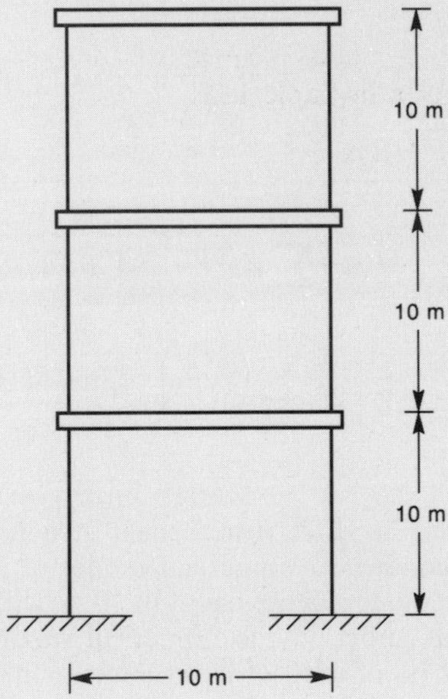

*Fig. 11.4. Three-storey shear structure*

*Determination of mean wind speeds 20 m and 30 m above the ground*

$$U(z) = 2{\cdot}5u_* \ln(z/z_0)$$

where

$$u_* = \sqrt{(k)}U(10)$$

Hence

$$u_* = \sqrt{(0{\cdot}015)} \times 25{\cdot}0 = 3{\cdot}062\,\text{m/s}$$
$$z_0 = 10 \times \exp\{-25{\cdot}0/2{\cdot}5 \times 3{\cdot}062\} = 0{\cdot}382\,\text{m}$$

Thus

$$U(10) = U_1 = 25{\cdot}000\,\text{m/s}$$
$$U(20) = U_2 = 2{\cdot}5 \times 3{\cdot}062 \ln(20{\cdot}0/0{\cdot}382) = 30{\cdot}299\,\text{m/s}$$
$$U(30) = U_3 = 2{\cdot}5 \times 3{\cdot}062 \ln(30{\cdot}0/0{\cdot}382) = 33{\cdot}403\,\text{m/s}$$

$$\mathbf{F}(10) = \mathbf{F}_1 = \tfrac{1}{2} \times 1{\cdot}226 \times 1{\cdot}3 \times 10{\cdot}0 \times 10{\cdot}0 \times 25{\cdot}000^2 = 49\,806{\cdot}250\,\text{N}$$
$$\mathbf{F}(20) = \mathbf{F}_2 = \tfrac{1}{2} \times 1{\cdot}226 \times 1{\cdot}3 \times 10{\cdot}0 \times 10{\cdot}0 \times 30{\cdot}299^2 = 73\,157{\cdot}763\,\text{N}$$
$$\mathbf{F}(30) = \mathbf{F}_3 = \tfrac{1}{2} \times 1{\cdot}226 \times 1{\cdot}3 \times 10{\cdot}0 \times 5{\cdot}0 \times 33{\cdot}403^2 = 44\,457{\cdot}474\,\text{N}$$

Hence

$$(F_1/U_1)^2 = (49\,806{\cdot}250/25{\cdot}000)^2 = 3{\cdot}81128 \times 10^6\,\text{N}^2\,\text{s}^2/\text{m}^2$$
$$(F_2/U_2)^2 = (73\,157{\cdot}763/30{\cdot}299)^2 = 5{\cdot}82994 \times 10^6\,\text{N}^2\,\text{s}^2/\text{m}^2$$
$$(F_3/U_3)^2 = (44\,457{\cdot}474/33{\cdot}403)^2 = 1{\cdot}77141 \times 10^6\,\text{N}^2\,\text{s}^2/\text{m}^2$$

The Kaimal spectrum values in $\text{m}^2/\text{s}$ for angular frequencies $\omega_i$ at heights H(z) are given in Table 11.1.

*Table 11.1. Example 11.3 data*

| $\omega_i$(rad/s) | $H = 10\,\text{m}$ | $H = 20\,\text{m}$ | $H = 30\,\text{m}$ |
|---|---|---|---|
| $\omega_1 = 4{\cdot}439$ | 8·10425 | 6·06436 | 5·02992 |
| $\omega_2 = 12{\cdot}446$ | 1·55880 | 1·13451 | 0·93013 |
| $\omega_3 = 18{\cdot}025$ | 0·85386 | 0·61838 | 0·50595 |

Determination of the reduced frequencies $\tilde{u}_i$, and aerodynamic admittance factors $A(\omega_i)$ corresponding to the natural angular frequencies $\omega_1$, $\omega_2$ and $\omega_3$, from the solid-line graph in Fig. 11.1, yields

$$\tilde{n}_1 = \frac{4{\cdot}439 \times 10{\cdot}0}{2\pi \times 25{\cdot}0} = 0{\cdot}2826\,\text{Hz}$$
$$\tilde{n}_2 = \frac{12{\cdot}446 \times 10{\cdot}0}{2\pi \times 25{\cdot}0} = 0{\cdot}7923\,\text{Hz}$$
$$\tilde{n}_3 = \frac{18{\cdot}025 \times 10{\cdot}0}{2\pi \times 25{\cdot}0} = 1{\cdot}1475\,\text{Hz}$$

and hence

$$A(\omega_1) = 0{\cdot}6732$$
$$A(\omega_2) = 0{\cdot}3398$$
$$A(\omega_3) = 0{\cdot}2371$$

The expression for the variance $\sigma_{q_i}^2$, which neglects the cross-spectral density function, is given by eq. (11.91) and implies the transposition of $\mathbf{Z}$ and the evaluation of $Z_{ji}^2$, which yield

$$\tilde{\mathbf{Z}}^{\mathrm{T}} = \begin{bmatrix} Z_{11}^2 & Z_{21}^2 & Z_{31}^2 \\ Z_{12}^2 & Z_{22}^2 & Z_{32}^2 \\ Z_{13}^2 & Z_{23}^2 & Z_{33}^2 \end{bmatrix} = \begin{bmatrix} 0{\cdot}8968 & 2{\cdot}9104 & 4{\cdot}5284 \\ 4{\cdot}5284 & 0{\cdot}9025 & 2{\cdot}9002 \\ 2{\cdot}9002 & 4{\cdot}5284 & 0{\cdot}9082 \end{bmatrix} \times 10^{-6}$$

Equation (11.91) may be written in matrix form. For the structure concerned, the aerodynamic admittance factors $A(\omega_i)$ are constants. Thus the variances $\sigma_{q2}^2$, $\sigma_{q2}^2$ and $\sigma_{q_3}^2$ may be calculated as follows

$$\sigma_{q1}^2 = \frac{1}{4{\cdot}439^3} \times \frac{0{\cdot}6732}{2 \times 0{\cdot}015} [0{\cdot}8968 \quad 2{\cdot}9104 \quad 4{\cdot}5284] \times \begin{bmatrix} 3{\cdot}81128 & 0 & 0 \\ 0 & 5{\cdot}82994 & 0 \\ 0 & 0 & 1{\cdot}77141 \end{bmatrix} \begin{bmatrix} 8{\cdot}10425 \\ 6{\cdot}06436 \\ 5{\cdot}02992 \end{bmatrix}$$
$$= 43{\cdot}8555\,\mathrm{m}^2$$

$$\sigma_{q2}^2 = \frac{1}{12{\cdot}446^3} \times \frac{0{\cdot}3398}{2 \times 0{\cdot}01} [4{\cdot}5284 \quad 0{\cdot}9025 \quad 2{\cdot}9002] \times \begin{bmatrix} 3{\cdot}81128 & 0 & 0 \\ 0 & 5{\cdot}82994 & 0 \\ 0 & 0 & 1{\cdot}77141 \end{bmatrix} \begin{bmatrix} 1{\cdot}55880 \\ 1{\cdot}13451 \\ 0{\cdot}93013 \end{bmatrix}$$
$$= 0{\cdot}3318\,\mathrm{m}^2$$

$$\sigma_{q_3}^2 = \frac{1}{18{\cdot}025^3} \times \frac{0{\cdot}2371}{2 \times 0{\cdot}01} [2{\cdot}9002 \quad 4{\cdot}5284 \quad 0{\cdot}9082] \times \begin{bmatrix} 3{\cdot}81128 & 0 & 0 \\ 0 & 5{\cdot}82994 & 0 \\ 0 & 0 & 1{\cdot}77141 \end{bmatrix} \begin{bmatrix} 0{\cdot}85386 \\ 0{\cdot}61838 \\ 0{\cdot}50595 \end{bmatrix}$$
$$= 0{\cdot}0538\,\mathrm{m}^2$$

*Determination of the generalized coordinates* $q_i = \kappa_i \sigma_i$

$$q_1 = \sqrt{[2\ln(4{\cdot}439 \times 3600)/2\pi]} + 0{\cdot}577/\sqrt{[2\ln(4{\cdot}439 \times 3600)/2\pi]} \times \sqrt{43{\cdot}8555} = 27{\cdot}190\,\mathrm{m}$$

$$q_2 = \sqrt{[2\ln(12{\cdot}446 \times 3600)/2\pi]} + 0{\cdot}577/\sqrt{[2\ln(12{\cdot}446 \times 3600)/2\pi]} \times \sqrt{0{\cdot}3318} = 2{\cdot}505\,\mathrm{m}$$

$$q_3 = \sqrt{[2\ln(18{\cdot}025 \times 3600)/2\pi]} + 0{\cdot}577/\sqrt{[2\ln(18{\cdot}025 \times 3600)/2\pi]} \times \sqrt{0{\cdot}0538} = 1{\cdot}028\,\text{m}$$

*Determination of maximum total displacements*

$$\boldsymbol{x} = \mathbf{K}^{-1}\mathbf{F} + \mathbf{Z}\boldsymbol{\kappa\sigma}$$

Given the shear stiffness of the columns between floors, the stiffness matrix for the structure is

$$\mathbf{K} = 12 \times 10^6 \begin{bmatrix} 2 & -1 & 0 \\ -1 & 2 & -1 \\ 0 & -1 & 1 \end{bmatrix} \text{N/m}$$

Hence

$$\mathbf{K}^{-1} = \frac{10^{-6}}{12} \times \begin{bmatrix} 1 & 1 & 1 \\ 1 & 2 & 2 \\ 1 & 2 & 3 \end{bmatrix} \text{m/N}$$

$$\begin{bmatrix} x_1 \\ x_2 \\ x_3 \end{bmatrix} = \frac{10^{-6}}{12} \times \begin{bmatrix} 1 & 1 & 1 \\ 1 & 2 & 2 \\ 1 & 2 & 3 \end{bmatrix} \begin{bmatrix} 49\,806{\cdot}250 \\ 73\,157{\cdot}763 \\ 44\,457{\cdot}474 \end{bmatrix} + \begin{bmatrix} 0{\cdot}947 & 2{\cdot}128 & 1{\cdot}703 \\ 1{\cdot}706 & 0{\cdot}950 & -2{\cdot}128 \\ 2{\cdot}128 & -1{\cdot}703 & 0{\cdot}953 \end{bmatrix} \begin{bmatrix} 27{\cdot}190 \\ 2{\cdot}505 \\ 1{\cdot}028 \end{bmatrix} \times 10^{-3}$$

Hence

$$\begin{bmatrix} x_1 \\ x_2 \\ x_3 \end{bmatrix} = \begin{bmatrix} 0{\cdot}0140 \\ 0{\cdot}0238 \\ 0{\cdot}0275 \end{bmatrix} + \begin{bmatrix} 0{\cdot}0328 \\ 0{\cdot}0466 \\ 0{\cdot}0546 \end{bmatrix} = \begin{bmatrix} 0{\cdot}0468 \\ 0{\cdot}0704 \\ 0{\cdot}0821 \end{bmatrix} \text{m}$$

Note that although all the three natural frequencies of the structure, i.e. $f_1 = 0{\cdot}7065$ Hz, $f_2 = 1{\cdot}9808$ Hz and $f_3 = 2{\cdot}8688$ Hz, lie within the part of the frequency spectrum in which the wind is considered to have a considerable amount of energy, the structure responds mainly in the first mode. It is therefore of interest to see to which extent the calculated displacements alter if it is assumed that the structure responds only in the first mode. This can easily be done by writing $q_2 = q_3 = 0$ in the transformation $\boldsymbol{x} = \mathbf{Z}$. Thus

$$\begin{bmatrix} x_1 \\ x_2 \\ x_3 \end{bmatrix} = \frac{10^{-6}}{12} \times \begin{bmatrix} 1 & 1 & 1 \\ 1 & 2 & 2 \\ 1 & 2 & 3 \end{bmatrix} \begin{bmatrix} 49\,806{\cdot}250 \\ 73\,157{\cdot}763 \\ 44\,457{\cdot}474 \end{bmatrix} + \begin{bmatrix} 0{\cdot}947 & 2{\cdot}128 & 1{\cdot}703 \\ 1{\cdot}706 & 0{\cdot}950 & -2{\cdot}128 \\ 2{\cdot}128 & -1{\cdot}703 & 0{\cdot}953 \end{bmatrix} \begin{bmatrix} 27{\cdot}190 \\ 0 \\ 0 \end{bmatrix} \times 10^{-3}$$

Hence

$$\begin{bmatrix} x_1 \\ x_2 \\ x_3 \end{bmatrix} = \begin{bmatrix} 0{\cdot}0140 \\ 0{\cdot}0238 \\ 0{\cdot}0275 \end{bmatrix} + \begin{bmatrix} 0{\cdot}0257 \\ 0{\cdot}0464 \\ 0{\cdot}0578 \end{bmatrix} = \begin{bmatrix} 0{\cdot}0397 \\ 0{\cdot}0702 \\ 0{\cdot}0854 \end{bmatrix} \text{m}$$

As can be seen, the differences, except in the case of the displacements $x_1$, are marginal and no greater than those that can be caused by uncertainties in the assumed values of damping ratios, and the degree of accuracy of the spectral density function used. In many cases, especially for buildings, it may therefore be sufficient—at least initially—to calculate the response in the first mode only in order to see if a further, more rigorous, investigation is required.

## Concluding remarks on the frequency domain method

This is a convenient method of predicting the dynamic response of structures. It is limited to the analysis of linear structures, although in practice it is also applied to some nonlinear structures by taking only the nonlinear response due to the mean wind speed component into account. When the frequency domain method is applied to determine the dynamic response of nonlinear structures such as cable roofs and guyed masts, whose stiffnesses and frequencies are functions of the degree of deformation, the natural frequencies should be determined for the deformed state due to the mean wind component and not for the case when there is no load on the structure.

Apart from the assumptions with respect to the statistical characteristics of wind, the main assumption made in order to make the method possible is that the amplitudes of the fluctuating component of the wind are sufficiently small compared to the mean wind speed the terms $\frac{1}{2}\rho C_d A u^2(t)$, $\frac{1}{2}\rho C_d \dot{x}^2(t)$ and $\rho C_d u(t)\dot{x}(t)$ in eq. (11.5) can be ignored. Generally this assumption is justified, but it may not be for sites in mountainous areas, where fluctuations of the same order of magnitude as the mean wind speed have been observed.

Finally, inspection of Fig. 11.3 and Table 10.3 makes it obvious that the degree of accuracy to which the dynamic response can be predicted by this method will vary with the type of spectral density function used. For important structures it may be advisable to construct spectral density functions from recordings at the site concerned.

## Vortex shedding of bluff bodies

So far only the along-wind response caused by the natural turbulence in the flow approaching the structure has been considered, not the different types of response due to the change of flow caused by the structure itself. Of these, the most important mechanism for wind-induced oscillations is the formation of vortices in the wake flow behind certain types of structure such as chimneys, towers, electrical transmission lines and suspended pipelines. Many failures due to vortex shedding have been reported.

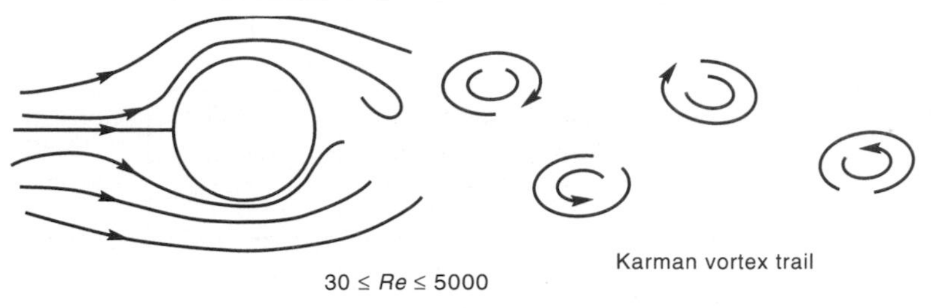


*Fig. 11.5. Regular periodic vortex shedding for flow past circular cylinder*

When bluff bodies are exposed to wind, vortices are shed from their sides, creating a pattern in their wake often referred to as the *Karman vortex trail* shown in Fig. 11.5. The frequency of the shedding depends on the shape of the body, the velocity of the flow, and to a lesser extent the surface roughness and the turbulence of the flow. The dominant frequency of vortex shedding is given by

$$n_{\mathrm{v}} = \frac{SU}{D} \tag{11.94}$$

where $S$ is a nondimensional constant referred to as the *Strouhal number*, $U$ is the mean wind velocity, and $D$ is the width of the bluff body. The manner in which vortices are formed is a function of the Reynolds number, which is given by

$$Re = \frac{UL}{\nu} \tag{11.95}$$

where $U$ is the mean velocity of the flow, $L$ is a representative dimension of the structural element, which in the case of members with circular cross-sections is equal to the diameter $D$, and $\nu$ is the kinematic viscosity, which for air is equal to $1{\cdot}5 \times 10^5$ m$^2$/s at 20°C.

The type of vortex shedding that is most important to civil engineers is when the shedding occurs regularly and alternates from side to side. For bodies with rectangular or square cross-sections the Strouhal number is nearly independent of the Reynolds number. For a body with a circular cross-section the Strouhal number varies with the rate of flow and hence with the Reynolds number. Three major regions are characterized by the Reynolds number: the subcritical region for $Re \leq 3 \times 10^5$, the supercritical region for $3 \times 10^5 \leq Re \leq 3 \times 10^6$, and the transcritical region for $Re \geq 3 \times 10^6$. Approximate values for the Strouhal number for circular and square sections are given in Table 11.2.

Vortex shedding will give rise to lift or across-wind forces forces, which as a first approximation per unit length may be written as

$$P_L(t) = \tfrac{1}{2}\rho D U^2 C_L(t) \tag{11.96}$$

where $C_L$ is a lift coefficient that fluctuates in a harmonic or random manner, and depends on the Reynolds number, the atmospheric turbulence and the surface roughness of the building. If the vortex shedding

*Table 11.2. Data for prediction of vortex-induced oscillations in turbulent flow (after Davenport, 1961)*

| Cross-section | Strouhal number $S$ | RMS lift coefficient $\sigma_L$ | Bandwidth $B$ | Correlation length (diameters) $L$ |
|---|---|---|---|---|
| Circular: region | | | | |
| Subcritical | 0·2 | 0·5 | 0·1 | 2·5 |
| Supercritical | Not marked | 0·14 | Not marked | 1·0 |
| Transcritical | 0·25 | 0·25 | 0·3 | 1·5 |
| Square: | | | | |
| Wind normal to face | 0·11 | 0·6 | 0·2 | 3·0 |

frequency $n_v$ coincides with the natural frequency of a structure, such as a chimney, quite large across-wind amplitudes of vibration will result unless sufficient damping is present. Values for lift coefficients and Strouhal numbers for different types of sections are given by ESDU (1978) and Simue & Scalan (1978).

If the vortex shedding is harmonic, eq. (11.96) may be written as

$$P_L(t) = P_0 \sin(\omega_v t) = \tfrac{1}{2}\rho D U^2 C_L \sin(2\pi n_v) \tag{11.97}$$

From eq. (2.8) the equivalent modal mass of a prismatic member is given by

$$M = \int_0^L m(x)\{\phi(x)\}^2 \, dx \tag{11.98}$$

and from eq. (2.26), assuming a constant wind profile, the equivalent modal force due to the fluctuating lift force given by eq. (11.97) is

$$P(t) = P_L \sin(2\pi n_v) = \tfrac{1}{2}\rho D U^2 \sin(2\pi n_v) \int_0^L C_L(x)\{\phi(x)\}dx \tag{11.99}$$

Since in eq. (4.14) $x_{st} = P_0/K = P_0/M\omega^2$, the maximum response of a one DOF system subjected to harmonic excitation may be written as:

$$x_{max} = \frac{P_L}{M\omega^2} \times \frac{1}{2\xi} \tag{11.100}$$

It follows that when the vortex shedding occurs with the same frequency as the natural frequency of the structure

$$x_{max} = \frac{\tfrac{1}{2}\rho D U^2 \int_0^L C_L\{\phi(x)\}dx}{\omega^2 \int_0^L m(x)\{\phi(x)\}^2 dx} \times \frac{1}{2\xi} \tag{11.101}$$

which can be simplified if it is assumed that the mass per unit length is constant, $m(x) = m$, and that the loss of spanwise correlation of the lift forces can be taken into account by assuming that the lift coefficient $C_L(x)$ is proportional to the modeshape, i.e.

$$C_L(x) = C_L\{\phi(x)\} \tag{11.102}$$

Substitution of the above expressions for $m(x)$ and $C_L(x)$ into eq. (11.101) yields

$$x_{\text{max}} = \frac{\frac{1}{2}\rho D U^2 C_L}{2\xi\omega^2} \tag{11.103}$$

From eq. (11.94) we have that

$$\omega_s{}^2 = \frac{4\pi^2 S^2 U^2}{D^2} \tag{11.104}$$

Substitution of the expression for $\omega^2$ into eq. (11.103), remembering that $\omega_v = \omega$ at resonance, yields for the maximum response

$$x_{\text{max}} = \frac{\rho D^3 C_L}{16\pi^2 S^2 m\xi} \tag{11.105}$$

For the first mode of a cantilever structure $x_{\text{max}}$ occurs at the tip. In higher modes this amplitude occurs where the resonance takes place. For circular cylinders a design value for $C_L$ is about $\sqrt{(2\sigma_L)}$. The maximum value for cylinders is 0·4. Approximate values for $\sigma_L$ are given in Table 11.2. Equation (11.105) may be used as a first estimate of likely response and will yield an upper bound solution.

**Example 11.4** A 20 m high industrial cable-stayed steel chimney has an external diameter of 1·0 m and a natural frequency of 2·4 Hz. The mass is 150 kg/m. The Strouhal number for the circular section of the chimney is $S = 0{\cdot}2$ and the root mean square value of the lift coefficient $\sigma_l = 0{\cdot}14$. Calculate the wind velocity that will cause vortex shedding with a frequency equal to the natural frequency of the chimney, the corresponding Reynolds number and, finally, the maximum first mode amplitude of response at the tip. The specific density of air $\rho = 1{\cdot}226\,\text{kg/m}^3$; the kinematic viscosity for air $\nu = 1{\cdot}5 \times 10^{-5}\,\text{m}^2/\text{s}$.

The velocity at which the frequency of vortex shedding is equal to the natural frequency of the chimney is

$$U = \frac{n_v D}{S} = \frac{2{\cdot}4 \times 1{\cdot}0}{0{\cdot}2} = \underline{12{\cdot}0\ \text{m/s}}$$

The Reynolds number for a flow of 12·0 m/s is

$$Re = \frac{UD}{\nu} = \frac{12{\cdot}0 \times 1{\cdot}0}{1{\cdot}5 \times 10^{-5}} = \underline{8{\cdot}0 \times 10^5}$$

The Reynolds number is therefore just at the lower end of the supercritical range.

The maximum amplitude of response at the tip of the chimney is given by

$$x_{\max} = \frac{\rho D^3 C_L}{16\pi^2 S^2 m\xi} = \frac{1{\cdot}226 \times 1{\cdot}0^3 \times \surd(2) \times 0{\cdot}14}{16 \times \pi^2 \times 0{\cdot}2^2 \times 150 \times 0{\cdot}01} = \underline{0{\cdot}0256\,\text{m}}$$

Even when the vortex shedding appears to be regular, the lift force and hence $C_L(t)$ are random rather than harmonic. From eq. (11.96) it follows that the spectral density function for the lift force per unit length can be expressed as

$$S_{P_L}(\omega) = \{\tfrac{1}{2}\rho D U^2\}^2 \times S_{C_L}(\omega) \tag{11.106}$$

where

$$\int_0^\infty S_{C_L}(\omega)\mathrm{d}\omega = \frac{1}{T}\int_0^T C_L(t) \times C_L(t)\mathrm{d}t = \sigma_L{}^2 \tag{11.107}$$

is the variance of the lift coefficient $C_L(t)$, and $S_{C_L}(\omega)$ is the spectral density function of $C_L(t)$.

Thus the spectral density function for response of a one DOF system assuming a correlation length $DL_C$, is given by

$$S_x(\omega) = \left\{\frac{1}{K}\right\}^2 \{\tfrac{1}{2}\rho D^2 L_c U^2\}^2 \times \mathrm{MF}^2(n) \times S_{C_L}(\omega) \tag{11.108}$$

From eq. (11.104)

$$U^2 = \frac{\omega_s{}^2 D^2}{4\pi^2 S^2} \tag{11.109}$$

Hence

$$S_x(\omega) = \left\{\frac{1}{K}\right\}^2 \left\{\frac{\omega_v{}^2 \rho D^4 L_c}{8\pi^2 S^2}\right\}^2 \times \mathrm{MF}^2(\omega) \times S_{C_L}(\omega)\mathrm{d}\omega \tag{11.110}$$

and thus

$$\begin{aligned}\sigma_x{}^2 &= \int_0^\infty S_x(\omega)\mathrm{d}\omega \\ &= \left\{\frac{1}{K}\right\}^2 \times \left\{\frac{\omega_v{}^2 \rho D^4 L_c}{8\pi^2 S^2}\right\}^2 \times \mathrm{MF}^2(\omega) \times \int_0^\infty S_{C_L}(\omega)\,\mathrm{d}\omega \end{aligned} \tag{11.111}$$

For weakly damped structures, eq. (11.111) may be written as

$$\sigma_x{}^2 = S_x(\omega)\Delta\omega = \left\{\frac{1}{K}\right\}^2 \times \left\{\frac{\omega_v{}^2 \rho D^4 L_c}{8\pi^2 S^2}\right\}^2 \times \mathrm{MF}^2(\omega) \times S_{C_L}(\omega)\Delta\omega \tag{11.112}$$

where

$$K = \omega_n M$$
$$MF(\omega) = 1/2\xi$$
$$\Delta\omega = \tfrac{1}{2}\xi\omega$$
$$\omega_v \approx \omega_n$$

$$\sigma_x{}^2 = S_x(\omega_n)\Delta\omega = \left\{\frac{1}{M}\right\}^2 \times \left\{\frac{\rho D^4 L_c}{8\pi^2 S^2}\right\}^2 \times \left\{\frac{\omega_n}{8\xi}\right\} \times S_{C_L}(\omega_n) \quad (11.113)$$

Approximate values for the correlation length $L$ in diameters are given in Table 11.2. The correlation length decreases with increasing turbulence intensity, increases with the ratio $2H/D$ (where $H$ is the height of the structure, and increases with the amplitude of the motion.

In the subcritical and transcritical range the energy of the lift force acting on circular cylinders is distributed closely on either side of the dominant shedding frequency, and can be represented by a Gaussian-type distribution curve. Harris (1988) and Lawson (1990) give the spectral density function for the lift coefficient for this type of distribution as

$$S_{C_L}(n) = \frac{\sigma_L{}^2}{n_s B\sqrt{(4\pi^3)}} \exp\left\{-\left(\frac{1 - n/n_v}{B}\right)^2\right\} \quad (11.114)$$

In the supercritical range the spectral density function is broad, and is given by Harris (1988) as

$$S_{C_L}(n) = 4{\cdot}8\sigma_L{}^2 \frac{1 + 682{\cdot}2(nD/U)^2}{\left[1 + 227{\cdot}4(nD/U)^2\right]^2} \times \frac{D}{U} \quad (11.115)$$

or

$$S_{C_L}(n) = 4{\cdot}8\sigma_L{}^2 \frac{1 + 682{\cdot}2(Sn/n_v)^2}{\left[1 + 227{\cdot}4(Sn/n_v)^2\right]^2} \times \frac{S}{n_v} \quad (11.116)$$

**Example 11.5** Use first eq. (11.114) and then eq. (11.116) to calculate the maximum transverse tip displacement of the 20 m high industrial steel chimney in example 11.4, which has an external diameter $D = 1{\cdot}0$ m, a natural frequency $n_n = 2{\cdot}4$ Hz and a mass $m = 150$ kg/m. The structural damping $\xi = 0{\cdot}01$. The Strouhal number for the circular section of the chimney $S = 0{\cdot}2$, the root mean square value of the lift coefficient $\sigma_l = 0{\cdot}14$, the bandwidth $B = 0{\cdot}1$, the correlation length $L = 2{\cdot}5D$, and the specific density of air $\rho = 1{\cdot}226$ kg/m$^3$. Use the same values for $S$ and $L$ when using eqs (11.114) and (11.116).

From eq. (11.113)

$$\sigma_x{}^2 = \left\{\frac{1}{M}\right\}^2 \times \left\{\frac{\rho D^4 L_c}{8\pi^2 S^2}\right\}^2 \times \left\{\frac{\omega_n}{8\xi}\right\} \times S_{C_L}(\omega_n)$$

where, from eq. (2.30)

$$M = (728/2835)mL = (728/2835) \times 150 \times 20 = 770{\cdot}37037\,\text{kg}$$

Hence

$$\sigma_x{}^2 = \left\{\frac{1}{770{\cdot}37}\right\}^2 \times \left\{\frac{1{\cdot}226 \times 1{\cdot}0^4 \times 2{\cdot}5}{8\pi^2 \times 0{\cdot}2^2}\right\}^2 \times \left\{\frac{2\pi \times 2{\cdot}4}{8 \times 0{\cdot}01}\right\} \times S_{C_L}(\omega_n)$$
$$= 3{\cdot}08235 \times 10^{-4} \times S_{C_L}(\omega_n)$$

The expression for the spectral density function for lift coefficients given by eq. (11.114) yields

$$S_{C_L}(\omega) = \frac{\sigma_L{}^2}{n_s B\sqrt{\pi}} \exp\left\{-\left(\frac{1 - n/n_s}{B}\right)^2\right\}$$
$$= \frac{0{\cdot}14^2}{2{\cdot}4 \times 0{\cdot}1\sqrt{\pi}} \exp\left\{-\left(\frac{1 - 2{\cdot}4/2{\cdot}4}{0{\cdot}1}\right)^2\right\} = 0{\cdot}0460754\,\text{m}^2\,\text{s/m}$$

Hence

$$\sigma_x{}^2 = 3{\cdot}08235 \times 10^{-4} \times 0{\cdot}0460754 = 14{\cdot}20207 \times 10^{-6}\,\text{m}^2$$
$$\sigma_x = 3{\cdot}76856 \times 10^{-3}\,\text{m}$$

Thus the maximum amplitude of lateral vibration due to vortex shedding, from eq. (11.114), is

$$x_{max} = \kappa\sigma_x$$
$$= \{\sqrt{2[\ln(2{\cdot}4 \times 3600)]} + 0{\cdot}577/\sqrt{[2\ln(2{\cdot}4 \times 3600)]}\} \times 3{\cdot}76856 \times 10^{-3}$$
$$= \underline{0{\cdot}0166\,\text{m}}$$

The expression for the spectral density function for lift coefficients given by eq. (11.116) yields

$$S_{C_L}(n) = 4{\cdot}8 \times 0{\cdot}14^2 \frac{1 + 682{\cdot}2(0{\cdot}2 \times 2{\cdot}4/2{\cdot}4)^2}{\left[1 + 227{\cdot}4(0{\cdot}2 \times 2{\cdot}4/2{\cdot}4)^2\right]^2} \times \frac{0{\cdot}2}{2{\cdot}4}$$
$$= 2{\cdot}1758 \times 10^{-3}\,\text{m}^2\,\text{s/m}$$

Hence

$$\sigma_x{}^2 = 3{\cdot}08235 \times 10^{-4} \times 2{\cdot}1758 \times 10^{-3} = 6{\cdot}70657\,\text{m}^2$$
$$\sigma_x = 0{\cdot}818936 \times 10^{-3}\,\text{m}$$

Thus the maximum amplitude of lateral vibration due to vortex shedding given by eq. (11.116) is

$$x_{max} = \kappa\sigma_x = \{\sqrt{[2\ln(2{\cdot}4 \times 3600)]} + 0{\cdot}577/\sqrt{[2\ln(2{\cdot}4 \times 3600)]}\}$$
$$\times\, 0{\cdot}818936 \times 10^{-3} = \underline{0{\cdot}0036\,\text{m}}$$

Comparison of the displacements calculated in examples 11.4 and 11.5 indicates that the response to vortex shedding in the supercritical region from eq. (11.116) is much less than in the subcritical region by eq. (11.114), and both equations lead to a smaller displacement than does eq. (11.105).

### The phenomenon of lock-in

The wind speed can be expressed in terms of a non-dimensional *reduced velocity* $U_r$ as

$$U = U_r n_n D \qquad (11.117)$$

where $N_n$ is the natural frequency of the structure and $D$ is the width of the structure. Combination of eqs (11.117) and (11.94) by elimination of $U$ yields

$$n_v = (Sn_n)U_r \qquad (11.118)$$

As both $S$ and $n_n$ are constants, it follows that the shedding frequency varies linearly with the reduced velocity. However, wind tunnel tests of flexible structural models have shown that in a region on either side of the reduced velocity, where this velocity is approximately equal to the inverse of the Strouhal number (i.e. where $U_r = 1/S$), the shedding frequency remains constant and is equal to the natural frequency of the structure. This phenomenon is referred to as a *lock-in*, because the shedding frequency is locked into the natural frequency of the structure. In steady flow the frequency of the structural vibration tends to be constant during a lock-in, with the greatest amplitude occurring when $n_v = n_n$.

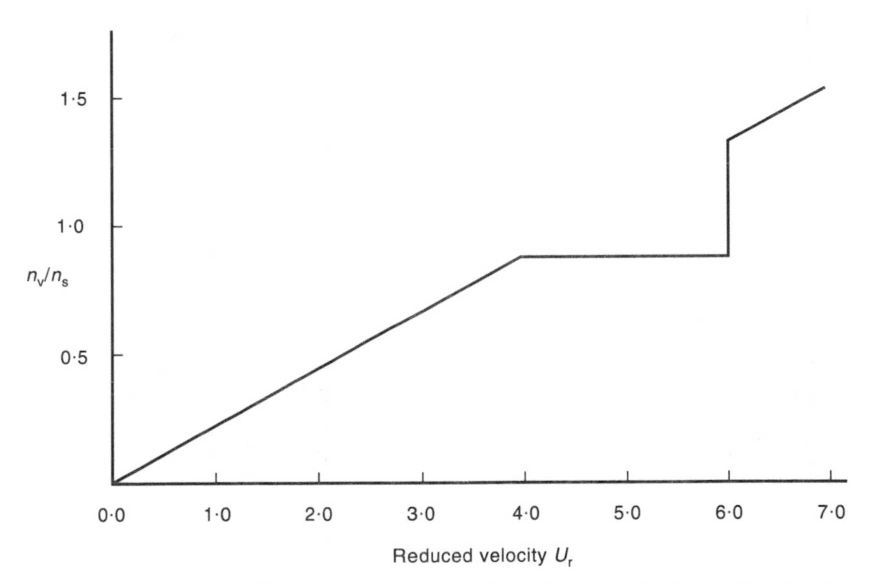


*Fig. 11.6. Variation of frequency ratio $n_v n_s$ with reduced wind velocity $U_r$, showing lock-in*

For a circular structure with Strouhal number $s = 0{\cdot}2$, the extent of lock-in can be seen in Fig. 11.6, where the ratio $n_v/n_n$ is plotted against the reduced velocity $U_r$.

The lock-in phenomenon can also be observed in the behaviour of real structures. In turbulent wind, however, the lock-in condition occurs only if the amplitudes of vibration are in excess of approximately 2% of the width of the building. The motion when this is the case will have large amplitude, and a regular frequency, at a reduced mean wind velocity equal to $1/S$. When the amplitudes of across-wind oscillations are smaller, the magnitude of the amplitude varies spasmodically, with a lock-in occurring from time to time. The response of horizontal structures such as the spans of pipeline bridges tends to correlate the vortex shedding along the span, and will therefore cause the transverse amplitude displacement to be increased further. Thus if either eq. (11.114) or eq. (11.116) yields a displacement greater than 2·0% of the diameter $D$, the calculations need to be repeated with a larger correlation length (ESDU, 1978; Harris, 1988).

**Example 11.6** Calculate the lengths of the 20 m high chimney in examples 11.3 and 11.4 likely to shed vortices if the lock-in of frequency shedding is assumed to last for wind velocities equal to ±20% of the one that will first cause across-wind vibration at the tip of the chimney. Assume roughness lengths of 0·2 m, 0·45 m and 0·9 m. Finally, assuming the effective correlation length $DL$ to be one-third of the above lengths, calculate the maximum tip displacements for each case using eq. (11.113) and (11.115). The specific density of air $\rho = 1{\cdot}226\,\text{kg/m}^3$.

The reduced velocity at which vortex shedding will occur

$$u_r = 1/S = 1/0{\cdot}2 = 5{\cdot}0$$

From eq. (11.95) the corresponding wind velocity is

$$U = U_r n_s D = 5{\cdot}0 \times 2{\cdot}4 \times 1{\cdot}0 = 12{\cdot}0\ \text{m/s}$$

Thus a lock-in will occur for wind velocities between

$$12{\cdot}20 \times 0{\cdot}8 = \underline{9{\cdot}8\ \text{m/s}} \text{ and } 12{\cdot}0 \times 1{\cdot}2 = \underline{14{\cdot}4\ \text{m/s}}$$

The maximum length of chimney shedding vortices will be found by assuming the maximum shedding velocity at the top of the chimney to be 14·4 m/s and the minimum velocity further down to be 9·6 m/s. In order to determine this distance it is first necessary to calculate the shear velocity corresponding to a the mean shedding velocity of 12·0 m/s.

When $z_0 = 0{\cdot}2$ m

$$u_* = \frac{12{\cdot}0}{2{\cdot}5\ln(20{\cdot}0/0{\cdot}2)} = 1{\cdot}0423068\ \text{m/s}$$

The corresponding height at which the velocity is 9·6 m/s is

$$z = z_0 e^{U(z)/2 \cdot 5 u_*} = 0 \cdot 2 e^{9 \cdot 6/2 \cdot 5 \times 1 \cdot 0423068} = \underline{7 \cdot 962 \text{ m}}$$

Thus the lock-in lengths, assuming the velocity fluctuations to be ±20% of the initiating shedding velocity, are

$$(\text{when } z_0 = 0 \cdot 20 \text{ m}) \ L_L = 20 \cdot 0 - 7 \cdot 962 = \underline{12 \cdot 038 \text{ m}}$$
$$(\text{when } z_0 = 0 \cdot 45 \text{ m}) \ L_L = 20 \cdot 0 - 9 \cdot 364 = \underline{10 \cdot 636 \text{ m}}$$
$$(\text{when } z_0 = 0 \cdot 90 \text{ m}) \ L_L = 20 \cdot 0 - 9 \cdot 364 = \underline{9 \cdot 243 \text{ m}}$$

The corresponding assumed correlation lengths are

$$(\text{when } z_0 = 0 \cdot 20 \text{ m}) \ L_C = 12 \cdot 038/3 \times 1 \cdot 0 = 4 \cdot 090 \text{ m}$$
$$(\text{when } z_0 = 0 \cdot 45 \text{ m}) \ L_C = 10 \cdot 636/3 \times 1 \cdot 0 = 3 \cdot 545 \text{ m}$$
$$(\text{when } z_0 = 0 \cdot 90 \text{ m}) \ L_C = 9 \cdot 243/3 \times 1 \cdot 0 = 3 \cdot 081 \text{ m}$$

When $z_0 = 0 \cdot 20$ m

$$\sigma_x{}^2 = 8 \cdot 2498974 \times 10^{-4} \times 0 \cdot 0460754 = 38 \cdot 01173 \times 10^{-6} \text{ m}^2$$
$$\sigma_x = 6 \cdot 16536 \times 10^{-3} \text{ m}$$
$$x_{\max} = \kappa \sigma_x = \{\surd[2 \ln(2 \cdot 4 \times 3600)] + 0 \cdot 577/\surd[2 \ln(2 \cdot 4 \times 3600)]\} \times 6 \cdot 1653600 \times 10^{-3} = \underline{0 \cdot 0271 \text{ m}}$$

When $z_0 = 0 \cdot 45$ m

$$\sigma_x{}^2 = 6 \cdot 1977551 \times 10^{-4} \times 0 \cdot 0460754 = 28 \cdot 556443 \times 10^{-6} \text{ m}^2$$
$$\sigma_x = 5 \cdot 3438229 \times 10^{-3} \text{ m}$$
$$x_{\max} = \kappa \sigma_x = \{\surd[2 \ln(2 \cdot 4 \times 3600)] + 0 \cdot 577/\surd[2 \ln(2 \cdot 4 \times 3600)]\} \times 5 \cdot 3438229 \times 10^{-3} = \underline{0 \cdot 0235 \text{ m}}$$

When $z_0 = 0 \cdot 90$ m

$$\sigma_x{}^2 = 4 \cdot 6815033 \times 10^{-4} \times 0 \cdot 0460754 = 21 \cdot 570214 \times 10^{-6} \text{ m}^2$$
$$\sigma_x = 4 \cdot 6443744 \times 10^{-3} \text{ m}$$
$$x_{\max} = \kappa \sigma_x = \{\surd[2 \ln(2 \cdot 4 \times 3600)] + 0 \cdot 577/\surd[2 \ln(2 \cdot 4 \times 3600)]\} \times 4 \cdot 6443744 \times 10^{-3} = \underline{0 \cdot 0204 \text{ m}}$$

A comparison of the above displacements with that of 0·0256 m calculated in example 11.5 using eq. (11.105), which yields an upper bound solution, indicates that the assumed correlation is not unreasonable.

## Random excitation of tapered cylinders by vortices

Tapered cylinders such as stacks also vibrate due to vortex shedding. However, less is known about the mechanism of excitation. Experience seems to indicate that the lift forces are narrow band random with a rather small correlation length, with the dominant frequency given by eq. (11.94). As the diameter varies, local resonance between $n_s$ and the natural frequency of the tapered cylinder takes place at different heights. As the wind speed increases the resonance first appears at the tip and then shifts downwards. The critical wind speed for each height occurs when $n_s$ is equal to $n_n$. An approximate method for calculating the mean standard deviation of tapered cylinders is given by Harris (1988).

## Suppression of vortex-induced vibration

Vortex shedding can be prevented by

- destroying the spanwise correlation of the vortices
- bleeding air into the near-wake region
- preventing the interaction of the two shear layers.

One method used to prevent vortex shedding of chimneys where the level of structural damping is insufficient is in the fitting of 'stakes'. The most efficient stake known for destroying the spanwise correlation of vortices is a three-start helix that makes one revolution in 5 diameters length of chimney and extends over the top third of the height. The disadvantage of stakes is that they increase the drag force.

Another method is the fitting of perforated shrouds. These prevent vortex shedding by bleeding air into the near-wake region. Shrouds tend to be heavier than stakes but increase the drag flow less.

A third method is the use of splitter plates. These are not generally a practical proposition because they need to be aligned in the direction of the wind. As they need to extend 4 diameters downwind they tend to be heavy. However, they have the advantage that they do not increase the drag forces to the same extent as stakes and perforated shrouds.

## Dynamic response to the buffeting of wind using time-integration methods

If spatially correlated wind histories can be generated, for example by the method presented in chapter 14, then the response of structures can be determined through step-by-step integration in the time domain. In chapter 6 three such integrated methods, based on the Newmark $\beta$- and Wilson $\theta$-equations, are presented. Experience seems to indicate that schemes employing the Newmark $[\beta = \frac{1}{4}]$-equations, i.e. assuming the accelerations to remain constant during the time step $\Delta t$, are the most efficient of these. For one DOF systems the response to wind can be calculated using eq. (6.51). The response of multi-DOF systems can be calculated using eq. (6.68), given as

$$\left[\mathbf{K} + \frac{2}{\Delta t}\mathbf{C} + \frac{4}{\Delta t^2}\mathbf{M} + \frac{4}{\Delta t}\mathbf{F}_\mathrm{d}(\mathbf{V} - \dot{\boldsymbol{x}})\right] =$$

$$2\mathbf{F}_\mathrm{d}(\mathbf{V} - \dot{\boldsymbol{x}})(\Delta\mathbf{V} + 2\dot{\boldsymbol{x}}) + 2\mathbf{C}\dot{\boldsymbol{x}} + \mathbf{M}\left(\frac{4}{\Delta t}\dot{\boldsymbol{x}} + 2\ddot{\boldsymbol{x}}\right) \qquad (11.119)$$

where **K**, **C** and **M** are the stiffness, damping and mass matrices for a structure, $\Delta\boldsymbol{x}$ is the change in displacement vector $\boldsymbol{x}$ during a time step $\Delta t$, $\dot{\boldsymbol{x}}$ is a velocity vector and $\ddot{\boldsymbol{x}}$ is a acceleration vector. Using the Newmark $[\beta = \frac{1}{4}]$-equations, from eqs (6.41)–(6.43) the $i$th elements in the displacement, velocity and acceleration vectors at time $(t + \Delta t)$ are

$$x_i(t + \Delta t) = x_i(t) + \Delta x_i \qquad (11.120a)$$

$$\dot{x}_i(t + \Delta t) = \frac{2}{\Delta t}\Delta x_i - \dot{x}_i(t) \qquad (11.120b)$$

$$\ddot{x}_i(t + \Delta t) = \frac{4}{\Delta t^2}\Delta x_i - \frac{4}{\Delta t}\dot{x}_i(t) - \ddot{x}_i(t) \qquad (11.120c)$$

The size of the time step $\Delta t$ is important, as over-large as well as over-small time steps will lead to inaccuracies in the calculated response. In the case of both wind and earthquakes most of the energy is contained within the part of the frequency spectrum that lies between 0 and 10 Hz. Thus the period of the smallest frequency component that needs to be considered is usually approximately 0·1 s. Experience has shown that frequency components of that order of magnitude can be sufficiently accurately modelled with time steps $\Delta t = 0{\cdot}1/10 = 0{\cdot}01$ s.

The forward integration process should be contained until the variance of response is constant. Experience indicates that this will occur after approximately 120 s of real time. The maximum response is found by multiplying the standard deviation of response by the peak factor $\kappa$.

**Problem 11.1** The tapering lattice tower shown in Fig. 11.6 supports a circular disc 40 m above the ground. The values of the lateral stiffness mass and damping coefficient of the equivalent mass–spring system of the tower are 323·723 kN/m, 7200 kg and 1030·44 N s/m respectively. The disc weighs 1·0 t and has a diameter of 3·0 m and a drag coefficient of 1·3. Determine the maximum displacement of the tower when the mean wind speed 10 m above the ground, averaged from 10 min recording, is 40 m/s. Assume the surface drag coefficient for the site to be equal to 0·006, and that the fluctuating component of the wind can be represented first by Davenport's, then by Harris's and finally by Kaimal's power spectrum.

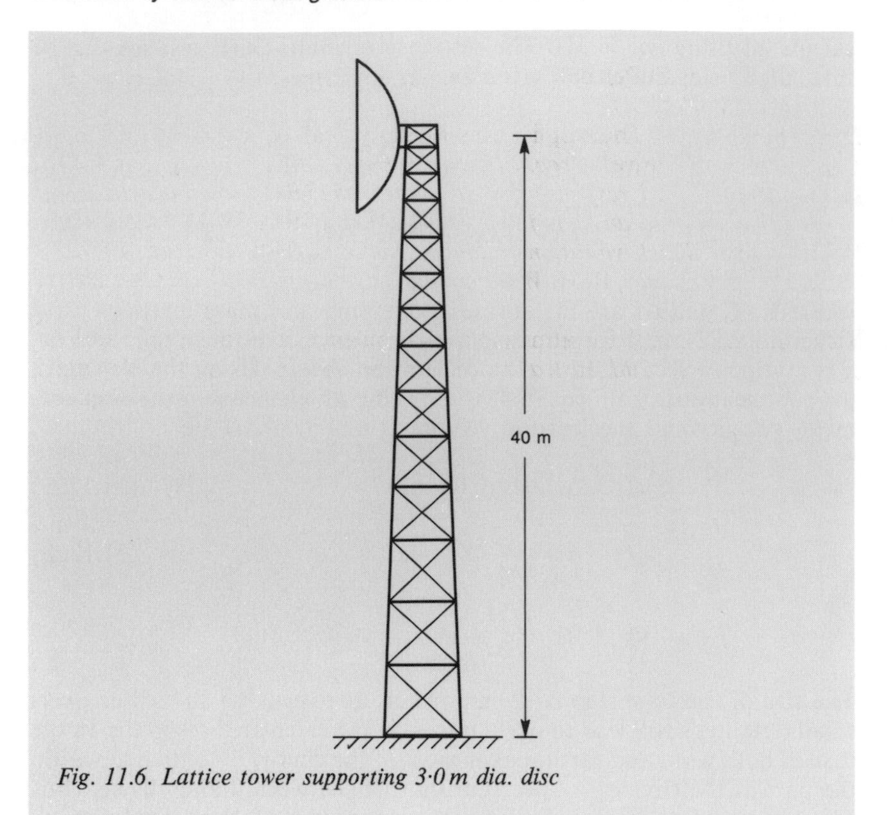


*Fig. 11.6. Lattice tower supporting 3·0 m dia. disc*

**Problem 11.2** Use Davenport's spectrum to calculate the response of the structure in example 11.3. Include the effect of aerodynamic damping, and comment on its effect on the calculated dynamic response.

**Problem 11.3** A cable-supported pipeline bridge has a span of 20 m. The mass and stiffness of an equivalent mass–spring system are 4000 kg and 39 478·418 N/m respectively. Make a preliminary estimate of the maximum across-wind response by assuming the modeshape of vibration to be similar to the deflected form of a built-in beam supporting a uniformly distributed load, and by further assuming that the response to random alternate vortex shedding will give rise to correlated vortex shedding along the span. The value of the Strouhal number for a circular section is $S = 0{\cdot}2$, the lift coefficient $C_l = 0{\cdot}3$ and the specific density of air $\rho = 1{\cdot}226\,\text{kg/m}^3$. Assume the first mode damping ratio to be 1% critical.

## References

Clough, R. W. & Penzien, J. *Dynamics of structures.* McGraw-Hill, London, 1975.

Davenport, A. G. The application of statistical concepts to the wind loading of structures. *Proc. Instn. Civ. Engrs*, 1961, 19, Aug., 449–472.

ESDU *Across-wind response due to vortex shedding isolated cylindrical structures in wind and gas flows.* ESDU Data Item 75011, Oct. 1978.

Harris, C. M. *Shock vibration*, 3rd edn. McGraw-Hill, London, 1988.

Simue, E. and Scalan, R. H. *Wind effects on structures.* Wiley, Chichester, 1978.

Vickery, B. J. Model for atmospheric turbulence for studies of wind on buildings. *Proc. 2nd Australasian Conf. on Hydraulics and Fluid Mech., Auckland*, 1965.

# 12. The nature and properties of earthquakes

## Introduction

Earthquakes are normally experienced as a series of cyclic movements of the earth's surface, and are the result of the fracturing or faulting of the earth's crust. The source of the vibratory energy is the release of accumulated strain energy resulting from sudden shear failures, which involve the slipping of the boundaries of large rock masses tens or even hundreds of kilometres beneath the earth's surface. On a global scale these large rock masses are continental in size and comprise the so-called *tectonic plates* into which the earth's crust is divided. The failure of the crust gives rise to propagation of two types of waves through the earth, pressure or primary waves and shear or secondary waves, referred to as P and S waves. The P wave travels faster than the S waves so that the waves arrive in alphabetical order. Thus if the velocities of the two types of waves are known, the distance from a focal point of observation can be calculated. Once P and S waves reach the surface, a surface wave is generated. Fig. 12.1 shows the principal geometrical terms used to describe earthquakes and the travel paths of P and S waves.

## Types and propagation of seismic waves

Only the pressure and shear waves are propagated within the earth's body. The P waves, as mentioned above, are the fastest of the two: their motion is the same as a sound wave that spreads out and alternatively compresses and dilates the rock. The P waves, like sound waves, can travel through solid rock and water. The S waves, which travel more slowly than the P waves, shear the rock sideways in a direction perpendicular to the direction of travel, and cannot propagate through water. Surface waves, as the name implies, travels only on the surface of the earth. Seismic surface waves are divided into two types, referred to as the *Love wave* and the *Rayleigh wave*. The motion of Love waves is essentially the same as that of S waves with no vertical components. They move from side to side on the earth's surface, in a direction normal to the direction of propagation. The Love waves are like rolling ocean waves, in which the disturbed material moves both vertically and horizontally in a vertical plane in the along direction of the quake. The surface waves travel more slowly than the P and S body waves, and generally the Love waves travel faster than the Rayleigh waves. The

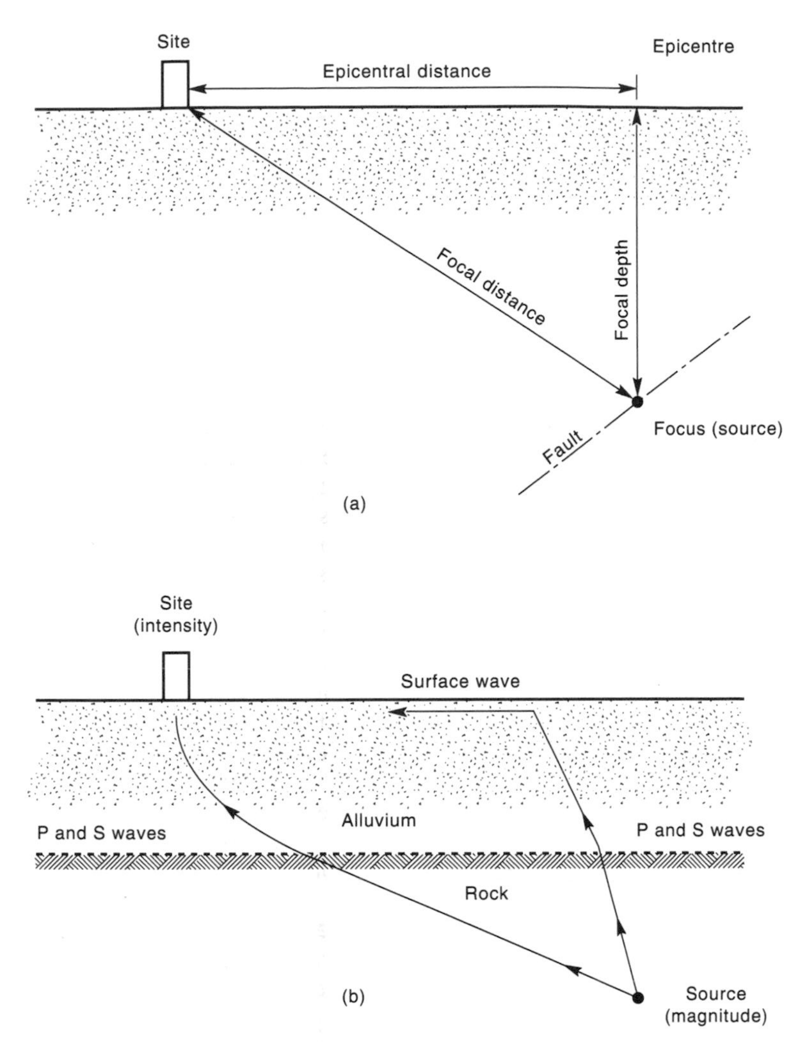


*Fig. 12.1. The principal terms used to describe earthquakes: (a) geometry; (b) transmission*

different forms of seismic waves described above are depicted in Fig. 12.2. When P and S waves are reflected or refracted at the interfaces between rock types, some of the wave energy can be converted to waves of the other types. Therefore on land and in strong earthquakes, after the first few shakes, two kinds of ground motion are usually felt simultaneously.

## Propagation velocity of seismic waves

The wave velocity within an elastic homogeneous isotropic solid can be defined by two constants $\kappa$ and $\mu$, where $\kappa$ is the modulus of incompressibility or bulk modulus and $\mu$ is the modulus of rigidity.

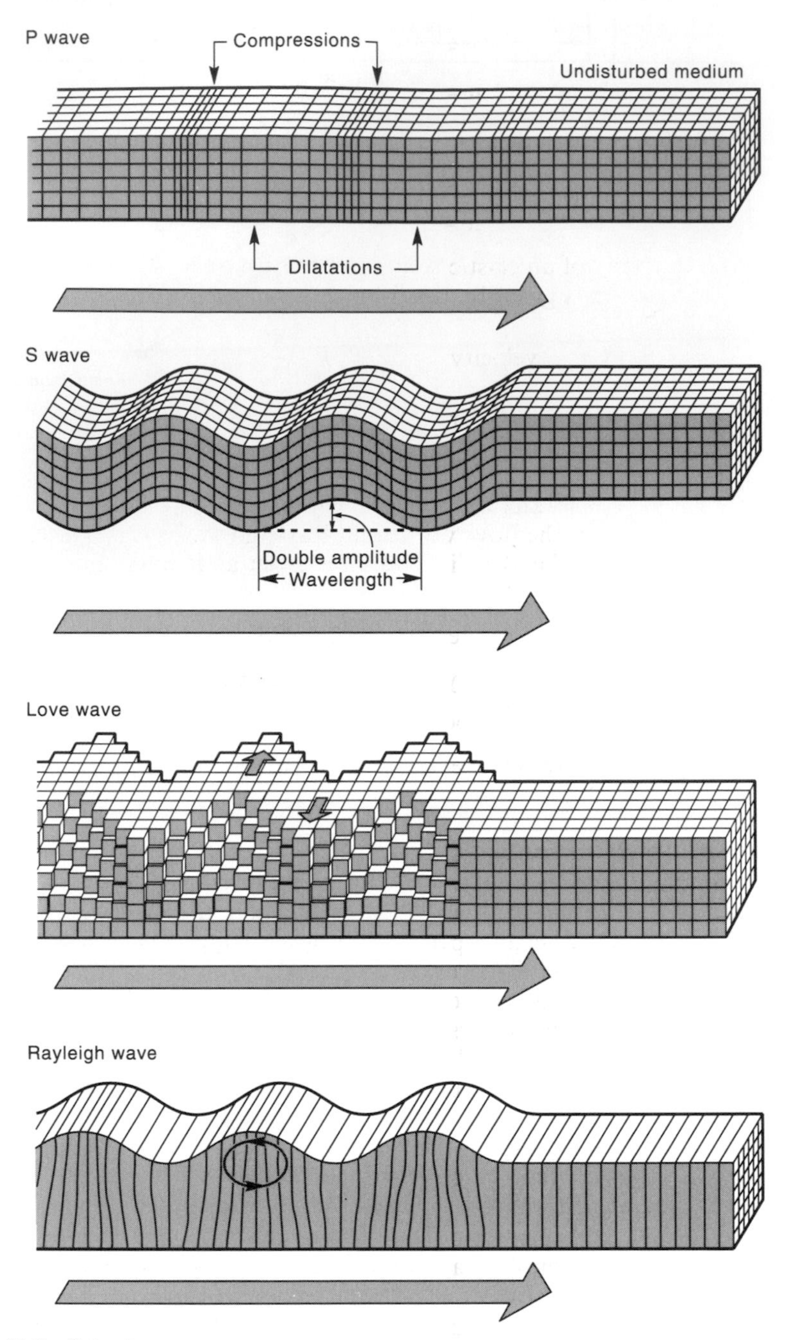


*Fig. 12.2. Seismic waves*

For granite

$$\kappa \approx 2{\cdot}7 \times 10^{10}\ \mathrm{N/m^2}$$
$$\mu = 1{\cdot}6 \times 10^{10}\ \mathrm{N/m^2}$$

For water

$$\kappa \approx 0{\cdot}2 \times 10^{10}\ \mathrm{N/m^2}$$
$$\mu = 0$$

Within the body of an elastic solid with density $\rho$ the velocity of pressure and shear waves is given by the following expression. For P waves

$$\text{velocity } \alpha = \sqrt{\left[\left(k + \frac{4}{3}\mu\right)/\rho\right]}$$

For granite, $\alpha = 5{\cdot}5\,\mathrm{km}/$; for water, $\alpha = 1{\cdot}5\,\mathrm{km/s}$. For S waves

$$\text{velocity } \beta = \sqrt{(\mu/\rho)}$$

For granite, $\beta = 3{\cdot}0\,\mathrm{km}$; for water, $\beta = 0{\cdot}0\,\mathrm{km/s}$.

The velocities of the Love waves and Rayleigh waves propagated along the surface of an elastic solid body are given as follows. For Rayleigh waves

$$\text{velocity } c_r < 0{\cdot}92\beta$$

where $\beta$ is the S wave velocity of the rock. For Love waves (layered solid)

$$\text{velocity } \beta_1 < c_L < \beta_2$$

where $\beta_1$ and $\beta_2$ are S wave velocities of the surface and deeper layers respectively.

## Recording of earthquakes

Ground motion is measured by seismometers that can detect very small vibrations but go off the scale in strong motion. Strong motion seismometers are usually set to operate only when triggered by accelerations above a certain level. The results from ordinary seismometers are used primarily in the study of earthquake mechanisms, while those from strong motion seismometers are of importance in establishing design critera and, when mounted on buildings, the behaviour of structures during earthquakes. The basic design concepts of seismometers are given in chapter 4.

## Magnitude and intensity of earthquakes

An earthquake disturbance, at its source, is measured by *magnitude* on the Richter scale, ranging from 0 to 8·9, which is the largest measured to date. The calculation of magnitude is based on seismometer measurements and is a measure of the strain released at source. The Richter scale is logarithmic, so that a magnitude = 5 event may be a minor one, while a magnitude = 6·5 event may be a major one with a release of energy at source 31·6 times that of an earthquake of magnitude 5. The determination of magnitude is shown in Fig. 12.3. To engineers the most important observation is that earthquakes less than 5 are not likely to cause any structural damage.

The effect of an earthquake diminishes with distance, so that the effect at a particular location is not defined by the magnitude. This is measured in terms of *intensity*, commonly on the modified *Mercalli scale*, although there are a number of other scales. The Mercalli scale is not a precise one, being based on subjective factors such as the type of building damage and whether the shock is felt by people in cars. The scale grades events from 1 (not felt) to 12 (damage nearly total).

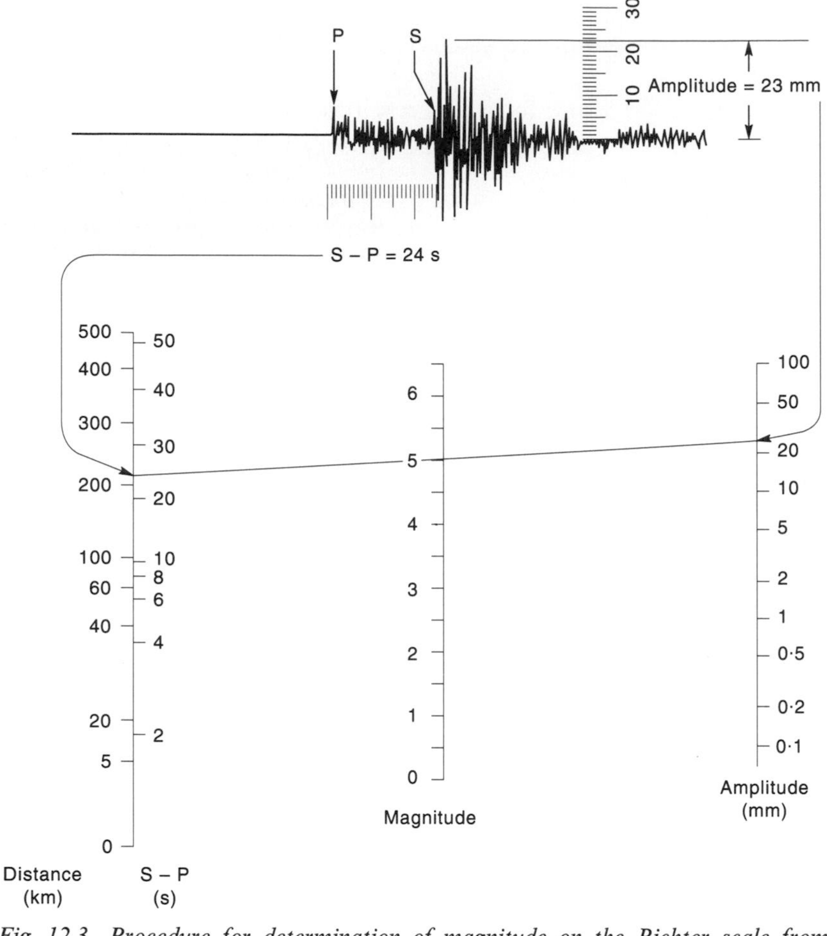


*Fig. 12.3. Procedure for determination of magnitude on the Richter scale from seismometer recording: first, the distance to the focus is measured using the time interval between the S and P waves ($S - P = 24$ s); then the height of the maximun wave is measured on the seismogram (23 mm); finally, a straight edge is placed between the appropriate points on the distance and amplitude scales to measure magnitude $M_L = 5{\cdot}0$*

## Influence of magnitude and surface geology on characteristics of earthquakes

An earthquake may have a duration of up to a minute or so, with the interaction of various types of waves depicted in Fig. 12.3 together with the effect of refraction at discontinuities producing extremely complex wave forms. Seismologists and engineers have developed formulae relating all the principal parameters of earthquake transmission such as duration, dominant period and attenuation. Earthquake waves, however, are affected by both soil conditions and topography, and practising engineers should bear in mind that real life results show a considerable amount of scatter on each side of these median rules. An extensive treatment of seismic risk is given by Lomnitz & Rosenbleuth (1976). However, the following nonquantitative rules are worth remembering

- the predominant period increases with increasing magnitude, distance and depth of alluvium (Figs 12.4 and 12.5)
- the peak acceleration increases with increasing magnitude and soil stiffness, and decreases with increasing distance (Figs 12.6 and 12.7)
- the duration increases with increasing magnitude.

The dominant frequency of the ground varies from site to site and from region to region, and is a function of the magnitude of the earthquake, the distance from the causative fault as shown in Fig. 12.4, and the depth

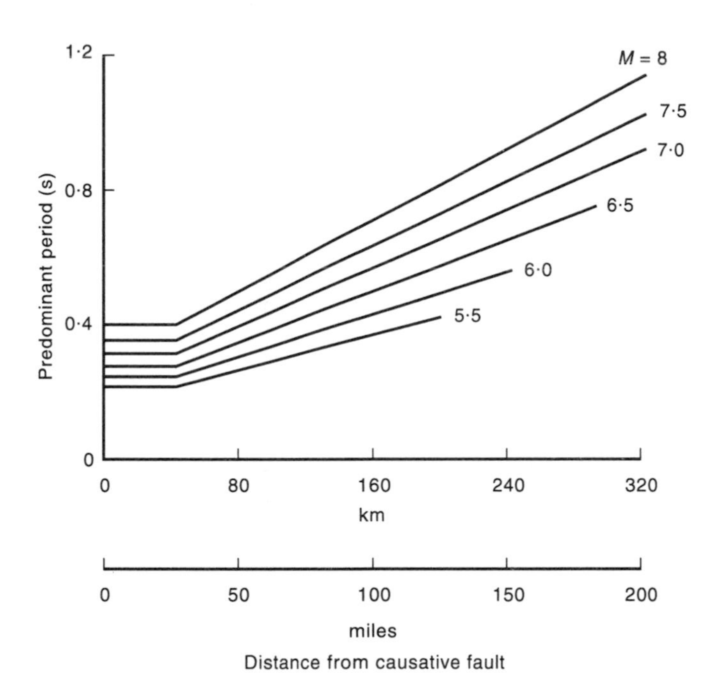


*Fig. 12.4. Predominant period–distance relationship for maximum acceleration in rock (after Seed, 1968)*

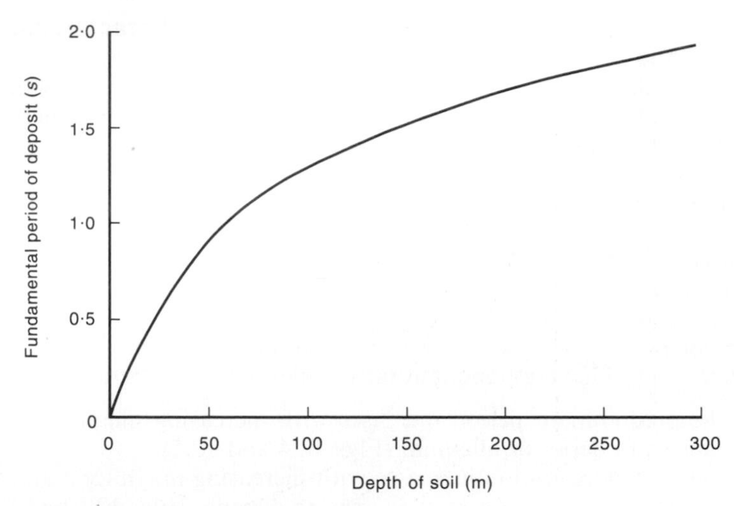


*Fig. 12.5. Relationship between the natural period of the soil and alluvium depth (after Seed, 1970)*

of alluvium as shown in Fig. 12.5. Soft surface material will behave similarly to jelly on a shaking table and can cause considerable amplification of the base rock motion. Thus an amplification factor of 20, as well as considerable modification of the predominant period, has been predicted for the San Francisco Bay mud. In California the dominant frequencies range from 3·3 to 4·0 Hz. As in the Caribbean, the dominant frequencies are lower and range from 2·5 to 2·8 Hz. Earthquakes with much lower frequencies have, however, been recorded. Thus the dominant frequency during the San Salvador earthquake in 1986 was 1·48 Hz, and that of the Mexico City earthquake in 1985 was as low as 0·41 Hz. This wide variation in the dominant frequency of the ground should be borne in mind by designers and writers of codes of practice, who for economical reasons attempt to simplify design proceedures.

Although very weak soils can produce substantial amplification of the base rock vibration for earthquakes of low intensity, in major shaking the effect is limited by shear failures in the soil. This produces an effective cut-off point in the transmission of large shocks. Weak soils have a bad reputation in earthquakes, but this is due to consolidation, liquefaction and other effects producing large displacements. The estimation of the effect of site geology on ground motion is complex and the literature is extensive.

In the same way that a building may be regarded as a dynamic system shaken at its base, the surface alluvium, extending in depth from a few metres to hundreds of metres, may be considered as a dynamic system shaken by the motion of the underlying rock. Obviously this argument can be extended to a combination of two dynamic systems. This is desirable in the case of most structures and necessary in the case of large rigid structures such as nuclear reactors. In the case of medium, relatively

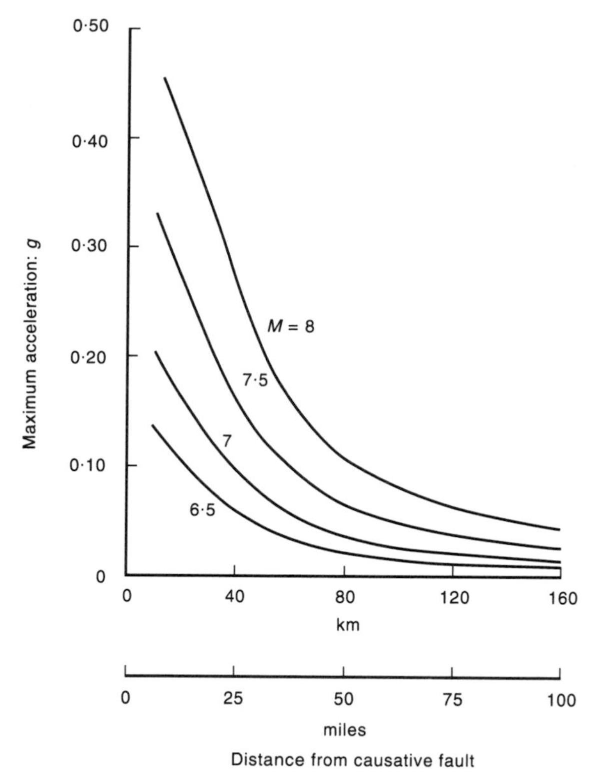


*Fig. 12.6. Acceleration–magnitude–distance relationship (after Seed et al., 1976)*

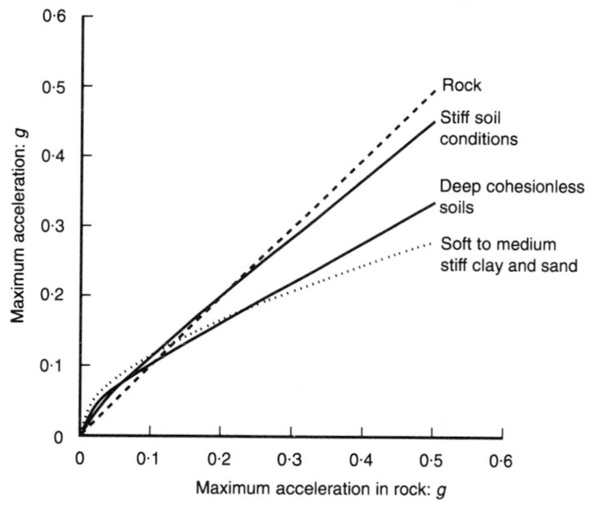


*Fig. 12.7. Effect of local soil conditions on peak acceleration (after Seed et al., 1976): the relationships shown are based on a ground acceleration of 0·3g and are extrapolated from a database*

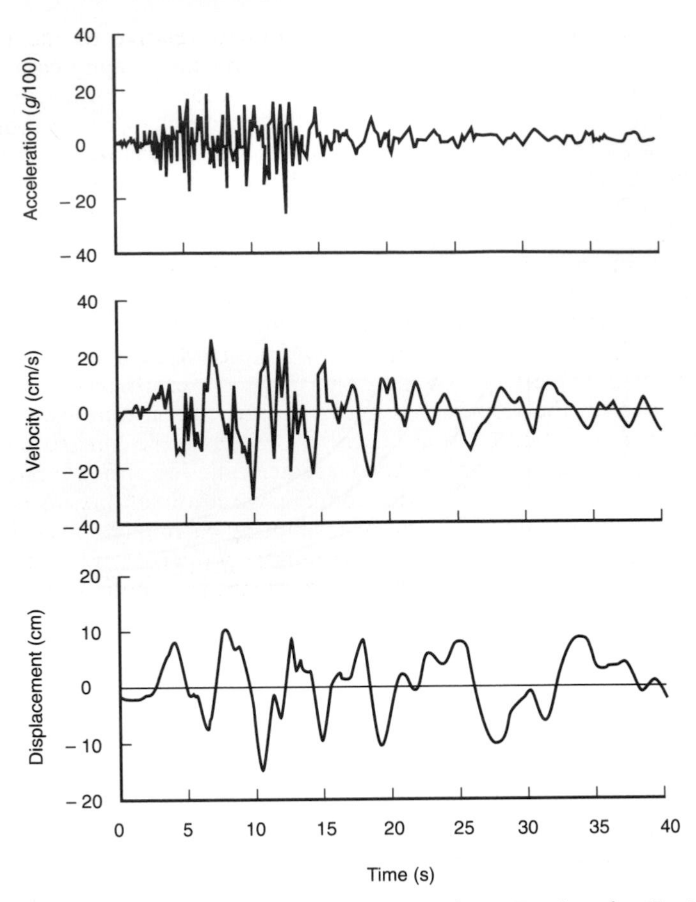


*Fig. 12.8. Strong motion earthquake records (from Earthquake Engineering Laboratory, 1980)*

flexible buildings, however, the dynamics of the soil and the building is usually considered separately. In the case of piled foundations, special considerations are needed, as the piles modify the surface response.

Although damage such as consolidation, liquefaction, landslides, avalanches and shear failures can be done to soils by earthquake motion, it is assumed in this book that the structures are sited on ground that will substantially retain its integrity during an earthquake.

## Representation of ground motion

At any point on the earth's surface, earthquake motion will comprise three translational components, two rocking components and one horizontal torsional component. Earthquakes are commonly classified by their intensity and peak acceleration, although these are only an approximate measure of their capacity for doing damage. Other important factors are the frequency content, duration, peak velocity and peak displacement. Of these, the frequency content relative to the natural frequencies of the structures is generally the most significant. Thus earthquakes with the main energy concentration in frequency bands corresponding to dominant structural frequencies can cause more damage than earthquakes with greater peak accelerations but with energy concentrated in different frequency bands from the structural ones.

Information about ground motion can be presented in the time domain in the form of acceleration, velocity and displacement histories (Fig. 12.8), and in the frequency domain in the form of response or power spectra. Response spectra are commonly used in design and form a convenient method of establishing suitable specifications for linear structures. Their values at any given frequency represent the peak response of a single-DOF oscillator to a specific earthquake record. In order to predict the response of nonlinear structures, time histories are needed. Strong motion histories, if not available, can be constructed from spectral density functions or autocovariance functions for ground accelerations, and require information on the variation of the variance of acceleration with time. Methods for generating earthquake histories and families of correlated earthquakes with similar properties are presented in chapter 13, together with methods for generating spatially correlated wind histories.

## References

Bolt, B. A. *Earthquakes, a primer*. W. H. Freeman, San Francisco, 1978.

Eiby, G. A. *Earthquakes*. Heinemann, London, 1980.

Key, D. E. *Earthquake design practice for buildings*. Thomas Telford, London, 1988.

Lomnitz, C. & Rosenbleuth, E. *Seismic risk and engineering decisions*. Elsevier, Amsterdam, 1976.

Seed, H. B. *Characteristics of rock motion during earthquakes*. University of California at Berkeley, Report EERC 63–5, 1868.

Seed, H. B. & Idriss, I. M. *Solid moduli and damping factors for dynamic response analysis*. University of California at Berkeley, Earthquake Engineering Research Center, Report EERC 70–10, 1970.

Seed. H. B., Murarka, R., Lysmer, J. & Idriss, I. M. Relationships of maximum acceleration, maximum velocity, distance from source, and local site conditions for moderately strong earthquakes. *Bull. Seismol. Soc. Am.*, 1976, **66**, No. 4, 1323–1342.

Seed, H. B. & Idriss, I. M. *Ground motion and ground liquefaction during earthquakes*. Earthquake Engineering Research Institute, Berkeley, 1982.

Earthquake Engineering Research Laboratory. *Earthquake strong motion records*. EERL, Pasadena, Report No. 80–01, 1980.

# 13. Dynamic response to earthquakes: frequency domain analysis

## Introduction

The most common form of data bank used in the design of structures to resist earthquakes is response spectra. As mentioned in chapter 12, a response spectrum is a curve that shows how the maximum response, velocity or acceleration of oscillators with the same damping ratio, but with different natural frequencies, respond to a specified earthquake. Another approach, which is gradually gaining ground, is the use of power spectral density functions for the ground acceleration caused by earthquakes. Their construction and use are similar to those of such functions as wind engineering. Both the above methods are of interest to the practising engineer because they are eminently very useful, as demonstrated below, in connection with the mode superposition method introduced in chapter 8.

Finally there are the time domain methods, which are generally only used in the case of nonlinear structures. One such method is described in chapter 6 and is briefly, for convenience and completeness, repeated in this chapter. The problem with time domain methods is that it is not sufficient to calculate the response to only one earthquake, as no two earthquakes at the same site are likely to be identical. It is therefore necessary to generate a family of earthquakes with properties appropriate for a given area. How many such earthquakes need to be included in any design calculation is a matter of experience and recommendations in design codes.

## Construction of response spectra

The linear acceleration method, Wilson $\theta$-method and the Newton $\beta$-method given in chapter 6, as well as the Duhamel's integral (Clough & Penzien, 1975; Key, 1988), may be used to calculate the maximum displacement, velocity and acceleration of a single oscillator for a given earthquake record such as that shown in Fig. 1.3 or Fig. 12.8.

From eq. (4.52), the equation of motion for a one DOF system, relative to the support, when subjected to a ground acceleration $\ddot{x}_g(t)$, can be written as

$$M\ddot{x} + C\dot{x} + Kx = M\ddot{x}_g(t) \tag{13.1}$$

or

$$\ddot{x} + 2\xi\omega_n\dot{x} + \omega_n{}^2x = \ddot{x}_g(t) \tag{13.2}$$

Thus by calculating the maximum response of oscillators with different frequencies, but with the same damping, it is possible to construct a response spectrum in the frequency domain for oscillators with the same damping ratio. By repeating this process for oscillators with different damping ratios, it is possible to construct a number of response spectra for the same record. An example of a response spectrum for the record is shown in Fig. 13.1.

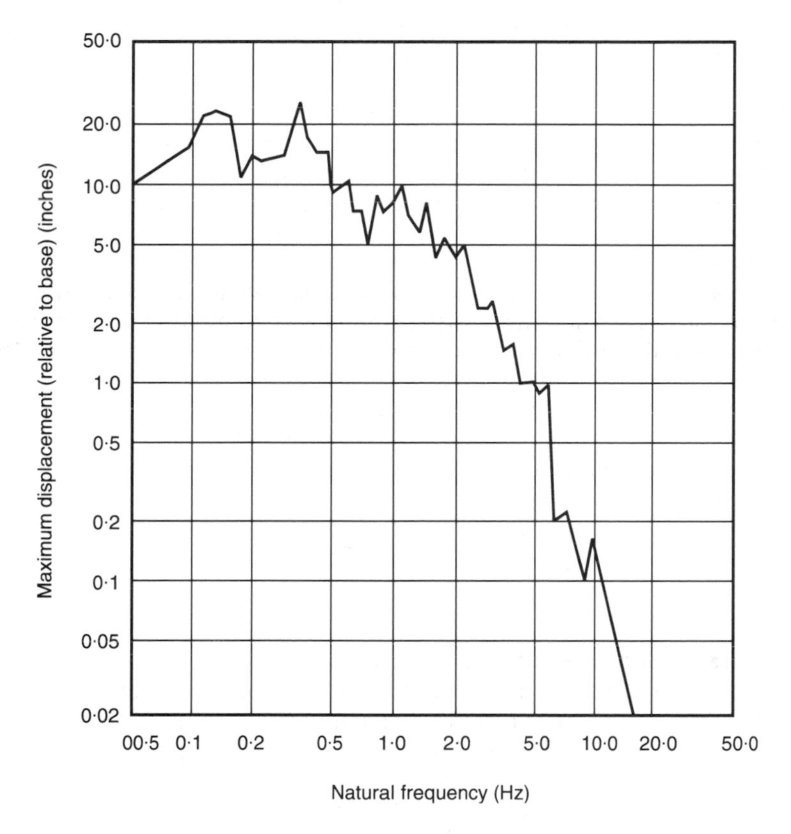


*Fig. 13.1. Displacement response spectra for elastic one DOF oscillator subjected to the ground motion of the 1940 El Centro earthquake (from Blum et al., 1961)*

**Tripartite response spectra**

Consider the single DOF mass–spring system shown in Fig. 13.2 when subjected to a support displacement $y_g$ and a corresponding support velocity $\dot{y}_g$. The equation of motion for this form of excitation is

$$M\ddot{y} + C(\dot{y} - \dot{y}_g) + K(y - y_g) = 0 \quad (13.3)$$

If the relative displacement and relative velocity are denoted $u$ and $\dot{u}$ respectively, then eq. (13.3) may be written as

$$M\ddot{y} + C\dot{u} + ku = 0 \quad (13.4)$$

and if the damping is ignored

$$M\ddot{y} + Ku = 0 \quad (13.5)$$

or

$$\ddot{y} + \omega_n{}^2 u = 0 \quad (13.6)$$

From eq. (13.6), the absolute acceleration is proportional to the relative displacement. Thus the maximum absolute acceleration $\ddot{y}_{max}$ is proportional to the maximum relative displacement $u_{max}$, i.e.

$$\ddot{y}_{max} = \omega_n{}^2 u_{max} \quad (13.7)$$

If damping is taken into account, and it is assumed that the relative velocity $\dot{u} = 0$ when the relative displacement is a maximum and equal to $u_{max}$, eq. (13.7) is again obtained. This expression for the maximum acceleration is, purely by coincidence, the same as for simple harmonic motion (SHM). The fictitious velocity associated with an apparent SHM is referred to as *pseudovelocity*. The maximum value of pseudovelocity is $\dot{u}_{max}$. Thus

$$\dot{u}_{max} = \omega_n u_{max} = 2\pi f_n u_{max} \quad (13.8)$$

$$\dot{u}_{max} = \ddot{y}_{max}/\omega_n = \ddot{y}_{max}/2\pi f_n \quad (13.9)$$

Taking the logarithm of both sides of eqs (13.8) and (13.9) yields

$$\log \dot{u}_{max} = \log f_n + \log(2\pi u_{max}) \quad (13.10)$$

$$\log \dot{u}_{max} = -\log f_n + \log(\ddot{y}_{max}/2\pi) \quad (13.11)$$

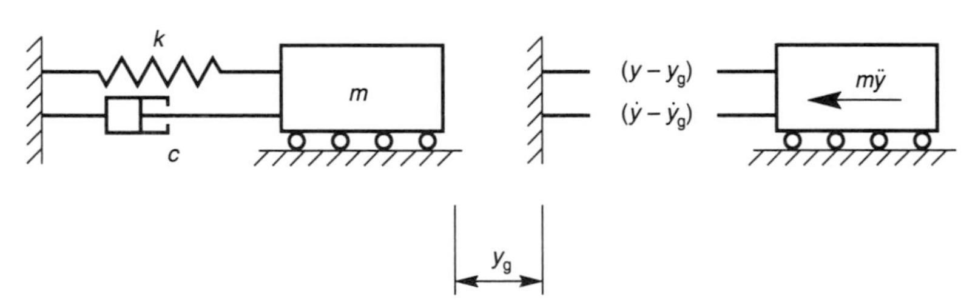


*Fig. 13.2. Single-DOF oscillator subjected to support motion*

For a constant value of $u_{\max}$ eq. (13.10) is a straight line plot of $\log \dot{u}_{\max}$ against $\log f_n$ with a slope of 45°, and for a constant value of $\ddot{y}_{\max}$ eq. (13.11) represents a straight line plot of $\log \dot{u}_{\max}$ against $\log f_n$ with a slope of 135°. Thus it is possible to plot the maximum spectral response $u_{\max}$, spectral acceleration $\ddot{y}_{\max}$ and spectral pseudovelocity $\dot{u}_{\max}$ on the same graph, as shown in Fig. 13.3. The graph shows the maximum predicted responses to the El Centro earthquake of oscillators with four levels of damping and with increasing natural frequencies.

The spectra shown in Fig. 13.3 are raw, and it is usual to smoothen them for design purposes, as it is highly unlikely that the duration, peak acceleration, frequency content and energy distribution of future earthquakes in the same area will be the same as those of previously recorded ones. Thus in design it is usual to employ consolidated response spectra normalized to a peak acceleration of $1{\cdot}0g$ with corresponding maximum values for ground displacement and velocity (Harris, 1988). One such set of spectra is shown in Fig. 13.4, where the maximum ground displacement is 36 inches and the maximum pseudo-ground velocity is 48 inches/s. These values are consistent with a motion that is more intense than those generally considered in earthquake engineering. They are, however, of proportional magnitudes deemed satisfactory for the design of most linear elastic structures.

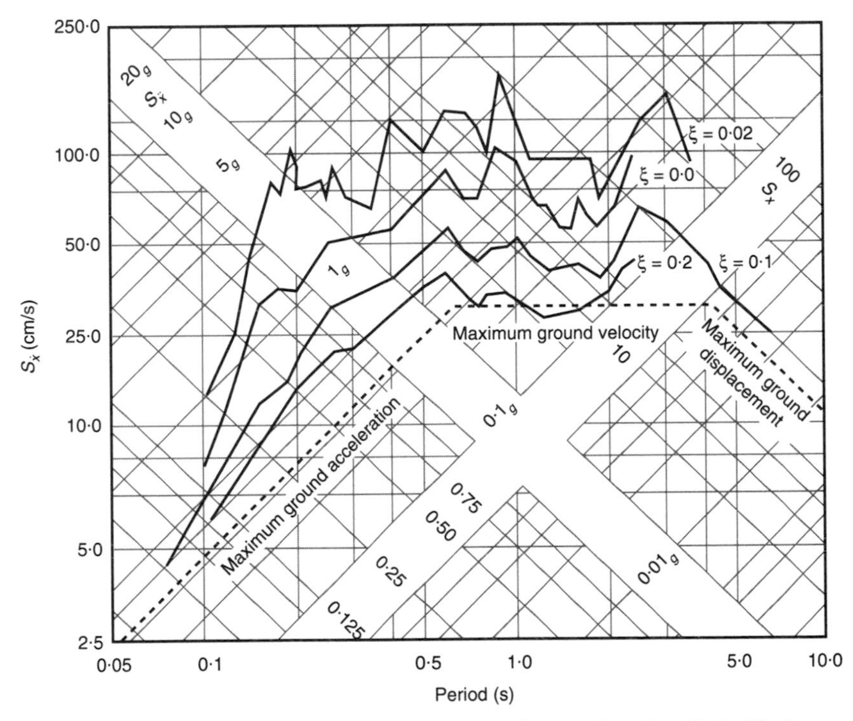


*Fig. 13.3. Response spectra for one DOF oscillators for the 1940 El Centro earthquake (from Blum et al., 1961)*

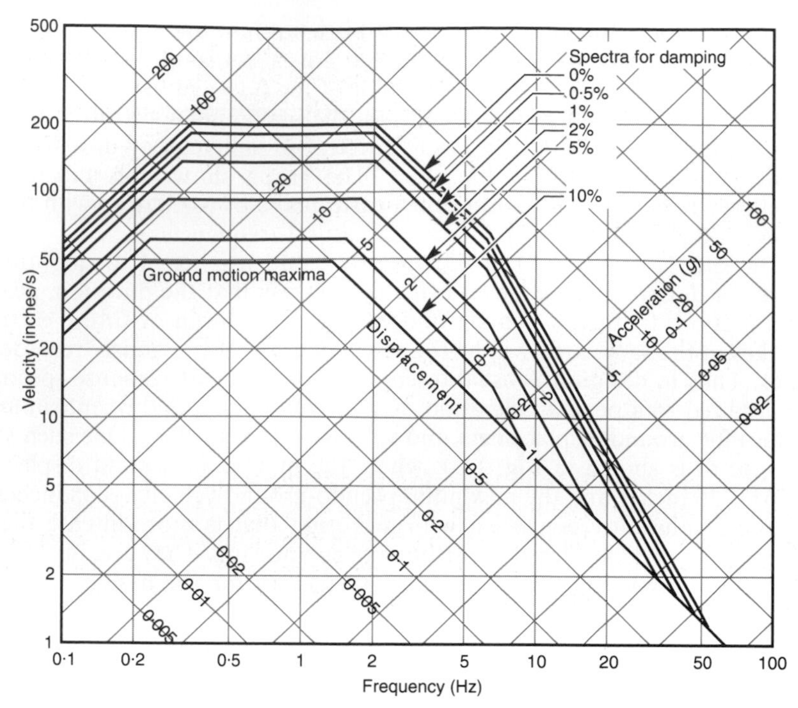


*Fig. 13.4. Basic design spectra normalized to 1·0g (from Newmark & Hall, 1973) (1 inch = 25·4 mm)*

Figure 13.4 shows an additional six curves to the one assumed for the ground motion. These are the corresponding response curves for single oscillators with damping ratios ranging from 0% to 10% of critical. For assumed peak ground accelerations different from 1·0*g* the values obtained from the graph need to be scaled linearly.

## Use of response spectra

Values for displacements, velocities and accelerations are obtained from Fig. 13.4 by taking the antilogarithm of the ratio of the coordinates of a variable (measured in millimetres, centimetres or inches) and the appropriate scaling factor. The coordinate magnitude for each variable is measured with the value 1·0 as origin. The position on the graph of a given frequency is found by taking the logarithm of the frequency and multiplying it by the scaling factor for frequencies.

The scaling factors for the variables in Fig. 13.4 (with the coordinates in cm) are determined as follows

frequencies: $\log 100 - \log 1 = 2 = 6{\cdot}20\,\text{cm}$, scaling factor $S_{\text{f}} = 3{\cdot}15$

displacements: $\log 100 - \log 1 = 2 = 4{\cdot}40\,\text{cm}$, scaling factor $S_{\text{d}} = 2{\cdot}20$

velocities: $\log 100 - \log 1 = 2 = 6{\cdot}10\,\text{cm}$, scaling factor $S_{\text{v}} = 3{\cdot}05$

accelerations: $\log 100 - \log 1 = 2 = 4{\cdot}50\,\text{cm}$, scaling factor $S_{\text{v}} = 2{\cdot}25$

**Example 13.1** The top of a tall building, which has a first natural frequency of 1·0 Hz and a first modal damping ratio of 1·0% of critical, is modelled as a mass–spring oscillator. Use the appropriate response spectrum in Fig. 13.4 to predict the maximum lateral displacement, pseudovelocity and acceleration of the roof of the structure that will be caused by an earthquake having an assumed peak acceleration of 0·3$g$.

$$x_{max} = 0{\cdot}3 \times 10^{3{\cdot}05/2{\cdot}20} = \underline{7{\cdot}30\text{ inches}}$$

$$\dot{x}_{max} = 0{\cdot}3 \times 10^{6{\cdot}73/3{\cdot}05} = \underline{48{\cdot}27\text{ inches/s}}$$

$$\ddot{x}_{max} = 0{\cdot}3 \times 10^{0{\cdot}86/2{\cdot}25} = \underline{0{\cdot}723g}$$

## Response of multi-DOF systems to earthquakes

Let a two-storey shear building subjected to a ground motion $x_g(t) = x_g$ be represented by the mass–spring system shown in Fig. 13.5, where the displacements $y_1$ and $y_2$ of the two masses $m_1$ and $m_2$ are relative to a fixed point. From Newton's law of motion

$$M_1\ddot{y}_1 + C_1(\dot{y}_1 - \dot{x}_g) + C_2(\dot{y}_1 - \dot{y}_2) + K_1(y_1 - x_g) + K_2(y_1 - y_2) = 0 \quad (13.12a)$$

$$M_2\ddot{y}_2 - C_2(\dot{y}_1 - \dot{y}_2) - K_2(y_1 - y_2) = 0 \quad (13.12b)$$

Now let

$$x_1 = y_1 - x_g$$
$$\dot{x}_1 = \dot{y}_1 - \dot{x}_g$$
$$\ddot{x}_1 = \ddot{y}_1 - \alpha\ddot{x}_g(t)$$
$$x_2 = y_2 - x_g$$
$$\dot{x}_2 = \dot{y}_2 - \dot{x}_g$$
$$\ddot{x}_2 = \ddot{y}_2 - \alpha\ddot{x}_g(t)$$

where $\ddot{x}_g(t)$ is the acceleration history of an earthquake normalized to a peak acceleration of 1·0$g$, and $\alpha$ is a constant that defines the magnitude

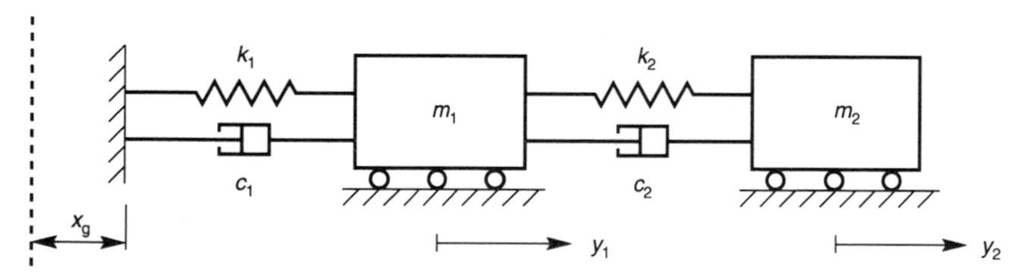


*Fig. 13.5. Mass–spring model of two-storey shear structure*

of the peak acceleration of the real quake. Substitution for $y$, $\dot{y}$, and $\ddot{y}$ in eqs (13.12a) and (13.12b) yields

$$M_1\ddot{x}_1 + (C_1 + C_2)\dot{x}_1 - C_2\dot{x}_2 + (K_1 + K_2)x_1 - K_2x_2 = M_1\alpha\ddot{x}_g(t) \tag{13.13a}$$

$$M_2\ddot{x}_2 - C_2\dot{x}_1 + C_2\dot{x}_2 - K_2x_1 + K_2x_2 = M_2\alpha\ddot{x}_g(t) \tag{13.13b}$$

Equations (13.13a) and (13.13b) may be written in matrix form as

$$\begin{bmatrix} M_1 & 0 \\ 0 & M_2 \end{bmatrix}\begin{bmatrix} \ddot{x}_1 \\ \ddot{x}_2 \end{bmatrix} + \begin{bmatrix} C_1 + C_2 & -C_2 \\ -C_2 & C_2 \end{bmatrix}\begin{bmatrix} \dot{x}_1 \\ \dot{x}_2 \end{bmatrix} + \begin{bmatrix} K_1 + K_2 & -K_2 \\ -K_2 & K_2 \end{bmatrix}\begin{bmatrix} x_1 \\ x_2 \end{bmatrix}$$
$$= \begin{bmatrix} M_1 & 0 \\ 0 & M_2 \end{bmatrix}\begin{bmatrix} \alpha\ddot{x}_g(t) \\ \alpha\ddot{x}_g(t) \end{bmatrix} \tag{13.14}$$

or as

$$\mathbf{M}\ddot{\boldsymbol{x}} + \mathbf{C}\dot{\boldsymbol{x}} + \mathbf{K}\boldsymbol{x} = \mathbf{M}\alpha\ddot{\boldsymbol{x}}_g(\boldsymbol{t}) \tag{13.15}$$

Equation (13.15) is obviously also the general form for the equation of motion for any linear multi-DOF structure subjected to a support motion $x_g(t) = x_g$ with acceleration $\alpha\ddot{x}_g(t)$.

In order to solve the system of equations given by eq. (13.15) and to predict the response of an $N$ DOF system to a given support motion let, as in chapters 8 and 11

$$\boldsymbol{x} = \boldsymbol{Zq}$$
$$\dot{\boldsymbol{x}} = \boldsymbol{Z}\dot{\boldsymbol{q}}$$
$$\ddot{\boldsymbol{x}} = \boldsymbol{Z}\ddot{\boldsymbol{q}}$$

where

$$\mathbf{Z} = [\mathbf{Z}_1, \mathbf{Z}_2, \ldots, \mathbf{Z}_i, \ldots, \mathbf{Z}_N]$$

is the normalized modeshape matrix associated with eq. (13.15). Substitution of the above expressions for $\boldsymbol{x}$, $\dot{\boldsymbol{x}}$ and $\ddot{\boldsymbol{x}}$ into eq. (13.15) and postmultiplication of each term in the equation by $\mathbf{Z}^{\mathrm{T}}$ yield

$$\mathbf{Z}^{\mathrm{T}}\mathbf{MZ}\ddot{\boldsymbol{q}} + \mathbf{Z}^{\mathrm{T}}\mathbf{CZ}\dot{\boldsymbol{q}} + \mathbf{Z}^{\mathrm{T}}\mathbf{KZ}\boldsymbol{q} = \mathbf{Z}^{\mathrm{T}}\mathbf{M}\alpha\ddot{\boldsymbol{x}}_{g(t)} \tag{13.16}$$

From the orthogonality properties of normalized eigenvectors considered in chapter 7

$$\mathbf{Z}^{\mathrm{T}}\mathbf{MZ} = \mathbf{I}$$
$$\mathbf{Z}^{\mathrm{T}}\mathbf{CZ} = 2\xi\omega$$
$$\mathbf{Z}^{\mathrm{T}}\mathbf{KZ} = \omega^2$$

Substitution of these expressions for the matrix products into eq. (13.16) will uncouple the equations of motion and yield

$$\ddot{q} + 2\xi\omega\dot{\boldsymbol{q}} + \omega^2\boldsymbol{q} = \mathbf{Z}^{\mathrm{T}}\mathbf{M}\alpha\ddot{\boldsymbol{x}}_g(\boldsymbol{t}) \tag{13.17}$$

where $2\xi\omega$ and $\omega^2$ are diagonal matrices. Equation (13.17) may also be written as

$$\begin{aligned}
\ddot{q}_1 + 2\xi_1\omega_1\dot{q}_1 + \omega_1{}^2 q_1 &= \mathbf{Z}_1{}^{\mathrm{T}}\mathbf{M}\boldsymbol{\alpha}\ddot{x}_{\mathrm{g}}(t) \\
\ddot{q}_2 + 2\xi_2\omega_2\dot{q}_2 + \omega_2{}^2 q_2 &= \mathbf{Z}_2{}^{\mathrm{T}}\mathbf{M}\boldsymbol{\alpha}\ddot{x}_{\mathrm{g}}(t) \\
&\cdots\cdots \\
&\cdots\cdots \\
\ddot{q}_i + 2\xi_i\omega_i\dot{q}_i + \omega_i{}^2 q_i &= \mathbf{Z}_i{}^{\mathrm{T}}\mathbf{M}\boldsymbol{\alpha}\ddot{x}_{\mathrm{g}}(t) \\
&\cdots\cdots \\
&\cdots\cdots \\
\ddot{q}_N + 2\xi_N\omega_N\dot{q}_N + \omega_N{}^2 q_N &= \mathbf{Z}_N{}^{\mathrm{T}}\mathbf{M}\boldsymbol{\alpha}\ddot{x}_{\mathrm{g}}(t)
\end{aligned} \tag{13.18}$$

where the product $\mathbf{Z}_i{}^{\mathrm{T}}\mathbf{M}$ is referred to as the $i$th *participation vector*. Since the mass in each of the above equivalent one DOF systems is unity, it follows that the equivalent ground acceleration in the generalized coordinate system $\ddot{q}_{\mathrm{g}_i}(t)$ is equal to the ground acceleration vector $\boldsymbol{\alpha}\ddot{x}_{\mathrm{g}}$ postmultiplied by the participation vector $\mathbf{Z}_i{}^{\mathrm{T}}\mathbf{M}$. The maximum value of $\ddot{q}_{\mathrm{g}_i}(t)$ occurs when $\ddot{x}_{\mathrm{g}}(t)$ is equal to $1{\cdot}0g$. Thus

$$\ddot{q}_{\mathrm{g}_i}(t) = \mathbf{Z}_i{}^{\mathrm{T}}\mathbf{M}\boldsymbol{\alpha}\ddot{x}_{\mathrm{g}}(t) \tag{13.19}$$

$$\ddot{q}_{\mathrm{g}_i,\,\max} = \mathbf{Z}_i{}^{\mathrm{T}}\mathbf{M}\boldsymbol{\alpha}\boldsymbol{g} \tag{13.20}$$

## Deterministic response analysis using response spectra

The $i$th generalized modal equation, eq. (13.18), can be considered as the equation of motion of a one DOF oscillator with unit mass subjected to a maximum ground acceleration $\ddot{q}_{\mathrm{g}_i,\max} = \mathbf{Z}_i{}^{\mathrm{T}}\mathbf{M}\boldsymbol{\alpha}\boldsymbol{g}$. The maximum response of this system can therefore be found by use of a response spectrum based on a damping ratio with value $\xi_i$ from which the response $\tilde{q}_{i,\max}$ corresponding to the frequency $\omega_i$ can be found. As the spectra in Fig. 13.4 are normalized to a peak acceleration of $1{\cdot}0g$, it follows that

$$q_{i,\,\max} = \mathbf{Z}_i{}^{\mathrm{T}}\mathbf{M}\boldsymbol{\alpha}\tilde{q}_{i,\max} \tag{13.21}$$

Thus

$$\boldsymbol{q}_{\max} = \mathbf{Z}_i{}^{\mathrm{T}}\mathbf{M}\boldsymbol{\alpha}\tilde{\boldsymbol{q}}_{\max} \tag{13.22}$$

and hence

$$\boldsymbol{x}_{\max} = \mathbf{Z}\boldsymbol{q}_{\max} = \mathbf{Z}\mathbf{Z}^{\mathrm{T}}\mathbf{M}\boldsymbol{\alpha}\tilde{\boldsymbol{q}}_{\max} \tag{13.23}$$

Equation (13.23) assumes that the maximum responses in each of the modes will occur simultaneously and relative to each other as in the modeshape matrix. As this is highly unlikely, the above expression for the maximum response vector is modified for design purposes, and each element in the response vector $\boldsymbol{x}_{\max}$ is recalculated as the square root of the sum of the squares of the contribution from each mode. Thus

$$x_{\mathrm{r}} = \left\{(Z_{\mathrm{r}1}q_1)^2 + (Z_{\mathrm{r}2}q_2)^2 + \ldots + (Z_{\mathrm{r}N}q_N)^2\right\}^{1/2} \tag{13.24}$$

or

$$x_{\mathrm{r}} = \sqrt{\left[\sum_{i=1}^{N}(Z_{\mathrm{r}i}q_i)^2\right]} \tag{13.25}$$

**Example 13.2** Use the response spectra in Fig. 90 to calculate the maximum and maximum modified displacements and accelerations of the floors in the three-storey shear structure shown in Fig. 2.14, if the building is subjected to an earthquake with a peak acceleration equal to 0·25$g$. Assume the damping in the first mode to be 2·0% of critical and that in the second and third mode to be 1·0% of critical. The stiffness matrix, mass matrix, natural frequencies and normalized modeshape matrix for the structure are as follows

$$\mathbf{K} = \begin{bmatrix} 114\,596{\cdot}3 & -49\,112{\cdot}7 & 0 \\ -49\,112{\cdot}7 & 81\,854{\cdot}5 & -32\,741{\cdot}8 \\ 0 & -32\,741{\cdot}8 & 32\,741{\cdot}8 \end{bmatrix} \mathrm{kN/m}$$

$$\mathbf{M} = \begin{bmatrix} 61{\cdot}16208 & 0 & 0 \\ 0 & 40{\cdot}77472 & 0 \\ 0 & 0 & 20{\cdot}38736 \end{bmatrix} \times 10^3\,\mathrm{kg}$$

$$\boldsymbol{\omega} = \begin{bmatrix} 18{\cdot}659513 \\ 41{\cdot}961902 \\ 58{\cdot}121498 \end{bmatrix} \mathrm{rad/s}$$

$$\boldsymbol{\omega}^2 = \begin{bmatrix} 348{\cdot}1774 \\ 1760{\cdot}8012 \\ 3378{\cdot}1085 \end{bmatrix} \mathrm{rad^2/s^2}$$

$$\mathbf{f} = \begin{bmatrix} 2{\cdot}9697537 \\ 6{\cdot}6784441 \\ 9{\cdot}6869163 \end{bmatrix} \mathrm{Hz}$$

$$\mathbf{Z} = \begin{bmatrix} 1{\cdot}7454 & 3{\cdot}0818 & 1{\cdot}9498 \\ 3{\cdot}3157 & 0{\cdot}4330 & -3{\cdot}6531 \\ 4{\cdot}2336 & -4{\cdot}4926 & 3{\cdot}3105 \end{bmatrix} \times 10^{-3}$$

From eq. (13.20), the peak acceleration vector in the generalized coordinate system is given by

$$\mathbf{Z}^{\mathrm{T}}\mathbf{M}\boldsymbol{\alpha}g = \begin{bmatrix} 1{\cdot}7454 & 3{\cdot}3157 & 4{\cdot}2336 \\ 3{\cdot}0818 & 0{\cdot}4330 & -4{\cdot}4926 \\ 1{\cdot}9498 & -3{\cdot}6531 & 3{\cdot}3105 \end{bmatrix} \begin{bmatrix} 61{\cdot}16208 & 0 & 0 \\ 0 & 40{\cdot}77472 & 0 \\ 0 & 0 & 20{\cdot}38736 \end{bmatrix} \begin{bmatrix} 0{\cdot}25g \\ 0{\cdot}25g \\ 0{\cdot}25g \end{bmatrix}$$

Hence

$$\mathbf{Z}^{\mathrm{T}}\mathbf{M}\boldsymbol{\alpha}g = \begin{bmatrix} 106{\cdot}7522 & 135{\cdot}1967 & 86{\cdot}3119 \\ 188{\cdot}4893 & 17{\cdot}6554 & -91{\cdot}5923 \\ 119{\cdot}2538 & -148{\cdot}9541 & 67{\cdot}4924 \end{bmatrix} \begin{bmatrix} 0{\cdot}25g \\ 0{\cdot}25g \\ 0{\cdot}25g \end{bmatrix}$$

$$= \begin{bmatrix} 82{\cdot}0652g \\ 28{\cdot}6381g \\ 9{\cdot}4480g \end{bmatrix}$$

Thus the decoupled equations of motion or generalized modal equations at the time when the peak acceleration occurs may be written as

$$\ddot{q}_1 + 2 \times 0{\cdot}02 \times 18{\cdot}659513\dot{q}_1 + 348{\cdot}1774q_1 = 82{\cdot}0652g$$
$$\ddot{q}_2 + 2 \times 0{\cdot}01 \times 41{\cdot}961902\dot{q}_2 + 1760{\cdot}8012q_2 = 26{\cdot}6381g$$
$$\ddot{q}_3 + 2 \times 0{\cdot}01 \times 58{\cdot}121498\dot{q}_3 + 3378{\cdot}1085q_3 = 9{\cdot}4480g$$

From the response spectra in Fig. 13.4 for oscillators with 1·0% and 2·0%, remembering that 1·0 inches = 0·0254 m, the following values are calculated for the generalized displacement coordinates

$$q_{1,\max} = 82{\cdot}0652 \times 10^{1{\cdot}80/2{\cdot}75} \times 0{\cdot}0254 = 9{\cdot}4089\,\mathrm{m}$$
$$q_{2,\max} = 28{\cdot}6381 \times 10^{0{\cdot}09/2{\cdot}75} \times 0{\cdot}0254 = 0{\cdot}7843\,\mathrm{m}$$
$$q_{3,\max} = 9{\cdot}4480 \times 10^{-1{\cdot}15/2{\cdot}75} \times 0{\cdot}0254 = 0{\cdot}0916\,\mathrm{m}$$

Thus

$$\begin{bmatrix} x_1 \\ x_2 \\ x_3 \end{bmatrix} = \begin{bmatrix} 1{\cdot}7454 & 3{\cdot}0818 & 1{\cdot}9498 \\ 3{\cdot}3157 & 0{\cdot}4330 & -3{\cdot}6531 \\ 4{\cdot}2336 & -4{\cdot}4926 & 3{\cdot}3105 \end{bmatrix} \begin{bmatrix} 9{\cdot}4089 \\ 0{\cdot}7843 \\ 0{\cdot}0916 \end{bmatrix} \times 10^{-3}$$

$$= \begin{bmatrix} 0{\cdot}0190 \\ 0{\cdot}0312 \\ 0{\cdot}0366 \end{bmatrix} \mathrm{m}$$

The corresponding modified displacement vector obtained applying eq. (13.25), used in structural design, is

$$\begin{bmatrix} x_1 \\ x_2 \\ x_3 \end{bmatrix} = \begin{bmatrix} 0{\cdot}0166 \\ 0{\cdot}0312 \\ 0{\cdot}0400 \end{bmatrix} \text{m}$$

Comparison of the elements in the two displacement vectors reveals that the relative displacement between the ground and first floor is greatest in the first vector, while the relative displacements between the first and second floor, and the second and third floor, are greater in the second one.

The maximum acceleration of each floor is found by again using the response spectra in Fig. 13.4. It should be noted that the acceleration is given in terms of acceleration due to gravity $g$, and not in inches/s$^2$. Thus

$$\ddot{q}_{1,\max} = 82{\cdot}0652g \times 10^{1{\cdot}70/2{\cdot}75} = 340{\cdot}675g$$

$$\ddot{q}_{2,\max} = 28{\cdot}6381g \times 10^{1{\cdot}80/2{\cdot}75} = 129{\cdot}268g$$

$$\ddot{q}_{3,\max} = 9{\cdot}4480g \times 10^{1{\cdot}45/2{\cdot}75} = 31{\cdot}814g$$

This yields the following acceleration vector for the floors

$$\begin{bmatrix} \ddot{x}_1 \\ \ddot{x}_2 \\ \ddot{x}_3 \end{bmatrix} = \begin{bmatrix} 1{\cdot}7454 & 3{\cdot}0818 & 1{\cdot}9498 \\ 3{\cdot}3157 & 0{\cdot}4330 & -3{\cdot}6531 \\ 4{\cdot}2336 & -4{\cdot}4926 & 3{\cdot}3105 \end{bmatrix} \begin{bmatrix} 340{\cdot}675 \\ 129{\cdot}268 \\ 31{\cdot}814 \end{bmatrix} \times 9{\cdot}81 \times 10^{-3}$$

$$= \begin{bmatrix} 10{\cdot}350 \\ 10{\cdot}490 \\ 9{\cdot}485 \end{bmatrix} \text{m/s}^2$$

The maximum modified acceleration vector, obtained by taking the square root of the sum of the squares of the contribution from each mode, yields

$$\begin{bmatrix} \ddot{x}_1 \\ \ddot{x}_2 \\ \ddot{x}_3 \end{bmatrix} = \begin{bmatrix} 7{\cdot}048 \\ 11{\cdot}153 \\ 15{\cdot}288 \end{bmatrix} \text{m/s}^2$$

In the case of accelerations, therefore, the modified solution leads to a much lower acceleration at the first floor level and a much greater one at the top level.

## Dynamic response to earthquakes using time domain integration methods

If earthquake histories are available, or can be generated, the maximum response of a structure can be determined through step-by-step integration in the time domain. In chapter 6 three such integration methods, based on the Newmark β- and Wilson θ-equations are presented. Of these, experience seems to indicate that schemes employing the Newmark $[\beta = \frac{1}{4}]$-equations are the most efficient. For one DOF systems the response to earthquakes can be calculated using eq. (6.54). The response of multi-DOF systems can be calculated using eq. (6.69), given as

$$\left[\mathbf{K} + \frac{2}{\Delta t}\mathbf{C} + \frac{4}{\Delta t^2}\mathbf{M}\right]\Delta \boldsymbol{x} = 2\mathbf{C}\dot{\boldsymbol{x}} + \mathbf{M}\left[\frac{\Delta \ddot{\boldsymbol{x}}_g + 4}{\Delta t}\dot{\boldsymbol{x}} + 2\ddot{\boldsymbol{x}}\right] \quad (13.26)$$

where **K**, **C** and **M** are the stiffness, damping and mass matrices for a structure, $\Delta \boldsymbol{x}$ is the change in the displacement vector $\boldsymbol{x}$ during a time step $\Delta t$, $\dot{\boldsymbol{x}}$ is a velocity vector and $\ddot{\boldsymbol{x}}$ is an acceleration vector. By use of the Newmark $[\beta = \frac{1}{4}]$-equations, from eqs (6.41)–(6.43) the $i$th elements in the displacement, velocity and acceleration vectors at time $(t + \Delta t)$ are

$$x_i(t + \Delta t) = x_i(t) + \Delta x_i \quad (13.27a)$$

$$\dot{x}_i(t + \Delta t) = \frac{2}{\Delta t}\Delta x_i - \dot{x}_i(t) \quad (13.27b)$$

$$\ddot{x}_i(t + \Delta t) = \frac{4}{\Delta t^2}\Delta x_i - \frac{4}{\Delta t}\dot{x}_i(t) - \ddot{x}_i(t) \quad (13.27c)$$

The size of the time step $\Delta t$ is important, as over-large as well as over-small time steps will lead to inaccuracies in the calculated response. In the case of both wind and earthquakes most of the energy is contained within the part of the frequency spectrum that lies between 0 and 10 Hz. Thus usually the period of the smallest frequency component that needs to be considered is approximately 0·1 s. Experience has shown that frequency components of this order of magnitude can be sufficiently accurately modelled with time steps $\Delta t = 0{\cdot}1/10 = 0{\cdot}01$ s.

As response spectra resulting from both recorded and generated earthquakes tend to be spiky, it is usually recommended to carry out a time domain analysis using different earthquakes normalized to the same peak acceleration to ensure that the combined spectra approximate a consolidated one. This, however, may prove to be tedious. It is therefore advantageous to generate only a suitable strong motion history, calculate the variance of response to this history and then multiply the resulting standard deviation of response with a suitable peak factor. This approach, however, is likely to require more research before being generally accepted.

## Power spectral density functions for earthquakes

The mean amplitude, variance and frequency content of earthquakes vary with time. Earthquakes are therefore not stationary processes. If divided into sufficiently small segments, the process within each segment may be considered to be approximately stationary. Each segmental process may be modelled mathematically by the summation of harmonic components. Thus, the acceleration of the ground motion may be expressed as

$$\ddot{x}_g(t) = \sum_{i=1}^{N} \ddot{x}_i \cos(\omega_i t + \phi_i) \tag{13.28}$$

where the values for $\ddot{x}_i$ and $\omega_i$ are found by Fourier analysis of real records, and $\phi_i$ is a phase angle that varies randomly between 0 and $2\pi$.

Power spectral density functions or power spectra for the strong motion part of earthquakes are constructed by plotting values lf $\ddot{x}_i^2/\omega_i$ against $\omega_i$, or values of $\ddot{x}_i^2/n_i$ against $n_i$, where $n_i = \omega_i/2\pi$. Such spectra, however, tend to be spiky and require adjustments if needed for design purposes. Power spectra used in design are, like wind spectra, averaged over a number of normalized earthquakes and smoothed out. Kanai and Taijiumi have proposed the following formulation for smoothed power spectra that are functions of the expected peak acceleration as well as the damping and natural frequency of the ground

$$S_{\ddot{x}_g}(\omega) = \frac{S_0\left[1 + 2\xi_g r)^2\right]}{(1 - r^2)^2 + (2\xi_g r)^2} \tag{13.29}$$

where

$$S_0 = \frac{0{\cdot}141\xi_g \ddot{x}_{g,\max}{}^2}{\omega_g\sqrt{(1 + 4\xi_g{}^2)}} \tag{13.30}$$

and $\ddot{x}_{g,\max}$ is the peak ground acceleration, $\omega_g$ is the natural angular frequency of ground, $r = \omega_i/\omega_g$, and $\xi_g$ is the damping ratio for ground. For firm ground Kanai has suggested the following values:
$\omega_g = 12{\cdot}7$ rad/s and $\xi_g = 0{\cdot}6$.

For low frequencies, i.e. when $\omega \to 0$, eq. (13.29) will lead to unbounded values for ground velocity and ground displacements. Clough & Penzien (1975) have therefore suggested the following modification of the spectral density function

$$S_{\ddot{x}_g}(\omega) = \frac{S_0\left[1 + (2\xi_g r)^2\right]}{(1 - r^2)^2 + (2\xi_g r)^2} \times \frac{r_1{}^4}{\left(1 - r_1{}^2\right)^2 + (2\xi r_1)^2} \tag{13.31}$$

where $r_1 = \omega/\omega_1$, and the frequency parameter $\omega_1$ and damping parameter $\xi_1$ are selected to give the spectral density function the desired characteristic. Suggested values for $\omega_1$ and $\xi_1$ are given in Key (1988) and Lin *et al.* (1989).

### Frequency domain analysis of single-DOF systems using power spectra for translational motion

In chapter 11 it is shown how frequency domain analysis can be used to predict the variance of dynamic response due to the buffeting of wind. In what follows it will be shown how the same approach can be extended to calculate the variance of the dynamic response to the strong motion part of an earthquake.

To obtain a relationship between the spectrum of the fluctuating force acting at a point on a structure due to the acceleration of the ground and the spectrum of the ground acceleration, let the frequency spans of both force and support motion be divided into unit frequency intervals, with each interval centred at the frequency ω. From eq. (4.53), the force acting on a mass $M$ due to support acceleration $\ddot{x}_g(t) = x_g\omega^2 \sin(\omega t)$ is

$$f_g(t) = M\ddot{x}_g(t) \tag{13.32}$$

If

$$\ddot{x}_g(t) = \ddot{x}_g \sin(\omega t) \tag{13.33}$$

then

$$f_g(t) = f_g \sin(\omega t) \tag{13.34}$$

since it is assumed that $f_g(t)$ varies linearly with $\ddot{x}_g(t)$. Substitution of the expressions for $\ddot{x}_g(t)$ and $f_g(t)$ into eq. (13.32) yield

$$f_g = M\ddot{x}_g \tag{13.35}$$

and hence

$$f_g{}^2 = M^2\ddot{x}_g{}^2 \tag{13.36}$$

As the coordinates of power spectra are proportional to the square of the amplitudes of the constituent harmonics and inversely proportional to their frequencies, it follows that

$$S_{f_g}(\omega) = M^2 S_{\ddot{x}_g}(\omega) \tag{13.37}$$

Having developed an expression for the force spectrum in terms of the ground acceleration spectrum, it remains to express the response spectrum in terms of the force spectrum. From the theory of forced vibrations of damped linear one DOF systems in chapter 4 (eq. (4.15)), the response to a force

$$f_g(t) = f_g \sin(\omega t) \tag{13.38}$$

is

$$x(t) = \frac{f_g}{K} \frac{1}{\sqrt{\left[(1 - r^2)^2 + (2\xi r)^2\right]}} \sin(\omega t - \alpha) \tag{13.39}$$

or

$$x(t) = \frac{f_g}{K} \mathrm{MF}(\omega) \sin(\omega t - \alpha) \tag{13.40}$$

Thus the maximum value of $x(t)$, which occurs when $\sin(\omega t - \alpha) = 1$, is

$$x = \frac{f_g}{K} \mathrm{MF}(\omega) \tag{13.41}$$

Squaring of each term in eq. (13.41) yields

$$x^2 = \frac{{f_g}^2}{K^2} M(\omega) \tag{13.42}$$

where $M(\omega) - MF^2(\omega)$, as in chapter 11, is the *mechanical admittance factor*. Because the coordinates of the power spectrum are proportional to the square of the amplitudes of the constituent harmonics, it follows that

$$S_x(\omega) = \frac{1}{K^2} M(\omega) S_{f_g}(\omega) \tag{13.43}$$

Finally, substitution of the expression for $S_{f_g}(\omega)$ given by eq. (13.37) into eq. (13.43) yields

$$S_x(\omega) = \frac{M^2}{K^2}(\omega) S_{\ddot{x}_g}(\omega) \tag{13.44}$$

hence

$$\sigma_x{}^2 = \int_0^\infty S_x(\omega) d\omega = \frac{M^2}{K^2} \int_0^\infty M(\omega) S_{\ddot{x}_g}(\omega) d\omega \tag{13.45}$$

For weakly damped structures, and since $\omega_n{}^2 = K/M$, the expression for $\sigma_x{}^2$ can be approximated to

$$\sigma_x{}^2 = \int_0^\infty S_x(\omega) d\omega \approx \frac{1}{\omega_n{}^4} M(\omega) S_{\ddot{x}_g}(\omega) \Delta\omega \tag{13.46}$$

where

$$\Delta\omega = \tfrac{1}{2}\xi\omega_n$$
$$M(\omega) = 1/4\xi^2$$

**Example 13.3** A tall building with a fundamental frequency of 1·0 Hz and a damping ratio of 1·0% of critical is submitted to an earthquake with a peak acceleration of 0·3*g*. Use a probabilistic method and Kanai's power spectrum to determine the mean standard deviation, or root mean square response, of the top of the building. Assume the dominant frequency of the ground to be 2·0 Hz and a ground damping ratio $\xi_g = 0{\cdot}6$. Finally, assuming the duration of the strong motion part of the earthquake to be 10 s, calculate the maximum response.

Because the structural damping is only 1·0% of critical, the expression for the variance of response given by eq. (13.46) may be used and written as

$$\sigma_x{}^2 = \frac{1}{8\omega_n{}^3} \times \frac{1}{\xi} S_{\ddot{x}_g}(\omega)$$

When

$$\xi_g = 0{\cdot}6$$
$$\omega_g = 2{\cdot}0 \times 2\pi \,\text{rad/s}$$
$$r = 1{\cdot}0/2{\cdot}0 = 0{\cdot}5$$
$$S_0 = \frac{0{\cdot}141 \times 0{\cdot}6 \times 0{\cdot}3^2 \times 9{\cdot}81^2}{2{\cdot}0 \times 2\pi\sqrt{(1 + 4 \times 0{\cdot}6^2)}} = 0{\cdot}0373289\,\text{m}^2/\text{s}^3$$
$$S_{\ddot{x}_g}(\omega_n) = \frac{0{\cdot}0373289 \times \left[1 + (2 \times 0{\cdot}6 \times 0{\cdot}5)^2\right]}{(1 - 0{\cdot}5^2)^2 + (2 \times 0{\cdot}6 \times 0{\cdot}5)^2} = 0{\cdot}0550323\,\text{m}^2/\text{s}^3$$

Hence

$$\sigma_x^{\,2} = \frac{1}{8 \times (1{\cdot}0 \times 2\pi)^3} \times \frac{1}{0{\cdot}01} 0{\cdot}0550323 = 2{\cdot}77324 \times 10^{-3}\,\text{m}^2$$
$$\underline{\sigma_x = 0{\cdot}0527\,\text{m}}$$

The maximum probable response is obtained by multiplication of the mean standard deviation by a peak factor which, since the structure is weakly damped, is given by

$$\kappa = \sqrt{[2\ln(\omega_n T/2\pi)]} + \frac{0{\cdot}577}{\sqrt{[2\ln(\omega_n T/2\pi)]}}$$

where $T$ is the assumed duration fo the strong motion part of the earthquake. Thus the maximum response is

$$x_{\max} = \kappa\sigma_x = \left\{\sqrt{[2\ln(1{\cdot}0 \times 10)]} + \frac{0{\cdot}577}{\sqrt{[2\ln(1{\cdot}0 \times 10)]}}\right\} \times 0{\cdot}0527$$
$$= \underline{0{\cdot}127\,\text{m}}$$

### Influence of the dominant frequency of the ground on the magnitude of structural response

In Fig. 13.6 the Kanai power spectrum is used to show how the root mean square response value or mean standard deviation of response of four different one DOF structures—with damping equal to 1·0% of critical but different natural frequencies—varies if the structures are sited on grounds with the same damping, but with increasing dominant frequencies, and shaken by earthquakes with peak accelerations equal to $0{\cdot}3g$. As can be seen from the graphs, the responses tend to increase as the dominant frequency decreases, and are greatest when the frequency of the structure and that of the dominant frequency of the ground coincide.

The level of ground damping will vary with the type of alluvium, and it can be shown that the response of a structure will increase with decreasing values of $\xi_g$, although the dominant frequency of the ground and the

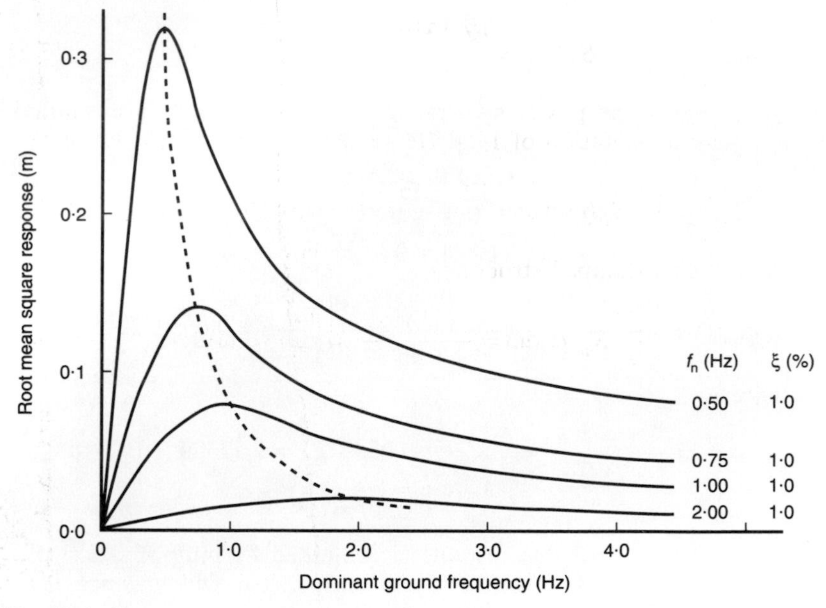


*Fig. 13.6. Root mean square response of one DOF structures sited on grounds with varying dominant frequencies and assumed damping equal to 60% of critical to earthquakes with peak acceleration $\ddot{x}_g = 0{\cdot}3g$*

peak acceleration of the earthquakes remain the same. Thus a $\xi_g$ value equal to 0·3 will result in root mean square responses of more than three times those shown in Fig. 13.6.

## Extension of the frequency domain method for translational motion to multi-DOF structures

It has previously been shown that the equations of motion for multi-DOF structures subjected to ground acceleration can be written in matrix notation as

$$\mathbf{M}\ddot{\boldsymbol{x}} + \mathbf{C}\dot{\boldsymbol{x}} + \mathbf{KX} = \mathbf{M}\boldsymbol{\alpha}\ddot{\boldsymbol{x}}_g(\boldsymbol{t}) \tag{13.47}$$

and that the decoupled equations of motion, obtained through the transformation $\boldsymbol{x} = \mathbf{Z}\boldsymbol{q}$, can be written in matrix notation as

$$\ddot{\boldsymbol{q}} + 2\boldsymbol{\xi\omega}\dot{\boldsymbol{q}} + \boldsymbol{\omega}^2\boldsymbol{q} = \mathbf{Z}^{\mathrm{T}}\mathbf{M}\boldsymbol{\alpha}\ddot{\boldsymbol{x}}_g(\boldsymbol{t}) \tag{13.48}$$

Thus the $i$th generalized modal equation is, as in eq. (13.18), given by

$$\ddot{q}_i + 2\xi_i\omega_i\dot{q}_i + \omega_i^{\,2}q_i = \mathbf{Z}_i^{\mathrm{T}}\mathbf{M}\boldsymbol{\alpha}\ddot{\boldsymbol{x}}_g(t) \tag{13.49}$$

where

$$\mathbf{Z}_i^{\mathrm{T}}\mathbf{M}\boldsymbol{\alpha}\ddot{\boldsymbol{x}}_g(\boldsymbol{t}) = [Z_{1i}M_1\alpha + Z_{2i}M_2\alpha + \ldots + Z_{Ni}M_N\alpha]\ddot{x}_g(t) \tag{13.50}$$

The spectral density function for $q_i$ is developed in exactly the same way as the spectral density function for the response of a one DOF system, and is given by

$$S_{q_i}(\omega) = \frac{(\mathbf{Z}_i^{\mathrm{T}}\mathbf{M}\boldsymbol{\alpha})^2}{\omega_i^4} M_i(\omega) S_{\ddot{x}_g}(\omega) \tag{13.51}$$

where $S_{\ddot{x}_g}(\omega)$ is the power spectral density function for an earthquake with a peak acceleration of $1{\cdot}0g$. Hence the variance of $q_i$ is given by

$$\sigma_{q_i}^2 = \int_0^\infty S_{q_i}(\omega)\mathrm{d}\omega = \frac{(\mathbf{Z}_i^{\mathrm{T}}\mathbf{M}\boldsymbol{\alpha})^2}{\omega_i^4} \int_0^\infty M_i(\omega) S_{\ddot{x}_g}(\omega)\mathrm{d}\omega \tag{13.52}$$

and for weakly damped structures by

$$\sigma_{q_i}^2 = \int_0^\infty S_{q_i}(\omega)\mathrm{d}\omega = \frac{(\mathbf{Z}_i^{\mathrm{T}}\mathbf{M}\boldsymbol{\alpha})^2}{\omega_i^4} M_i(\omega) S_{\ddot{z}_g}(\omega)\Delta\omega \tag{13.53}$$

where

$$\Delta\omega = \tfrac{1}{2}\xi_i\omega_i$$

$$M_i(\omega) = 1/4\xi_i^2$$

Hence

$$q_i = \kappa_i\sigma_{q_i} \tag{13.54}$$

where, from eq. (10.48)

$$\kappa_i = \sqrt{[2\ln(\omega_i T/2\pi)]} + \frac{0{\cdot}577}{\sqrt{[2\ln(\omega_i T/2\pi)]}} \tag{13.55}$$

Finally

$$\boldsymbol{x} = \mathbf{Z}\boldsymbol{q} \tag{13.56}$$

**Example 13.4** Let the structure shown in Fig. 2.14 be situated in an area where the dominant ground frequency is 2·0 Hz and the ground damping is assumed to be 60% of critical. Calculate the structural response to an earthquake, the strong motion part of which can be represented by Kanai's spectrum, if the peak acceleration is $0{\cdot}25g$. Assume the duration of the strong motion to be 10 s. The damping in the first mode is 2·0% and in the second and third modes is 1·0% of critical. The mass matrix, angular frequencies and the normalized modeshape matrix for the structure are as follows

$$\mathbf{M} = \begin{bmatrix} 600 & 0 & 0 \\ 0 & 400 & 0 \\ 0 & 0 & 200 \end{bmatrix} \times \frac{10^3}{g}\ \mathrm{kg}$$

$$\boldsymbol{\omega} = \begin{bmatrix} 18{\cdot}850 \\ 42{\cdot}390 \\ 56{\cdot}518 \end{bmatrix} \mathrm{rad/s}$$

$$\mathbf{Z} = \begin{bmatrix} 1{\cdot}7454 & 3{\cdot}0818 & 1{\cdot}9498 \\ 3{\cdot}3157 & 0{\cdot}4330 & -3{\cdot}6531 \\ 4{\cdot}2336 & -4{\cdot}4926 & 3{\cdot}3105 \end{bmatrix} \times 10^{-3}$$

From eq. (13.50), the generalized peak mode acceleration is given by

$$\ddot{\boldsymbol{q}}_{\mathrm{g,\,max}} = 0{\cdot}25\mathbf{Z}^{\mathrm{T}}\mathbf{M}\boldsymbol{\alpha}\boldsymbol{g}$$

Hence

$$\ddot{\boldsymbol{q}}_{\mathrm{g,\,max}} = \begin{bmatrix} 1{\cdot}7454 & 3{\cdot}3157 & 4{\cdot}2336 \\ 3{\cdot}0818 & 0{\cdot}4330 & -4{\cdot}4926 \\ 1{\cdot}9498 & -3{\cdot}6531 & 3{\cdot}3105 \end{bmatrix} \begin{bmatrix} 600 & 0 & 0 \\ 0 & 400 & 0 \\ 0 & 0 & 200 \end{bmatrix} \begin{bmatrix} 0{\cdot}25 \\ 0{\cdot}25 \\ 0{\cdot}25 \end{bmatrix}$$

$$= \begin{bmatrix} 805{\cdot}06 \\ 280{\cdot}94 \\ 92{\cdot}69 \end{bmatrix} \mathrm{m/s^2}$$

For weakly damped structures the variance of the generalized mode response $q$ is given by eq. (13.53). Thus

$$\sigma_q^{\,2} = \frac{1}{8\omega^3} \times \frac{1}{\xi} S_{\ddot{q}_{\mathrm{g,\,max}}}(\omega)$$

where from eq. (13.29)

$$S_{\ddot{q}_{\mathrm{g}}}(\omega_i) = \frac{S_0\left[1 + (2\xi_{\mathrm{g}} r_i)^2\right]}{\left(1 - r_i^{\,2}\right)^2 + (2\xi_{\mathrm{g}} r_i)^2}$$

and from eq. (13.30)

$$S_0 = \frac{0{\cdot}141\xi_{\mathrm{g}}\ddot{q}_{\mathrm{g,\,max}}^{\;2}}{\omega_{\mathrm{g}}\sqrt{(1 + 4\xi_{\mathrm{g}}^{\,2})}}$$

For $\omega_1 = 18{\cdot}850$ rad/s

$$S_0 = \frac{0{\cdot}141 \times 0{\cdot}6 \times 805{\cdot}06^2}{2{\cdot}0 \times 2\pi\sqrt{(1 + 4 \times 0{\cdot}6^2)}} = 2783{\cdot}3290\ \mathrm{m^2/s^3}$$

For $\omega_2 = 42{\cdot}390$ rad/s

$$S_0 = \frac{0{\cdot}141 \times 0{\cdot}6 \times 280{\cdot}94^2}{2{\cdot}0 \times 2\pi\sqrt{(1 + 4 \times 0{\cdot}6^2)}} = 342{\cdot}5934\ \mathrm{m^2/s^3}$$

For $\omega_3 = 56{\cdot}718$ rad/s

$$S_0 = \frac{0{\cdot}141 \times 0{\cdot}6 \times 92{\cdot}69^2}{2{\cdot}0 \times 2\pi\sqrt{(1 + 4 \times 0{\cdot}6^2)}} = 37{\cdot}0281\ \mathrm{m^2/s^3}$$

In the expression for $S_{\ddot{q}_g}(\omega_i)$ the values for $r_i$ are

$$r_1 = 18{\cdot}850/2{\cdot}0 \times 2\pi = 1{\cdot}50004$$

$$r_2 = 42{\cdot}390/2{\cdot}0 \times 2\pi = 3{\cdot}37329$$

$$r_3 = 56{\cdot}718/2{\cdot}0 \times 2\pi = 4{\cdot}52348$$

Hence

$$S_{\ddot{q}_g}(18{\cdot}850) = \frac{2783{\cdot}3290\left[1 + (2 \times 0{\cdot}6 \times 1{\cdot}50004)^2\right]}{(1 - 1{\cdot}50004^2)^2 + (2 \times 0{\cdot}6 \times 1{\cdot}50004)^2}$$

$$= 2457{\cdot}1857\,\mathrm{m^2/s^3}$$

$$S_{\ddot{q}_g}(42{\cdot}390) = \frac{342{\cdot}5934\left[1 + (2 \times 0{\cdot}6 \times 3{\cdot}37329)^2\right]}{(1 - 3{\cdot}37329^2)^2 + (2 \times 0{\cdot}6 \times 3{\cdot}37329)^2}$$

$$= 47{\cdot}9915\,\mathrm{m^2/s^3}$$

$$S_{\ddot{q}_g}(56{\cdot}718) = \frac{37{\cdot}0281\left[1 + (2 \times 0{\cdot}6 \times 4{\cdot}52348)^2\right]}{(1 - 4{\cdot}52348^2)^2 + (2 \times 0{\cdot}6 \times 4{\cdot}52348)^2}$$

$$= 2{\cdot}7633\,\mathrm{m^2/s^3}$$

Thus the variances and mean standard deviations of response in the generalized coordinate system are

$$\sigma_{q_1}{}^2 = \frac{1}{8} \times \frac{1}{18{\cdot}850^3} \times \frac{1}{0{\cdot}02} \times 2457{\cdot}1857 = 2{\cdot}2928938\,\mathrm{m^2}$$

$$\sigma_{q_1} = 1{\cdot}5142304\,\mathrm{m}$$

$$\sigma_{q_2}{}^2 = \frac{1}{8} \times \frac{1}{42{\cdot}390^3} \times \frac{1}{0{\cdot}01} \times 47{\cdot}9915 = 0{\cdot}0078756\,\mathrm{m^2}$$

$$\sigma_{q_2} = 0{\cdot}0887446\,\mathrm{m}$$

$$\sigma_{q_3}{}^2 = \frac{1}{8} \times \frac{1}{56{\cdot}718^3} \times \frac{1}{0{\cdot}01} \times 2{\cdot}7633 = 0{\cdot}0001893\,\mathrm{m^2}$$

$$\sigma_{q_3} = 0{\cdot}0137586\,\mathrm{m}$$

Thus

$$q_1 = \left\{\sqrt{[2\ln(18{\cdot}850 \times 10/2\pi)]} + \frac{0{\cdot}577}{\sqrt{[2\ln(18{\cdot}850 \times 10/2\pi)]}}\right\} \times 1{\cdot}5142304$$

$$= 4{\cdot}2843\,\mathrm{m}$$

$$q_2 = \left\{ \sqrt{[2\ln(42{\cdot}390 \times 10/2\pi)]} + \frac{0{\cdot}577}{\sqrt{[2\ln(42{\cdot}390 \times 10/2\pi)]}} \right\} \times 0{\cdot}0887446$$

$$= 0{\cdot}2752\,\text{m}$$

$$q_3 = \left\{ \sqrt{[2\ln(56{\cdot}718 \times 10/2\pi)]} + \frac{0.577}{\sqrt{[2\ln(56{\cdot}718 \times 10/2\pi)]}} \right\} \times 0{\cdot}0137586$$

$$= 0{\cdot}0439\,\text{m}$$

$$\begin{bmatrix} x_1 \\ x_2 \\ x_3 \end{bmatrix} = \begin{bmatrix} 1{\cdot}7454 & 3{\cdot}0818 & 1{\cdot}9498 \\ 3{\cdot}3157 & 0{\cdot}4330 & -3{\cdot}6531 \\ 4{\cdot}2336 & -4{\cdot}4926 & 3{\cdot}3105 \end{bmatrix} \begin{bmatrix} 4{\cdot}2843 \\ 0{\cdot}2752 \\ 0{\cdot}0439 \end{bmatrix} \times 10^{-3}$$

$$= \begin{bmatrix} 0{\cdot}0084 \\ 0{\cdot}0142 \\ 0{\cdot}0170 \end{bmatrix} \text{m}$$

A comparison with the displacements calculated in example 13.2 reveals that the use of the Kanai's power spectrum leads to a much smaller response than does the use of the Newmark response spectra. The main reason for this is that the variances of the underlying power spectrum values for the Newmark response spectra have been found to be much greater than those of Kanai's spectra.

The modified displacement vector used in structural design, and calculated by taking the root of the sum of the squares of the contribution from each mode, as expressed by eq. (13.25) is

$$\begin{bmatrix} x_1 \\ x_2 \\ x_3 \end{bmatrix} = \begin{bmatrix} 0{\cdot}0075 \\ 0{\cdot}0142 \\ 0{\cdot}0182 \end{bmatrix} \text{m}$$

Finally, it is of interest to calculate the response on the assumption that the structure responds mainly in the first mode. This yields

$$\begin{bmatrix} x_1 \\ x_2 \\ x_3 \end{bmatrix} = \begin{bmatrix} 1{\cdot}7454 & 3{\cdot}0818 & 1{\cdot}9498 \\ 3{\cdot}3157 & 0{\cdot}4330 & -3{\cdot}6531 \\ 4{\cdot}2336 & -4{\cdot}4926 & 3{\cdot}3105 \end{bmatrix} \begin{bmatrix} 4{\cdot}2843 \\ 0 \\ 0 \end{bmatrix} \times 10^{-3}$$

$$= \begin{bmatrix} 0{\cdot}0075 \\ 0{\cdot}0142 \\ 0{\cdot}0181 \end{bmatrix} \text{m}$$

Thus, the first mode response vector is nearly identical to the modified one calculated above.

## Response of one DOF structures to rocking motion

So far only the response of structures to the translational motion of earthquakes has been considered. However, earthquakes also contain rocking components about the two horizontal axes and one torsional component about the vertical axis, of which the former are caused by the shear waves and Rayleigh waves shown in Fig. 12.2. Modern codes require that the effect of these components be taken into account, although very little information on their amplitudes and frequencies is available.

Consider the column shown in Fig. 13.7, in which the rotational moment of inertia of the equivalent lumped mass at the top is assumed to be zero. Let the base of the column be subjected to a rocking motion

$$\theta_g(t) = \theta_g \sin(\omega t) \tag{13.57}$$

The translational equation of motion for the lumped mass is

$$M\ddot{y} + C(\dot{y} - \dot{x}_g(t)) + K(y - x_g(t)) = 0 \tag{13.58}$$

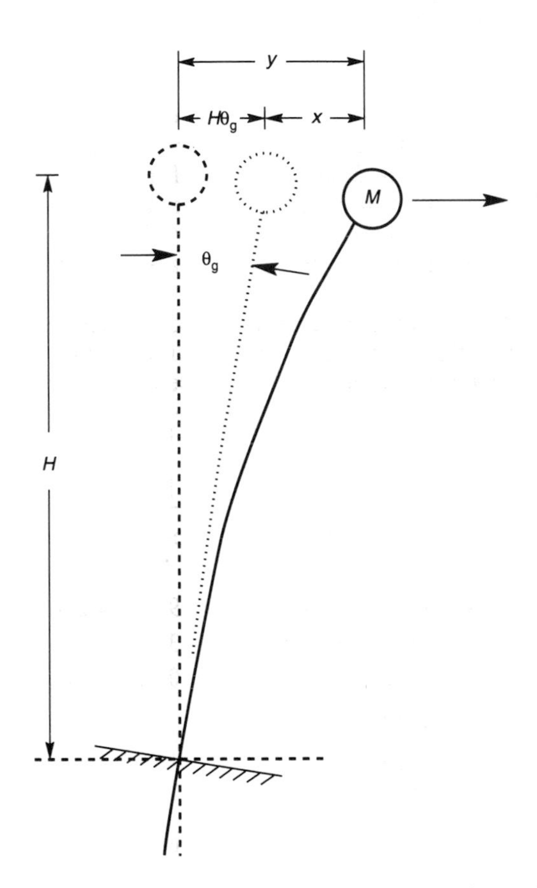


*Fig. 13.7. Column considered as a one DOF system subjected to rocking excitation*

where

$$\ddot{y} = \ddot{x} + \ddot{x}_g(t)$$
$$\dot{y} = \dot{x} + \dot{x}_g(t)$$
$$y = x + x_g(t)$$

Hence

$$M\ddot{x} = C\dot{x} + Kx = -M\ddot{x}_g(t) \tag{13.59}$$

From eq. (13.57) it follows that

$$\ddot{x}_g(t) = H\ddot{\theta}_g(t) = -H\theta_g\omega^2 \sin(\omega t) \tag{13.60}$$

and finally that

$$M\ddot{x} + C\dot{x} + Kx = MH\theta_g\omega^2 \sin(\omega t) \tag{13.61}$$

Thus the response at the top relative to the position of the rotated but undeformed column, resulting from the sinusoidal rocking motion $\theta_g(t) = \theta_g \sin(\omega t)$, is

$$x = \frac{MH\theta_g\omega^2}{K} \times \frac{1}{\sqrt{\left[(1 - r^2)^2 + (2\xi r)^2\right]}} \sin(\omega t - \alpha) \tag{13.62}$$

$$x_{\max} = \frac{MH\theta_g\omega^2}{K} \times \frac{1}{2\xi} \tag{13.63}$$

## Frequency domain analysis of single-DOF systems using power spectra for rocking motion

It has previously been shown how frequency domain analysis can be used to predict the variance of response due to the strong motion translational components of earthquakes. In the following the same approach is extended to include the calculation of the variance of response due to rocking components.

To obtain a relationship between the spectrum of the fluctuating force acting at a point on a structure due to the angular acceleration of the ground and the spectrum of the angular ground acceleration, let the frequency spans of both force and rocking motion be divided into unit frequency intervals, with each interval centred at the angular frequency $\omega$. From eqs (13.59) and (13.60), the force acting on a mass $M$ due to support acceleration $\ddot{\theta}_g\omega^2$ is

$$f_g(t) = MH\ddot{\theta}_g(t) \tag{13.64}$$

If

$$\ddot{\theta}_g(t) = \ddot{\theta}_g \sin(\omega t) \tag{13.65}$$

then

$$f_g(t) = f_g \sin(\omega t) \tag{13.66}$$

since it is assumed that $f_g(t)$ varies linearly with $\ddot{\theta}_g(t)$. Substitution of the expressions for $\ddot{\theta}_g(t)$ and $f_g(t)$ into eq. (13.64) yields

$$f_g = MH\ddot{\theta} \tag{13.67}$$

and hence

$$f_g^2 = M^2H^2\ddot{\theta}_g^{\,2} \tag{13.68}$$

As the coordinates of power spectra are proportional to the square of the amplitudes of the harmonic components and inversely proportional to their frequencies, it follows that

$$S_{f_g}(\omega) = M^2H^2S_{\ddot{\theta}_g}(\omega) \tag{13.69}$$

From eq. (13.43)

$$S_x(\omega) = \frac{1}{K^2}M(\omega)S_{f_g}(\omega) \tag{13.70}$$

Substitution of the expression for $S_{f_g}(\omega)$ given by eq. (13.69) into eq. (13.70) yields

$$S_x(\omega) = \frac{M^2H^2}{K^2}M(\omega)S_{\ddot{\theta}_g}(\omega) \tag{13.71}$$

Hence

$$\sigma_x^{\,2} = \int_0^{\infty} S_x(\omega)\mathrm{d}\omega = \frac{M^2H^2}{K^2}\int_0^{\infty} M(\omega)S_{\ddot{\theta}_g}(\omega)\mathrm{d}\omega \tag{13.72}$$

For weakly damped structures, and since $\omega_n^{\,2} = K/M$, the expression for $\sigma_x^{\,2}$ can be approximated to

$$\sigma_x^{\,2} = \int_0^{\infty} S_x(\omega)\mathrm{d}\omega \approx \frac{H^2}{\omega_n^{\,4}}M(\omega)S_{\ddot{\theta}_g}(\omega)\Delta\omega \tag{13.73}$$

where

$$\Delta\omega = \tfrac{1}{2}\xi\omega_n$$
$$M(\omega) = 1/4\xi^2$$

## Assumed power spectral density function for rocking motion used in examples

At the time of writing the author is not aware of any spectral density function for rocking motion. In order to present an example to illustrate the above theory it is therefore necessary to construct a function that will yield reasonable response values. In Eurocode 8, part 3, response spectra for rocking acceleration about the *x*-axis, *y*-axis and *z*-axis have been proposed in terms of the lateral response acceleration spectra. Thus

$$\ddot{\theta}_x(\omega) = 0{\cdot}85\omega\ddot{x}(\omega)/\beta \qquad (13.74a)$$

$$\ddot{\theta}_y(\omega) = 0{\cdot}85\omega\ddot{y}(\omega)/\beta \qquad (13.74a)$$

$$\ddot{\theta}_z(\omega) = 1{\cdot}00\omega\ddot{z}(\omega)/\beta \qquad (13.74a)$$

where $\beta$ is the shear wave velocity in m/s, and $\omega$ is the frequency under consideration. By definition, the coordinates of a spectral density function are equal to the square of the amplitudes of the constituent frequency components divided by the frequencies of the same component. Thus the value of the spectral density function for rocking acceleration about the $y$-axis at an angular frequency $\omega_i$ is

$$S_{\ddot{\theta}_{gy}}(\omega_i) = \frac{\ddot{\theta}_{gy}{}^2(\omega_i}{\omega_i} = \frac{0{\cdot}85^2\omega_i{}^2\ddot{x}_g{}^2(\omega_i)}{\omega_i\beta^2} = \frac{0{\cdot}85^2\omega_i{}^2}{\beta^2}S_{\ddot{x}_g}(\omega_i) \qquad (13.75)$$

Finally, substitution of the expression for Kanai's power spectrum given by eqs (13.30) and (13.31) into eq. (13.75) yields

$$S_{\ddot{\theta}_{gy}}(\omega_i) = \frac{0{\cdot}1019r_i{}^2\omega_g\xi_g\ddot{x}_p{}^2}{\beta^2\sqrt{(1+4\xi_g{}^2)}} \times \frac{1+2\xi_g r_i)^2}{(1-r_i{}^2)^2+(2\xi_g r_i)^2} \qquad (13.76)$$

**Example 13.5** Let the 40·0 m tall tapering lattice tower shown in Fig. 11.6 be subjected to an earthquake with a peak acceleration of 0·3$g$. Calculate the acceleration and response at the top of the tower due to the translation and rocking of the ground if the natural frequency of the mast is 2·0 Hz, the equivalent mass of a mass–spring system depicting the movement of the disc is 8200 kg, the structural damping is 2·0% of critical, the dominant frequency of the ground is 2·0 Hz and the ground damping is 60·0% of critical. Assume the duration of the strong motion of the earthquake to be 10 s, and the velocity of the shear waves to be 500 m/s.

For weakly damped structures, from eq. (13.73) the variance of response to rocking motion is given by

$$\sigma_x{}^2 = \int_0^\infty S_x(\omega)\mathrm{d}\omega \approx \frac{H^2}{\omega_n{}^4}M(\omega)S_{\ddot{\theta}_{gy}}(\omega)\Delta\omega$$

where

$$\Delta\omega = \tfrac{1}{2}\xi\omega_n$$

$$M(\omega) = 1/4\xi^2$$

Hence

$$\sigma_x{}^2 = \frac{40{\cdot}0^2}{8\times(2\pi\times 2{\cdot}0)^3}\times\frac{1}{0{\cdot}02}\times S_{\ddot{\theta}_{gy}}(\omega) = 5{\cdot}0393023 S_{\ddot{\theta}_{gy}}(\omega)$$

where from eq. (13.76), since $r = 2{\cdot}0/2{\cdot}0 = 1{\cdot}0$

$$S_{\ddot{\theta}_{gy}}(\omega_i) = \frac{0{\cdot}1019 \times 1{\cdot}0^2 \times (2\pi \times 2{\cdot}0) \times 0{\cdot}6 \times (0{\cdot}3 \times 9{\cdot}81)^2}{500{\cdot}0^2\sqrt{(1+4\times 0{\cdot}6^2)}}$$

$$\times \frac{1+(2\times 0{\cdot}6\times 1{\cdot}0)^2}{(1-1{\cdot}0^2)^2+(2\times 0{\cdot}6\times 1{\cdot}0)^2}$$

$$S_{\ddot{\theta}_{gy}}(\omega_i) = 28{\cdot}87408 \times 10^{-6}\,\mathrm{m^2/s^3}$$

Hence

$$\sigma_x{}^2 = 5{\cdot}0393023 \times 28{\cdot}87408 \times 10^{-6} = 145{\cdot}50522 \times 10^{-6}\,\mathrm{m}^2$$

$$\sigma_x = 12{\cdot}062554 \times 10^{-3}\,\mathrm{m}$$

Thus

$$x_{\text{max, rocking}} = \{\sqrt{[2\ln(2{\cdot}0\times 10)]} + 0{\cdot}577/\sqrt{[2\ln(2{\cdot}0\times 10)]}\}$$

$$\times\ 12{\cdot}062554 \times 10^{-3} = \underline{0{\cdot}0324}\,\mathrm{m}$$

The corresponding variance and hence displacement due to translational motion of the ground is found by applying eq. (13.46), which on substitution of the expressions for $M(\omega)$ and $\Delta\omega$ is

$$\sigma_x{}^2 = \int_0^\infty S_x(\omega)\mathrm{d}\omega \approx \frac{1}{8\omega_n{}^3} \times \frac{1}{\xi} \times S_{x_g}(\omega)$$

Hence

$$\sigma_x{}^2 = \frac{1}{8\times(2\pi\times 2{\cdot}0)^3} \times \frac{1}{0{\cdot}02} \times S_{\ddot{x}_g}(\omega) = 3{\cdot}14956 \times 10^{-3} S_{\ddot{x}_g}(\omega)$$

where, from eqs (13.29) and (13.30)

$$S_{\ddot{x}_g}(\omega) = \frac{0{\cdot}141 \times 0{\cdot}6 \times (0{\cdot}3\times 9{\cdot}81)^2}{2\pi\times 2{\cdot}0\sqrt{(1+4\times 0{\cdot}6^2)}} \times \frac{1+(2\times 0{\cdot}6\times 1{\cdot}0)^2}{(1-1{\cdot}0^2)^2+(2\times 0{\cdot}6\times 1{\cdot}0)^2}$$

$$= 0{\cdot}0632518\,\mathrm{m^2/s^3}$$

Hence

$$\sigma_x{}^2 = 3{\cdot}14956 \times 10^{-3} \times 0{\cdot}0632518 = 1{\cdot}99215 \times 10^{-4}\,\mathrm{m}^2$$

$$\sigma_x = 0{\cdot}0141143\,\mathrm{m}$$

Thus

$$x_{\text{max, translation}} = \{\sqrt{[2\ln(2{\cdot}0\times 10)]} + 0{\cdot}577/\sqrt{[2\ln(2{\cdot}0\times 10)]}\}$$

$$\times\ 0{\cdot}0141143 = \underline{0{\cdot}0379\,\mathrm{m}}$$

Therefore, with the assumed spectral density function for the rocking motion and the assumed velocity of the shear waves, the rocking contributes approximately 46% to the lateral motion at the top of the tower.

**Extension of the frequency domain method for rocking motion to multi-DOF structures**

It has previously been shown that the equations of motion for multi-DOF structures subjected to translational ground acceleration can be written in matrix notation as

$$\mathbf{M}\ddot{x} + \mathbf{C}\dot{x} + \mathbf{KX} = \mathbf{M}\alpha\ddot{x}_g(t) \tag{13.77}$$

With reference to eqs (13.59) and (13.60), it is clear that the equations of motion for structures subjected to rocking as well as translational support motion can be established simply by addition of the force vector $\mathbf{MH}\ddot{\boldsymbol{\theta}}_g(t)$ to the right-hand side of eq. (13.77), where $\ddot{\theta}_g(t)$ is the history of the angular acceleration of the rocking motion corresponding to a translational motion with a peak acceleration of $\alpha g$ m/s$^2$, and $\alpha$ is a factor that defines the magnistude of the peak translational accleration. Thus

$$\mathbf{M}\ddot{\mathbf{x}} + \mathbf{C}\dot{\mathbf{x}} + \mathbf{KX} = \mathbf{M}\alpha\ddot{\mathbf{x}}_g(t) + \mathbf{MH}\dot{\boldsymbol{\theta}}_g(t) \tag{13.78}$$

where $\mathbf{H}\ddot{\boldsymbol{\theta}}_g(t)$ is an acceleration vector in which the element $H_i$ is the height of mass $M_i$ above the ground. Diagonalization of the above equations, achieved through the transformation $\boldsymbol{x} = \mathbf{Z}\boldsymbol{q}$ and premultiplication of each term by $\mathbf{Z}^{\mathrm{T}}$, yields

$$\ddot{\boldsymbol{q}} + 2\xi\omega\dot{\boldsymbol{q}} + \omega^2\boldsymbol{q} = \mathbf{Z}^{\mathrm{T}}\mathbf{M}\alpha\ddot{x}_g(t) + \mathbf{Z}^{\mathrm{T}}\mathbf{MH}\ddot{\boldsymbol{\theta}}_g(t) \tag{13.79}$$

Considering the rocking motion only, the $i$th generalized modal equation is given by

$$\ddot{q}_i + 2\xi_i\omega_i\dot{q}_i + \omega_i^2 q_i = \mathbf{Z}^{\mathrm{T}}\mathbf{MH}\ddot{\boldsymbol{\theta}}_g(t) \tag{13.80}$$

where

$$\mathbf{Z}_i^{\mathrm{T}}\mathbf{MH}\ddot{\boldsymbol{\theta}}_g(t) = [Z_{1i}M_1H_1 + Z_{2i}M_2H_2 + \ldots + Z_{Ni}M_NH_N]\ddot{\theta}_g(t) \tag{13.81}$$

The spectral density function for the generalized coordinate $q_i$ is not found by following exactly the same procedure as used for the one DOF system, and yields

$$S_{q_i}(\omega) = \frac{(\mathbf{Z}_i^{\mathrm{T}}\mathbf{MH})^2}{\omega_i^4} M_i(\omega) S_{\ddot{\theta}_g}(\omega) \tag{13.82}$$

where $S_{\ddot{\theta}_g}(\omega)$ is the spectral density function for a rocking motion with a peak angular acceleration of 1·0 rad/s. Thus the variance of $q_i$ is given by

$$\sigma_{q_i}^2 = \int_0^\infty S_{q_i}(\omega)\mathrm{d}\omega = \frac{(\mathbf{Z}_i^{\mathrm{T}}\mathbf{MH})^2}{\omega_i^4}\int_0^\infty M_i(\omega)S_{\ddot{\theta}_g}(\omega)\mathrm{d}\omega \tag{13.83}$$

For weakly damped structures

$$\sigma_{q_i}^2 = \int_0^\infty S_{q_i}(\omega)\mathrm{d}\omega = \frac{(\mathbf{Z}_i^{\mathrm{T}}\mathbf{MH})^2}{\omega_i^4} M_i(\omega)S_{\ddot{\theta}_g}(\omega)\Delta\omega \tag{13.84}$$

where

$$\Delta\omega = \tfrac{1}{2}\xi_i\omega_i$$
$$M_i(\omega) = 1/4\xi_i^2$$

Hence

$$q_i = \kappa_i\sigma_{q_i} \tag{13.85}$$

where from eq. (10.48)

$$\kappa_i = \sqrt{[2\ln(\omega_i T/2\pi)]} + \frac{0{\cdot}577}{\sqrt{[2\ln(\omega_i T/2\pi)]}} \tag{13.86}$$

Finally

$$\boldsymbol{x} = \mathbf{Z}\boldsymbol{q} \tag{13.87}$$

**Example 13.6** Let the shear structure shown in Fig. 11.4 be subjected to an earthquake with a peak acceleration of 0·3$g$. The dominant frequency, damping ratio of the ground, duration of the strong motion, and velocity of the shear waves may be taken as 2·0 Hz, 0·6, 10 s and 500 m/s respectively. The mass of each equivalent floor is 120 000 kg. The modal damping ratios in the first, second and third modes are respectively, 0·03, 0·02 and 0·01. The natural frequencies and normalized modeshape matrix for the structure are given below. Assume that the foundation supporting the structure behaves as a rigid plate, and calculate the response due to rocking motion.

$$\omega = \begin{bmatrix} 4{\cdot}439 \\ 12{\cdot}446 \\ 18{\cdot}025 \end{bmatrix} \text{rad.s}$$

$$\omega^2 = \begin{bmatrix} 19{\cdot}70 \\ 155{\cdot}40 \\ 324{\cdot}90 \end{bmatrix} \text{rad}^2/\text{s}^2$$

$$\mathbf{Z} = \begin{bmatrix} 0{\cdot}947 & 2{\cdot}128 & 1{\cdot}703 \\ 1{\cdot}706 & 0{\cdot}950 & -2{\cdot}128 \\ 2{\cdot}128 & -1{\cdot}703 & 0{\cdot}953 \end{bmatrix} \times 10^{-3}$$

From eq. (13.79) the decoupled equations for rocking motion are given by the matrix equation

$$\ddot{\boldsymbol{q}} + 2\xi\omega\dot{\boldsymbol{q}} + \omega^2\boldsymbol{q} = \mathbf{Z}^{\mathrm{T}}\mathbf{M}\mathbf{H}\ddot{\boldsymbol{\theta}}_{\mathrm{g}}(t)$$

where, from eq. (13.84), the variance of $q_i$ for weakly damped structures is given by

$$\sigma_{q_i}{}^2 = \int_0^\infty S_{q_i}(\omega)\mathrm{d}\omega = \frac{(\mathbf{Z}_i{}^{\mathrm{T}}\mathbf{NH})^2}{8\omega_i{}^3} \times \frac{1}{\xi} S_{\ddot{\theta}_g}(\omega)$$

The three values for $\mathbf{Z}_i{}^{\mathrm{T}}\mathbf{MH}$ are determined through evaluation of the matrix product

$$\mathbf{Z}^{\mathrm{T}}\mathbf{MH} = \begin{bmatrix} 0{\cdot}947 & 1{\cdot}706 & 2{\cdot}128 \\ 2{\cdot}128 & 0{\cdot}950 & -1{\cdot}703 \\ 1{\cdot}703 & -2{\cdot}128 & 0{\cdot}953 \end{bmatrix} \begin{bmatrix} 120{\cdot}0 & 0 & 0 \\ 0 & 120{\cdot}0 & 0 \\ 0 & 0 & 120{\cdot}0 \end{bmatrix} \begin{bmatrix} 10{\cdot}0 \\ 20{\cdot}0 \\ 30{\cdot}0 \end{bmatrix}$$

$$= \begin{bmatrix} 12\,891{\cdot}60 \\ -1297{\cdot}20 \\ 650{\cdot}76 \end{bmatrix} \text{kg m}$$

The assumed expression for $S_{\ddot{\theta}_g}(\omega)$ is given by eq. (13.76). Thus, when

$$r_1 = \omega_1/\omega g = 4{\cdot}439/2\pi \times 2{\cdot}0 = 0{\cdot}3532$$

$$S_{\ddot{\theta}_{gy}}(\omega_1) = \frac{0{\cdot}1019 \times 0{\cdot}3532^2 \times (2\pi \times 2{\cdot}0) \times 0{\cdot}6 \times (0{\cdot}3 \times 9{\cdot}81)^2}{500^2\sqrt{(1 + 0{\cdot}6^2)}} \times \frac{1 + (2 \times 0{\cdot}6 \times 0{\cdot}3532)^2}{(1 - 0{\cdot}3532^2)^2 + (2 \times 0{\cdot}6 \times 0{\cdot}3532)^2}$$

Hence

$$S_{\ddot{\theta}_{gy}}(\omega_1) = 2{\cdot}12579 \times 10^{-6} \times 1{\cdot}2473694 = 2{\cdot}6516454 \times 10^{-6}\,\text{rad}^2/\text{s}$$

and when

$$r_2 = \omega_2/\omega_g = 12{\cdot}466/2\pi \times 2{\cdot}0 = 0{\cdot}9920$$

$$S_{\ddot{\theta}_{gy}}(\omega_2) = \frac{0{\cdot}1019 \times 0{\cdot}9920^2 \times (2\pi \times 2{\cdot}0) \times 0{\cdot}6 \times (0{\cdot}3 \times 9{\cdot}81)^2}{500^2\sqrt{(1 + 0{\cdot}6^2)}} \times \frac{1 + (2 \times 0{\cdot}6 \times 0{\cdot}9920)^2}{(1 - 0{\cdot}9920^2)^2 + (2 \times 0{\cdot}6 \times 0{\cdot}9920)^2}$$

Hence

$$S_{\ddot{\theta}_{gy}}(\omega_2) = 16{\cdot}768813 \times 10^{-6} \times 1{\cdot}7073847 = 28{\cdot}597277 \times 10^{-6}\,\text{rad}^2/\text{s}$$

and when

$$r_3 = \omega_3/\omega_g = 18{\cdot}025/2\pi \times 2{\cdot}0 = 1{\cdot}4344$$

$$S_{\ddot{\theta}_{gy}}(\omega_3) = \frac{0{\cdot}1019 \times 1{\cdot}4344^2 \times (2\pi \times 2{\cdot}0) \times 0{\cdot}6 \times (0{\cdot}3 \times 9{\cdot}81)^2}{500^2\sqrt{(1 + 0{\cdot}6^2)}} \times \frac{1 + (2 \times 0{\cdot}6 \times 1{\cdot}4344)^2}{(1 - 1{\cdot}4344^2)^2 + (2 \times 0{\cdot}6 \times 1{\cdot}4344)^2}$$

Hence

$$S_{\ddot{\theta}_{gy}}(\omega_3) = 35{\cdot}060614 \times 10^{-6} \times 0{\cdot}9710095 = 34{\cdot}044192 \times 10^{-6}\,\mathrm{rad^2/s}$$

Substitution of the given values for $\omega_i$ and $\xi_i$ and the calculated values for $\mathbf{Z}_i^{\mathrm{T}}\mathbf{MHW}$ and $S_{\ddot{\theta}_{gy}}(\omega_i)$ into eq. (13.84) yields

$$\sigma_{q_1}^2 = \frac{12\,891{\cdot}60^2}{8 \times 4{\cdot}429^3} \times \frac{1}{0{\cdot}03} \times 2{\cdot}6516454 \times 10^{-6} = 21{\cdot}13493\,\mathrm{m}^2,$$

hence $\sigma_{q_1} = 4{\cdot}59727\,\mathrm{m}$

$$\sigma_{q_2}^2 = \frac{1297{\cdot}20^2}{8 \times 12{\cdot}466^3} \times \frac{1}{0{\cdot}02} \times 28{\cdot}597277 \times 10^{-6} = 0{\cdot}15525\,\mathrm{m}^2,$$

hence $\sigma_{q_2} = 0{\cdot}39402\,\mathrm{m}$

$$\sigma_{q_3}^2 = \frac{650{\cdot}76^2}{8 \times 18{\cdot}025^3} \times \frac{1}{0{\cdot}01} \times 34{\cdot}044192 \times 10^{-6} = 0{\cdot}03077\,\mathrm{m}^2,$$

hence $\sigma_{q_3} = 0{\cdot}17541\,\mathrm{m}$

Thus

$$q_1 = \left\{\sqrt{[2\ln(4{\cdot}429 \times 10/2\pi)]} + \frac{0{\cdot}577}{\sqrt{[2\ln(4{\cdot}429 \times 10/2\pi)]}}\right\} \times 4{\cdot}59727$$

$$= 10{\cdot}42780\,\mathrm{m}$$

$$q_2 = \left\{\sqrt{[2\ln(12{\cdot}466 \times 10/2\pi)]} + \frac{0{\cdot}577}{\sqrt{[2\ln(12{\cdot}466 \times 10/2\pi)]}}\right\} \times 0{\cdot}39402$$

$$= 1{\cdot}05618\,\mathrm{m}$$

$$q_3 = \left\{\sqrt{[2\ln(18{\cdot}025 \times 10/2\pi)]} + \frac{0{\cdot}577}{\sqrt{[2\ln(18{\cdot}025 \times 10/2\pi)]}}\right\} \times 0{\cdot}17541$$

$$= 0{\cdot}49354\,\mathrm{m}$$

$$\begin{bmatrix} x_1 \\ x_2 \\ x_3 \end{bmatrix} = \begin{bmatrix} 0{\cdot}947 & 2{\cdot}128 & 1{\cdot}703 \\ 1{\cdot}706 & 0{\cdot}950 & -2{\cdot}128 \\ 2{\cdot}128 & -1{\cdot}703 & 0{\cdot}953 \end{bmatrix} \begin{bmatrix} 10{\cdot}42780 \\ 1{\cdot}05618 \\ 0{\cdot}49354 \end{bmatrix} \times 10^{-3} = \begin{bmatrix} 0{\cdot}013 \\ 0{\cdot}018 \\ 0{\cdot}021 \end{bmatrix} \mathrm{m}$$

## Torsional response to seismic motion

Torsional response of buildings to ground motion is due to the nature of the motion itself, lack of symmetry in the structure, and/or lack of symmetry in the distribution of the total mass about the shear centre of the building. In order to determine the contribution of the torsional vibration to the lateral, vertical and rocking responses it is necessary to analyse the structure as a three-dimensional one. This requires the assembly of three-dimensional stiffness and mass matrices. The construction of the former is given by Coates *et al.* (1972), and that of the latter by Clough & Penzien (1975). The total response may then be calculated either in the frequency domain by first calculating the natural frequencies and modes of vibration and then applying the method of mode superposition, using either translational and rotational response spectra or power spectra, or in the time domain using either real or generated earthquake histories. For nonsymmetric structures with large overhangs, such as cable-stayed cantilever roofs (see Fig. 1.1) and cantilevered cranes, the torsional response modes could be the more significant ones. Even for symmetric structures some codes require that the effect of possible vibration in torsional modes be taken into account by assuming the position of the centre of gravity of the structural mass to be eccentric to that of the structure's shear centre. Most cases, except the very simplest ones, will require the use of a computer. As an introduction to torsional vibration one such problem is considered in example 13.7, where the translational and rotational response of a platform considered as a shear structure is calculated using Kanai's power spectrum (eqs (13.30) and (13.31)).

The expression for the torsional response spectrum of a one-DOF spectrum due to a torsional moment

$$T(t) = P(t) \times e = M\ddot{x}_g(t) \times e \tag{13.88}$$

where $M$ is the equivalent mass of the structure, $\ddot{x}_g(t)$ is the ground acceleration at time $t$, and $e$ is the eccentricity of the mass relative to the shear centre of the structure, can be developed in exactly the same manner as that for the translational one, and can be shown to be

$$S_\theta(\omega) = \frac{M^2e^2}{K_t^{\,2}} M(\omega) S_{\ddot{x}_g}(\omega) \tag{13.89}$$

Hence the variance of torsional response is

$$\sigma_\theta^{\,2} = \int_0^\infty S_\theta(\omega)\mathrm{d}\omega = \frac{M^2e^2}{K_t^{\,2}} \int_0^\infty M(\omega) S_{\ddot{x}_g}(\omega)\mathrm{d}\omega \tag{13.90}$$

For weakly damped structures, and since $\omega_n^{\,2} = K_t/I_p = K_t/Mk^2$, where $k$ is the radius of gyration, the expression for $\sigma_\theta^{\,2}$ can be approximated to

$$\sigma^2 = \int_0^\infty S_\theta(\omega)\mathrm{d}\omega \approx \frac{1}{\omega_n^{\,4}} \times \frac{e^2}{k^4} M(\omega) S_{\ddot{x}_g}(\omega)\Delta\omega \tag{13.91}$$

where

$$\Delta\omega = \tfrac{1}{2}\xi\omega_n$$

$$M(\omega) = 1/4\xi^2$$

Thus, for weakly damped structures the expression for the variance of torsional response is

$$\sigma_\theta^{\,2} = \int_0^\infty S_\theta(\omega)\mathrm{d}\omega \approx \frac{1}{8} \times \frac{1}{\omega_n^{\,3}} \times \frac{e^2}{K^4} \times \frac{1}{\xi} S_{\ddot{x}_g}(\omega) \tag{13.92}$$

**Example 13.7** Calculate the translational and rotational response of the platform structure shown in Fig. 4.15 when subjected to an earthquake with a peak acceleration of $0{\cdot}25g$. Assume the dominant frequency of the ground to be 0·565 Hz, and the ground damping ratio to be 60·0% of critical. The equivalent mass of the structure at platform level is $4{\cdot}722 \times 10^6$ kg. The eccentricity of the centre of gravity of the mass relative to the shear centre of the structure measured perpendicular to the direction of the quake is assumed to be 1·0 m. The polar moment of inertia of the mass is $1\,361{\cdot}2421 \times 10^6$ kg m$^2$, the translational stiffness is $135{\cdot}748 \times 10^3$ kN/m and the rotational stiffness is $45\,475{\cdot}523 \times 10^3$ kN/rad. The translational and rotational frequencies are 0·8533 Hz and 0·9199 Hz respectively. The damping in both the translational and the rotational mode may be assumed to be 2·0% of critical, and the duration of the strong motion 10 s.

From eq. (13.46), the variance of the translational response of a weakly damped one DOF system can be written as a function of an earthquake acceleration spectrum

$$\sigma_x{}^2 = \frac{1}{8} \times \frac{1}{\omega_n{}^3} \times \frac{1}{\xi} S_{\ddot{x}_g}(\omega)$$

where

$$\omega_n = 2\pi \times 0{\cdot}8533 = 5{\cdot}361442 \text{ rad/s}$$

Thus

$$\sigma_x{}^2 = \frac{1}{8} \times \frac{1}{5{\cdot}361442^3} \times \frac{1}{0{\cdot}02} S_{\ddot{x}_g}(\omega) = 0{\cdot}0405541 S_{\ddot{x}_g}(\omega)$$

Similarly, from eq. (13.91), the variance of the rotational response of weakly damped one DOF systems can be written as

$$\sigma_\theta{}^2 = \frac{1}{8} \times \frac{1}{\omega_n{}^3} \times \frac{e^2}{k^4} \times \frac{1}{\xi} S_{\ddot{x}_g}(\omega)$$

where

$$\omega_n = 2\pi \times 0{\cdot}9199 = 5{\cdot}7799022 \text{ rad/s}$$

$$k = \sqrt{(1\,361{\cdot}2421 \times 10^6 / 4{\cdot}722 \times 10^6)} = 16{\cdot}97871 \text{ m}$$

Thus

$$\sigma_\theta{}^2 = \frac{1}{8} \times \frac{1}{5{\cdot}7799022^3} \times \frac{1{\cdot}0^2}{16{\cdot}97871^4} \times \frac{1}{0{\cdot}02} S_{\ddot{x}_g}(\omega)$$
$$= 0{\cdot}389493 \times 10^{-6} S_{\ddot{x}_g}(\omega)$$

From eqs (13.29) and (13.30)

$$S_{\ddot{x}_g}(\omega) = \frac{S_0\left[1 + (2\xi_g r)^2\right]}{(1 - r^2)^2 + (2\xi r)^2}$$

where

$$S_0 = \frac{0{\cdot}141\xi_g \ddot{x}_{g,\,max}{}^2}{\omega_g\sqrt{(1+4\xi_g{}^2)}}$$

Thus

$$S_0 = \frac{0{\cdot}141 \times 0{\cdot}6 \times 0{\cdot}25^2 \times 9{\cdot}81^2}{2\pi \times 0{\cdot}920\sqrt{(1+4\times 0{\cdot}6^2)}} = 0{\cdot}0563541\,\mathrm{m^2\,s^{-4}\,rad^{-1}}$$

When $f_n = 0{\cdot}8533$ Hz

$$r = 0{\cdot}8533/0{\cdot}920 = 0{\cdot}9275$$

$$S_{\ddot{x}_g}(\omega) = \frac{0{\cdot}0563541\left[1+(2\times 0{\cdot}6\times 0{\cdot}9275)^2\right]}{(1-0{\cdot}9275^2)^2 + (2\times 0{\cdot}6\times 0{\cdot}9275)^2}$$

$$= 0{\cdot}1002655\,\mathrm{m^2\,s^{-4}\,rad^{-1}}$$

Hence

$$\sigma_x{}^2 = 0{\cdot}0405541 \times 0{\cdot}1002655 = 4{\cdot}06617\times 10^{-3}$$

$$\sigma_x = 0{\cdot}0637665\,\mathrm{m}$$

The corresponding peak factor is

$$\kappa = \sqrt{[2\ln(0{\cdot}8533\times 10)]} + 0{\cdot}577\sqrt{[2\ln(0{\cdot}8533\times 10)]} = 2{\cdot}3493672$$

Thus the maximum translational amplitude is

$$x_{max} = \kappa\sigma_x = 2{\cdot}3493672 \times 0{\cdot}0637665 = \underline{0{\cdot}1498\,\mathrm{m}}$$

When $f_n = 0{\cdot}9199$ Hz

$$r = 0{\cdot}9199/0{\cdot}920 = 0{\cdot}9998913$$

$$S_{\ddot{x}_g}(\omega) = \frac{0{\cdot}0563541\left[1+(2\times 0{\cdot}6\times 0{\cdot}9998913)^2\right]}{(1-0{\cdot}9998913^2)^2 + (2\times 0{\cdot}6\times 0{\cdot}9998913)^2}$$

$$= 0{\cdot}0954973\,\mathrm{m^2\,s^{-4}\,rad^{-1}}$$

Hence

$$\sigma_\theta{}^2 = 0{\cdot}389493\times 10^{-6} \times 0{\cdot}0954973 = 0{\cdot}0371955\times 10^{-6}$$

$$\sigma_\theta = 0{\cdot}1928614\times 10^{-3}\,\mathrm{rad}$$

The corrresponding peak factor is

$$\kappa = \sqrt{[2\ln(0{\cdot}9199\times 10)]} + 0{\cdot}577\sqrt{[2\ln(0{\cdot}9199\times 10)]} = 2{\cdot}380589$$

Thus the maximum angular rotational amplitude is

$$\theta_{max} = \kappa\sigma_\theta = 2{\cdot}380589 \times 0{\cdot}1928614\times 10^{-3} = \underline{0{\cdot}4591327\times 10^{-3}\,\mathrm{rad}}$$

The corresponding movements at the corners of the platform are therefore

$$x_t = 0{\cdot}4591327\times 10^{-3} \times \sqrt{(20^2+20^2)} = \underline{0{\cdot}0123\,\mathrm{m}}$$

## Reduction of dynamic response

The two most common techniques used for reducing the vibration caused by earthquakes are isolation and energy absorption. A third method involves active control in which feedback from sensors recording the vibration of the structure is utilized to control the behaviour of the structure.

Isolation involves the installation at the base of a structure, mainly in order to limit the amount of horizontal ground acceleration transmitted to the building. Isolators are therefore essentially soft springs, usually constructed from lamination of steel and rubber. Their effectiveness depends on correct anticipation of the frequency contents of future earthquakes. In practice they are often designed to reduce the dominant frequency of a structure to 0·5 Hz or less.

Elastic absorption involves linking dampers between points with relative displacements, either within the structure or between the structure and the ground. Such dampers may be fabricated in the form of hydraulic dash pots, but are more generally designed to behave elastically up to a given maximum permitted relative displacement, above which they will yield. For further information about isolation and energy absorption the reader is referred to Harris (1988), Key (1988) and Warburton (1992).

As a first attempt to establish the spring stiffness of isolators, a building may be reduced to a one DOF structure constructed on a base which again is supported on isolators. Such a system, containing an internal energy absorber, is shown in Fig. 13.8.

The equation of motion for a one DOF shear structure with the base plate supported on isolators, as shown in Fig. 13.8, when subjected to a ground acceleration $\ddot{x}_g(t)$ is

$$\mathbf{M}\ddot{\boldsymbol{x}} + \mathbf{C}\dot{\boldsymbol{x}} + \mathbf{K}\boldsymbol{x} = \mathbf{M}\ddot{x}_g(t) \qquad (13.93)$$

or

$$\begin{bmatrix} M_1 & \\ 0 & M_2 \end{bmatrix}\begin{bmatrix} \ddot{x}_1 \\ \ddot{x}_2 \end{bmatrix} + \begin{bmatrix} (C_1 + C_2) & -C_2 \\ -C_2 & C_2 \end{bmatrix}\begin{bmatrix} \dot{x}_1 \\ \dot{x}_2 \end{bmatrix} + \begin{bmatrix} (K_1 + K_2 & -K_2 \\ -K_2 & K_2 \end{bmatrix}\begin{bmatrix} x_1 \\ x_2 \end{bmatrix}$$
$$= \begin{bmatrix} M_1 & 0 \\ 0 & M_2 \end{bmatrix}\begin{bmatrix} \ddot{x}_g(t) \\ \ddot{x}_g(t) \end{bmatrix} \qquad (13.94)$$

If one ignores the effect of damping, the corresponding eigenvalue equation is

$$\mathbf{KX} - \omega^2\mathbf{MX} = 0 \qquad (13.95)$$

which is satisfied when

$$\begin{vmatrix} (K_1 + K_2 - \omega^2 M_1) & -K_2 \\ -K_2 & (K_2 - \omega^2 M_2) \end{vmatrix} = 0 \qquad (13.96)$$

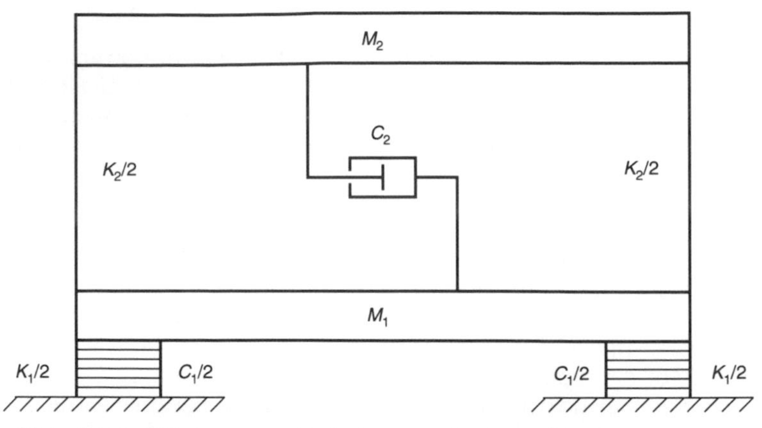


*Fig. 13.8. One DOF structure with isolators and energy absorber*

Evaluation of the determinant and solving the resulting characteristic equation with respect to $K_1$ yields the following expression for the shear stiffness of the isolators in terms of the required frequency

$$K_1 = \frac{\omega^2 K_2(M_1 + M_2) - \omega^4 M_1 M_2}{K_2 i\omega^2 M_2} \tag{13.97}$$

**Example 13.8** The first natural frequency of the shear structure in example 2.5 (Fig. 2.14) obtained from an eigenvalue analysis is 0·7419 Hz. The shear stiffness of the columns at ground level is $25{\cdot}0 \times 10^6$ kN/m. Reduce the structure to a one DOF system and calculate the shear stiffness of isolators required to reduce the first natural frequency to 0·5 Hz. Assume the mass of the base slab supporting the columns at ground level to be $1700{\cdot}0 \times 10^3$ kg.

The equivalent mass of the one DOF system is

$$M_2 = \frac{K_2}{\omega_1{}^2} = \frac{2500{\cdot}0 \times 10^6}{(2\pi \times 0{\cdot}7419)^2} = 115\,051{\cdot}77 \times 10^3 \text{ kg}$$

Thus, from eq. (2.92) the required combined shear stiffness of the isolators is

$$K_1 = \Big\{(2\pi \times 0{\cdot}5)^2 \times 2500{\cdot}0 \times 10^6(1500{\cdot}0 + 115\,051{\cdot}77) \times 10^3$$
$$-(2\pi \times 0{\cdot}5)^4 \times 1500{\cdot}0 \times 115\,051{\cdot}77 \times 10^6\Big\}/$$
$$\Big\{2500{\cdot}0 \times 10^6 - (2\pi \times 0{\cdot}5)^2 \times 115\,051{\cdot}77 \times 10^3\Big\}$$

$$\underline{K_1 = 2095{\cdot}28 \times 10^6 \text{ N/m}}$$

## Soil–structure interaction

When one is constructing numerical dynamic models of structures, it is necessary to consider the flexibility of the soil and also to what extent the weight of the structure is likely to reduce the dominant frequency of the soil above the bedrock. Fortunately the weight of most structures is very small compared with the amount of soil, and will not significantly alter the dynamic characteristics of the latter. Such changes usually need to be considered only in the case of exceptionally heavy rigid structures such as nuclear containment buildings, when the weight of the structure can affect the surface ground motion below and adjacent to the foundations. This may be better appreciated by considering the lumped mass model of soil shown in Fig. 13.9. When modelling such structures it is necessary to incorporate the supporting soil down to the rock base. This type of analysis is sophisticated and specialized, and outside the scope of this book. For detailed work on the subject the reader is referred to Wolf.

As stated above, the weight of most buildings will not alter the characteristics of the supporting ground, but the flexibility of the soil will tend to reduce the overall stiffness of the structure and thus reduce its frequencies and modify its modal response, as well as generating additional damping through energy dissipation. At resonance the surrounding layers of certain types of soil, such as wet clays, will also tend to vibrate in phase with the structure in the same manner as water and air will, and therefore add to the amount of the vibrating mass. For the

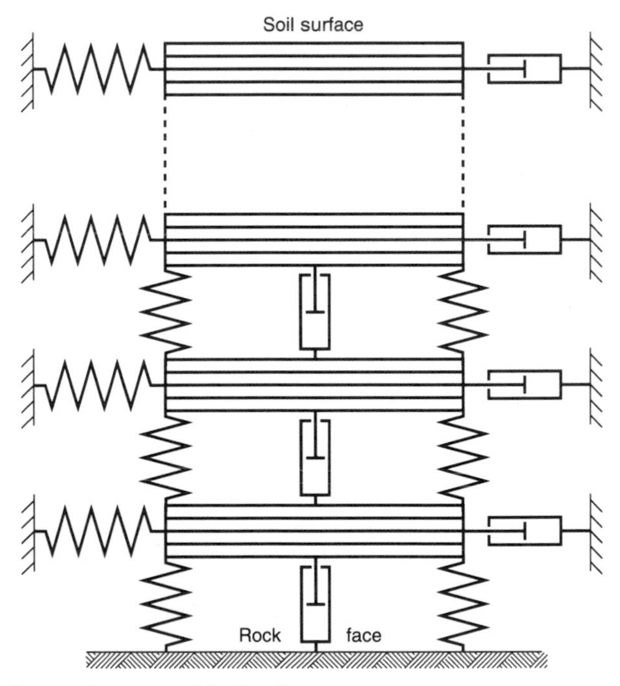


*Fig. 13.9. Lumped mass model of soil*

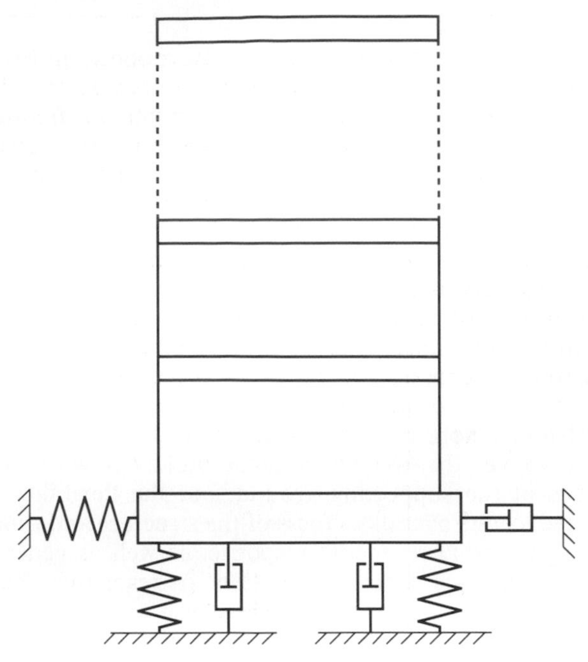

*Fig. 13.10. Numerical modelling of the stiffness and damping of soil by equivalent elastic springs and viscous dampers*

purpose of analysis the stiffness and damping properties of soil can be modelled as springs and equivalent viscous dampers, as indicated in Fig. 13.10. The modelling of such springs and dampers is considered to be outside the scope of this book. The reader is referred to Key (1988).

**Problem 13.1** A prestressed concrete bridge, which can be considered as a simply supported beam, spans 40 m. The flexural rigidity and the mass of the bridge are $8{\cdot}15887 \times 10^{10}\,\mathrm{N\,m^2}$ and 35 t/m respectively. The structural damping is 2·0% of critical. Use the response spectra shown in Fig. 13.4 to calculate the maximum first mode response due to an earthquake with a peak acceleration of $0{\cdot}3g$.

**Problem 13.2** The bridge in problem 13.1 is sited at a point where the depth of soil is approximately 21 m. For this depth the dominant frequency of the ground is estimated to be approximately 2·0 Hz. As the ground is firm, the ground damping is assumed to be 60% of critical. Use Kanai's power spectrum to estimate the maximum first mode response due to an earthquake with a peak acceleration of $0{\cdot}3g$.

**Problem 13.3** The two-storey shear structure shown in Fig. 13.11 is to be erected in a seismic zone on a site where the dominant frequency is unknown and therefore conservatively assumed to be equal to the first natural frequency of the building. Use first the response spectra shown in Fig. 13.4 and then Kanai's power spectrum to calculate the maximum response of the structure to an earthquake with a peak acceleration of $0{\cdot}35g$. Assume the damping in the first and second modes to be 1·5% and 1·0% of critical respectively, the damping in the ground to be 60% of critical, and the duration of the strong motion part of the quake to be 10 s. The mass matrix, natural angular frequencies and normalized modeshape matrix for the structure are as follows

$$\mathbf{M} = \begin{bmatrix} 6{\cdot}0 & 0 \\ 0 & 6{\cdot}0 \end{bmatrix} \times 10^4 \text{ kg}$$

$$\boldsymbol{\omega} = \begin{bmatrix} 9{\cdot}4248 \\ 24{\cdot}6743 \end{bmatrix} \text{rad/s}$$

$$\mathbf{Z} = \begin{bmatrix} 0{\cdot}2146 & 0{\cdot}4024 \\ 0{\cdot}3473 & -0{\cdot}0687 \end{bmatrix} \times 10^{-2}$$

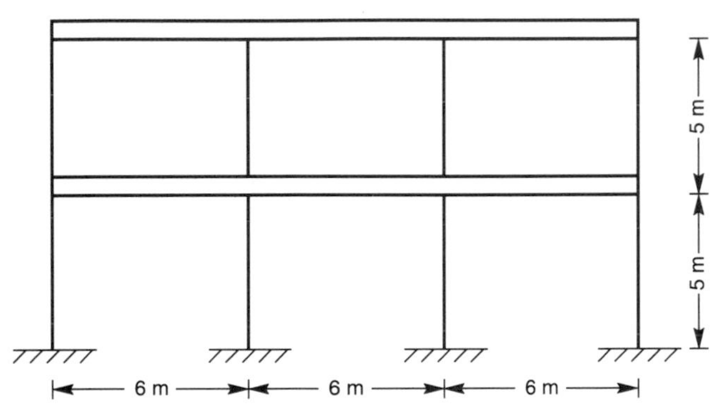


*Fig. 13.11. Two-storey shear structure*

**Problem 13.4** Calculate the response of the stepped tower in problem 7.4 to the lateral and rocking motions of an earthquake with a peak acceleration of $0{\cdot}35g$, if the duration of the strong motion part of the quake is 10 s. Assume the dominant frequency of the ground to be 2·0 Hz, the damping in the ground to be 60% of critical, and the shear velocity of the ground to be 500 m/s. The structural damping may be taken as 3·0% of critical in the first mode and as 2·0% in the second mode. The stiffness matrix, mass matrix, natural frequencies and normalized damping matrix for the tower are given below. The given stiffness and mass matrices have been obtained by eliminating the rotational degrees of freedom at stations 10 and 20 m above the ground.

$$\tilde{\mathbf{K}} = \begin{bmatrix} 27\,568{\cdot}761 & -7657{\cdot}989 \\ -7657{\cdot}989 & 3063{\cdot}186 \end{bmatrix} \text{kN/m}$$

$$\tilde{\mathbf{M}} = \begin{bmatrix} 20\,066{\cdot}47 & 2795{\cdot}75 \\ 2795{\cdot}75 & 4659{\cdot}62 \end{bmatrix} \text{kg}$$

$$\boldsymbol{\omega} = \begin{bmatrix} 25{\cdot}133 \\ 119{\cdot}098 \end{bmatrix} \text{rad/s}$$

$$\mathbf{Z} = \begin{bmatrix} 3{\cdot}443 & 6{\cdot}521 \\ 10{\cdot}109 & -10{\cdot}753 \end{bmatrix} \times 10^{-3}$$

## References

Blum, J. A., Newmark, N. M. & Corning, L. H. *Design of multistorey reinforced building for earthquake motions.* Portland Cement Association, Chicago, 1961.

Bolt, B. A. *Earthquakes, a primer.* W.H. Freeman, San Francisco, 1978.

Clough, R. W. & Penzien, J. *Dynamics of structures.* McGraw-Hill, London, 1975.

Coates, R. C., Coutie, M. G. & Kong, F. K. *Structural analysis.* Nelson, London, 1972.

Craig, R. R. Jr. *Structural dynamics.* Wiley, Chichester, 1981.

Eiby, G. A. *Earthquakes.* Heinemann, London, 1980.

Harris, C. M. *Shock vibration handbook*, 3rd edn. McGraw-Hill, London, 1988.

Kanai, K. Semi-empirical formula for the seismic characteristics of the ground. *University of Tokyo Bull. Earthquake Research Institute*, 1957, **35**, 309–325.

Key, D. E. *Earthquake design practice for buildings.* Thomas Telford, London, 1988.

Lin, B. C., Tadjbasksh, I. G., Papageorgiu, A. A. and Ahmadi, G. Response of base-isolated buildings to random excitation described by Clough–Penzien spectral model. *Earthquake Engng Struct. Dyn.* 1989, **18**, 49–62.

Lomnitz, F. & Rosenbleuth, E. *Seismic risk and engineering decisions.* Elsevier, Amsterdam, 1976.

Newmark, N. M. and Hall, W. J. *Earthquake spectra and design.* Earthquake Engineering Research Institute, Berkeley, 1982.

Paz, M. *Structural dynamics.* Van Nostrand Reinhold, New York, 1980.

Seed, H. B. & Idriss, I. M. *Ground motion and ground liquefaction during earthquakes.* Earthquake Engineering Research Institute, Berkeley, 1982.

Tajimi, H. A statistical method of determining the maximum response of building structures during an earthquake. *Proc. 2nd Int. Conf. Earthquake Engineering*, Tokyo and Kyoto, 1960, **II**, 781–798.

Warburton, G. B. *Reduction of vibrations.* Wiley, London, 1992.

Wolf, J. H. *Dynamic soil–structure interaction.* Prentice-Hall, Englewood Cliffs.

# 14. Generation of wind and earthquake histories

## Introduction

In chapter 6 a time domain method is presented for predicting the linear and nonlinear response of one DOF systems to wind and earthquakes, and to multi-DOF systems in general. The equations developed are based on the incremental equation of motion, and arise from various assumptions with respect to the change in acceleration during a time step $\Delta t$. Other time domain methods, which are particularly suitable for highly nonlinear structures such as guyed masts, cable and membrane roofs, are those in which equilibrium of the dynamic forces at the end of each time step is sought by minimization of the gradient vector of the total potential dynamic energy by use of the Newton-Raphson or conjugate gradient method, and where increased convergency and stability are achieved through scaling and the calculation of a steplength in the descent direction to a point where the energy is a minimum (Buchholdt, 1985; Buchholdt *et al.*, 1986). The prediction of response using any of the above methods requires the ability to generate earthquake histories and single and spatially correlated wind histories. The problem with using recorded earthquake histories is that no two earthquakes are the same. For the purpose of design it is therefore necessary to calculate the response to a family of simulated earthquakes compatible with a given site. Because wind histories can be considered as stationary stochastic processes, they are simpler to generate the earthquake ones. Therefore methods for simulating wind histories are presented first.

## Generation of single wind histories by a Fourier series

Shinozuka and Jan (1952) have shown that it is possible to express the fluctuating velocity component $u(t)$ of wind at any time $t$ as

$$u(t) = \sqrt{(2)} \sum_{i=1}^{n} \sqrt{[(s_u(n_i) \times \Delta n)]} \cos(2\pi n_i + \phi_i) \qquad (14.1)$$

where $S_u(n)$ is the value of the power spectral density function for the fluctuating component of wind at the frequency $n$, $\Delta n = n_{i+1} - n_i$, and $\phi_i$ is the phase angle with a uniform probability distribution function that varies randomly between 0 and $2\pi$.

The frequency band in eq. (14.1), which has been divided into $N$ parts, must contain all the significant natural frequencies of the structure. For nonlinear structures the frequency step $\Delta n$ needs to be small, as the natural frequencies of such structures vary with the amplitude of response.

### Generation of wind histories by the autoregressive method

Another method for generating single wind histories that yields variances of response similar to those of real wind is the autoregressive (AR) one. It is computationally more efficient that the Fourier series (FS) method given by eq. (14.1), and can also be used to generate earthquake histories. The AR method filters white noise and transforms it into a signal with a specified variance and autocovariance function.

Mathematically the method for transforming white noise may be expressed as

$$u(t) = \gamma(B) \times a(t) \tag{14.2}$$

where $u(t)$ is the stochastic process to be generated, $a(t)$ is the input white noise with zero mean and variance $\sigma_a^2$ and $\gamma(B)$ is a transfer function or filter. The white noise $a(t)$ may also be expressed as

$$a(t) = \sigma_{Nu} \times N(t) \tag{14.2}$$

where $N(t)$ are random shocks with zero mean and unit variance. Substitution of this expression for $a(t)$ into eq. (14.2) yields

$$u(t) = \gamma(B) \times \sigma_{Nu} \times N(t) \tag{14.3}$$

Thus the white noise process $a(t)$ is transformed into the process $u(t)$ by the filter or transfer function $\gamma(B)$. One type of filter that has proved to be very suitable for modelling wind and earthquakes is the so called autoregressive filter, which regressively weights and sums previous values.

In an autoregressively simulated process of order $p$, the instantaneous values of $u(t)$ are expressed as a finite linear aggregate of the previous values of $u(t)$, plus a random impulse with zero mean and variance $\sigma_{Nu}^2$. Thus the expression for $u(t)$ may be written as

$$u(t) = \sum_{s=1}^{p} \phi_s u(t - s\Delta t) + \sigma_{Nu} N(t) \tag{14.4}$$

where $\phi$ is an autoregressive parameter, $N(t)$ is a random impulse with zero mean and unit variance, and

$$\sigma_{Nu}^2 = \frac{1}{T} \int_0^T \sigma_{Nu} N(t) \times u(t) \mathrm{d}t \tag{14.5}$$

Alternatively, eq. (14.4) may be written as

$$u(t) = \sum_{s=1}^{p} \phi_s B_s u(t) + \sigma_{Nu} N(t) \tag{14.6}$$

where $B_s$ is a back shift operator, which is defined through the equality

$$B_s u(t) = u(t + s\Delta t) \tag{14.7}$$

Solving eq. (14.6) with respect to $u(t)$ yields

$$u(t) = \frac{1}{1 - \sum_{s=1}^{p} \phi_s B_s} \times \sigma_{Nu} N(t) \tag{14.8}$$

Comparing eq. (14.8) with eq. (14.3) yields the following expression for an autoregressive filter of order $p$

$$\gamma(B) = \frac{1}{\sum_{s=1}^{p} \phi_s B_s} \tag{14.9}$$

In order to obtain expressions for determining the values for the parameters $\phi$ and the variance $\sigma_{Nu}{}^2$, both sides of eq. (14.4) are multiplied by $u(t - kt)$, where $k = 1, 2, \ldots p$. Integration and averaging over time $T$ yields

$$\frac{1}{T}\int_0^T u(t) \times u(t - kt)\mathrm{d}t = \sum_{s=1}^{p} \frac{1}{T}\int_0^T \phi_s u(t - s\Delta t)u(t - k\Delta t)\mathrm{d}t + \frac{1}{T}\int_0^T \sigma_{Nu} N(t)u(t - k\Delta t)\mathrm{d}t \tag{14.10}$$

When $k > 0$, eq. (14.10) yields

$$C_u(f\Delta t) = C_u(-k\Delta t) = \sum_{s=1}^{p} \phi_s C_u[(k - s)\Delta t], \quad k = 1, 2, \ldots p \tag{14.11}$$

because of the symmetry of the auto-covariance function and the randomness of the process $N(t)$. When $k = 0$, eq. (14.10) yields

$$\sigma_u{}^2 = \sum_{s=1}^{p} \phi_s C_u(s\Delta t) + \sigma_{Nu}{}^2 \tag{14.12}$$

Division of all the elements in eqs (14.11) and (14.12) by $\sigma_u{}^2$ yields

$$c_u(k\Delta t) = \sum_{s=1}^{p} \phi_s c_u[(k - s)\Delta t] \quad k = 1, 2, \ldots p \tag{14.13}$$

$$\sigma_{Nu}{}^2 = \sigma_u{}^2 \left\{ 1 - \sum_{s=1}^{p} \phi_s c_u(s\Delta t) \right\} \tag{14.14}$$

where $c_u(k\Delta t)$ is the autocovariance coefficient at time lag $\tau = k\Delta t$ corresponding to the power spectral density functions $S_u(n)$, and $c_u[(k - s)\Delta t] = 1{\cdot}0$ when $k = s$.

The values of the autocovariance coefficient $c_u$ are determined, given an expression for the power spectrum $S_u(n)$, by dividing both sides of eq. (10.27) by $\sigma_u^2$ and completing the integration. Thus

$$c_u(k\Delta t) = \frac{1}{\sigma_u^2}\int_0^\infty S_u(n)\cos(2\pi nk\Delta t)\mathrm{d}n \tag{14.15}$$

The values of the autocovariance function having been determined using eq. (14.11), the autoregressive parameters $\phi$ can be determined by use of eq. (14.13) and the variances of the impulses $\sigma_{Nu}^2$ from eq. (14.14).

Unlike the Fourier series model, the autoregressive model is not unconditionally stationary, and tends to become nonstationary when a short time step is chosen. In this case, the right-hand side of eq. (14.11) may become negative. Another problem is concerned with the number of parameters $\phi$ to be used in generation of the autoregressive model. Both problems have been dealt with by Box and Jenkins (1977) in terms of the so-called partial autocorrelation function, which for a process suitable for simulation by an autoregressive method of order $q$ is nearly zero when $p > q$.

Figure 14.1 shows the variation of the partial autocorrelation function with the order of the model and with the size of the time step. The curves are based on Kaimal's spectrum (eq. (10.34)), with $V(10) = 30$ m/s and $z_0 = 0{\cdot}1$ m, and indicate that for the data used a suitable number of $\phi$ parameters is 3–5, and that the size of the time step should not be less than 0·1 s. Thus if $p = 3$ and $\Delta t = 0{\cdot}1$ s, from eqs (14.13) and (14.14)

$$\begin{bmatrix}\phi_1\\ \phi_2\\ \phi_3\end{bmatrix} = \begin{bmatrix} c_u(0{\cdot}0) & c_u(0{\cdot}1) & c_u(0{\cdot}2)\\ c_u(0{\cdot}1) & c_u(0{\cdot}0) & c_u(0{\cdot}2)\\ c_u(0{\cdot}2) & c_u(0{\cdot}1) & c_u(0{\cdot}0)\end{bmatrix}^{-1}\begin{bmatrix}c_u(0{\cdot}1)\\ c_u(0{\cdot}2)\\ c_u(0{\cdot}3)\end{bmatrix} \tag{14.16}$$

$$\sigma_{Nu}^2 = \sigma_u^2\{1 - \phi_1 c_u(0{\cdot}1) - \phi_2 c_u(0{\cdot}2) - \phi_3 c_u(0{\cdot}3)\} \tag{14.17}$$

It should be noted that a time step of $\Delta t = 0{\cdot}1$ s is of the order of ten times the size of the time step used in the forward integration method when the acceleration is assumed to remain constant during the time step. Thus it is necessary to interpolate to obtain the wind velocities required at any time $t$ during the dynamic analysis.

As the process $u(t)$ is generated with $\phi$ parameters that are functions of the autocovariance coefficients, the simulated histories need to be multiplied by the ratio $\sigma_u/\sigma_{gu}$, where $\sigma_u$ is the standard deviation of the required history and $\sigma_{gu}$ is the standard deviation of the generated process.

Of the two methods for generating wind histories, the FS method is more expensive in terms of computer time. This is shown in Fig. 14.2, where the time taken to generate wind histories by the two methods is compared. An example of an FS model and AR models generated with the same power spectrum and autocovariance functions as those of a recorded history is given by Buchholdt *et al.* (1986).

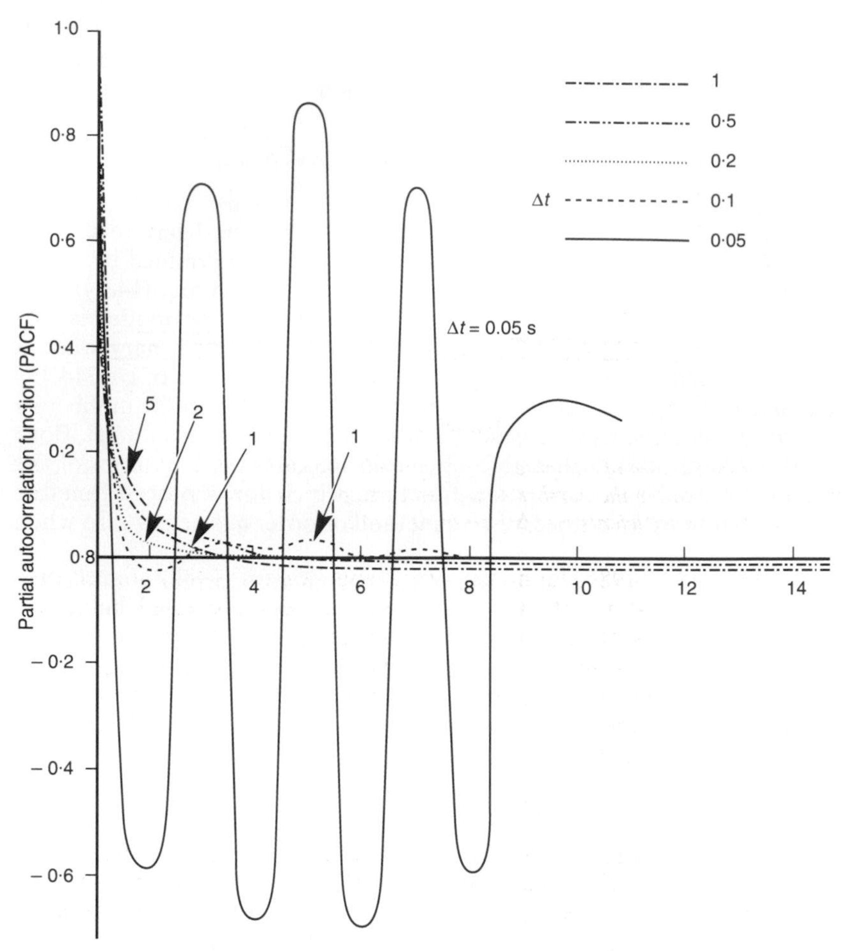


*Fig. 14.1. Variation of the partial autocorrelation model with number of parameters* $\phi$ *and size of time step*

## Generation of spatially correlated wind histories

In chapter 10 the correlation between the fluctuating wind velocities at two points in space is expressed in terms of the cross-covariance function (eq. (10.37)) and the cross-spectrum (eq. (10.41)), the latter being expressed as a function of the square root of the product of the power spectra of the individual histories and the coherence function (eq. (10.42)). In chapter 11 it is shown how power spectra and cross-spectra are used to establish model force spectra for calculating the variance of response of multi-DOF structures in the frequency domain (eqs (11.52)–(11.59)). The interdependence of the velocity fluctuations in space must also be included when generating spatial wind fields. This can be achieved in different ways. Spinelli devised a method with correlation at time lag

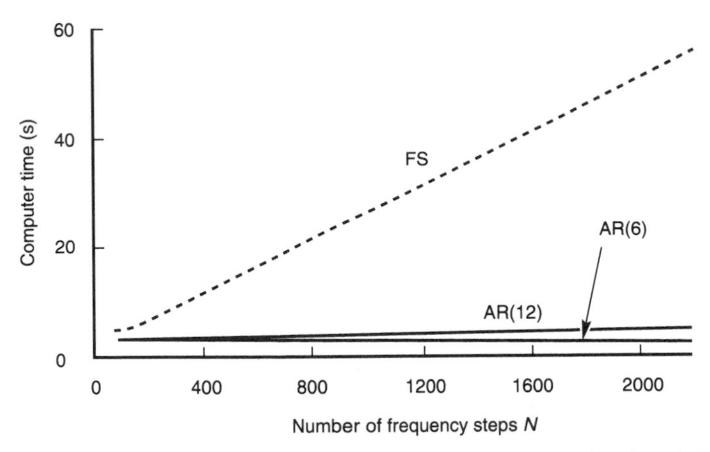


*Fig. 14.2. Comparison of computer time required to generate single FS and AR wind models: for the latter the number of frequency steps N is used only to compute the autocovariance coefficients $c_u(k\Delta t)$ from eq. (14.8)*

$\tau = 0$ (Buchholdt, 1985; Iannuzzi, 1987), and Iwatani (1982) devised one with correlation at $\tau \geq 0$. A third method is now presented based an eigenvalue analysis of the cross-covariance matrix $\mathbf{C}_{uv}(0)$at$\tau = 0$.

In chapter 10 it is shown that the wind velocity vector at any time $t$ may be considered to consist of a steady-state component $\boldsymbol{U}(z)$, whose element can be determined using either eq. (10.4) or eq. (10.10), and a fluctuating component $\boldsymbol{u}(z, t)$. Thus

$$\boldsymbol{U}(z, t) = \boldsymbol{U}(z) + \boldsymbol{u}(z, t) \tag{14.18}$$

The vector $\boldsymbol{u}(z, t)$ may further be expressed as $\mathbf{D} \times \boldsymbol{v}(z, t)$, where $\mathbf{D}$ is a correlation matrix whose elements are evaluated from the cross-covariance of the elements in $\boldsymbol{u}(z, t)$ at $\tau = 0$ and $\boldsymbol{v}(z, t)$ is a fluctuating velocity vector in which the elemental time histories are uncorrelated and can be modelled as either FS or AR series. Equation (14.11) may therefore be written as

$$\boldsymbol{U}(z, t) = \boldsymbol{U}(z) + \mathbf{D} \times \boldsymbol{v}(z, t) \tag{14.19}$$

The elements in matrix $\mathbf{D}$ are determined as follows. From eqs (10.37) and (10.38), it is shown that the cross-covariance matrix at time lag $\tau = 0$ is

$$\mathbf{C}_{u(z,t),u(z,t)}(0) - \mathbf{C}_u(0) = \frac{1}{T}\int_0^T \boldsymbol{u}(z, t) \times \boldsymbol{u}(z, t)^{\mathrm{T}}\, \mathrm{d}t = [\sigma_{ij}] \tag{14.20}$$

where $[\sigma_{ij}]$ is a square symmetrical matrix in which the elements on the leading diagonal $\sigma_{ij} = \sigma_i^2$ are the variances, and the off-diagonal elements are the cross-variances of the elemental processes in $\mathbf{u}(\mathbf{z}, \mathbf{t})$. Thus the elements on the leading diagonal may be calculated by use of eq. (10.28), and the off-diagonal elements by use of eq. (10.40). To proceed, the eigenvalues and the normalized eigenvectors of $\mathbf{C}_u(0)$ must be determined. Let the corresponding eigenvalue equation be

$$\mathbf{C}_u(0)\mathbf{X} = \boldsymbol{\lambda}\mathbf{I}\mathbf{X} \tag{14.21}$$

whose $\boldsymbol{\lambda} = \text{diag.}\{\lambda_1, \lambda_2, \ldots \lambda_n\}$ is the eigenvalue matrix and $\mathbf{X} = [\mathbf{X}_1, \mathbf{X}_2, \ldots \mathbf{X}_n]$ is the eigenvector matrix. In order to normalize an eigenvector $\boldsymbol{X}_i$, let

$$\boldsymbol{X}_i^{\mathrm{T}}\mathbf{I}\boldsymbol{X}_i = L_i^2 \tag{14.22}$$

The normalized eigenvector $\boldsymbol{Z}_i$ is now found by dividing each element in $\mathbf{X}_i$ by $L_i$. Thus

$$\boldsymbol{Z}_i = \mathbf{X}_i/L_i \tag{14.23}$$

$$\boldsymbol{Z}_i^{\mathrm{T}}\mathbf{I}\boldsymbol{Z}_i = 1 \tag{14.24}$$

Writing the eigenvalue eq. (14.21) in terms of the eigenvalue $\lambda_i$ and the normalized eigenvector $\boldsymbol{Z}_i$ and postmultiplication of each term by $\boldsymbol{Z}_i^{\mathrm{T}}$ yields

$$\boldsymbol{Z}_i^{\mathrm{T}}\mathbf{C}_u(0)\boldsymbol{Z}_i = \boldsymbol{\lambda}_i\boldsymbol{Z}_i^{\mathrm{T}}\mathbf{I}\boldsymbol{Z}_i = \boldsymbol{\lambda}_i \tag{14.25}$$

Hence

$$\boldsymbol{Z}^{\mathrm{T}}\mathbf{C}_u(0)\boldsymbol{Z} = \boldsymbol{\lambda} \tag{14.26}$$

$$\mathbf{C}_u(0) = \boldsymbol{Z}^{-\mathrm{T}}\boldsymbol{\lambda}\mathbf{Z}^{-1} \tag{14.27}$$

Because $\mathbf{C}_u(0)$ is a symmetric positive definite matrix, $\boldsymbol{Z}^{-1} = \boldsymbol{Z}^{\mathrm{T}}$ and $\boldsymbol{Z}^{-\mathrm{T}} = \boldsymbol{Z}$. Hence

$$\mathbf{C}_u(0) = \boldsymbol{Z}\boldsymbol{\lambda}\boldsymbol{Z}^{\mathrm{T}} \tag{14.28}$$

Thus if the uncorrelated wind histories in $\boldsymbol{v}(\boldsymbol{t})$ in eq. (14.19) are generated with power spectra having p7 variances, $\lambda_1, \lambda_2, \ldots \lambda_n$, then

$$\mathbf{C}_{v(z,\,y)v(z,\,t)}(0) = \mathbf{C}_v(0) = \frac{1}{T}\int_0^T \boldsymbol{v}(z, \boldsymbol{t}) \times \boldsymbol{v}(z, \boldsymbol{t})\mathrm{d}\boldsymbol{t} = \boldsymbol{\lambda} \tag{14.29}$$

From eqs (14.19) and (14.20) it follows that

$$\frac{1}{T}\int_0^T \mathbf{u}(\mathbf{z}, \mathbf{t}) \times \mathbf{u}(\mathbf{z}, \mathbf{t})^{\mathrm{T}}\,\mathrm{d}t = \left\{\frac{1}{T}\int_0^T \mathbf{D}\boldsymbol{v}(z, \boldsymbol{t}) \times [\mathbf{D}\boldsymbol{v}(z, \boldsymbol{t})]^{\mathrm{T}}\,\mathrm{d}\boldsymbol{t}\right\} \tag{14.30}$$

or

$$\frac{1}{T}\int_0^T \mathbf{u}(\mathbf{z}, \mathbf{t}) \times \mathbf{u}(\mathbf{z}, \mathbf{t})^{\mathrm{T}}\,\mathrm{d}t = \mathbf{D}\left\{\frac{1}{T}\int_0^T \boldsymbol{v}(z, \boldsymbol{t}) \times \boldsymbol{v}(z, \boldsymbol{t})^{\mathrm{T}}\,\mathrm{d}\boldsymbol{t}\right\}\mathbf{D}^{\mathrm{T}} \tag{14.31}$$

Hence

$$\mathbf{C}_u(0) = \mathbf{D}\boldsymbol{\lambda}\mathbf{D}^{\mathrm{T}} \tag{14.32}$$

Comparison of eqs (14.28) and (14.32) reveals that $\mathbf{D} = \mathbf{Z}$, from which it follows that

$$U(z,t) = U(z) + Z \times v(z,t) \tag{14.33}$$

where the histories in $v(t)$ are generated with different sets of random numbers and with variances $\lambda_1$, $\lambda_2$, $\ldots\lambda_n$, and where $\lambda_1$ to $\lambda_n$ are the eigenvalues of the cross-correlation matrix $\mathbf{C}_u(0) = [\sigma_{ij}]$, whose elements are defined by eq. (14.20).

Examples of the use of spatially correlated wind histories to determine the dynamic response of guyed masts are given by Buchholdt *et al.* (1986), Iannuzzi (1987) and Ashmawy (1991).

## Generation of earthquake histories

If the response to seismic excitation can be considered to be linear, then the analysis can be undertaken in the frequency domain, and the input excitation for the site under consideration can be prescribed in the form of response or power spectra as shown in chapter 13. If, on the other hand, the structure is likely to behave nonlinearly, the analysis should be carried out in the time domain and the input prescribed in the form of earthquake accelerograms. Fig. 14.3 shows the accelerograms and Fig. 14.4 the power spectral density functions for three different earthquakes. The former are of relatively short duration and the amplitude and hence the variance vary with time. The latter show the distribution of the square of the amplitudes of the acceleration histories of the frequency components in frequency domain, and also quite clearly the values of the dominant frequencies of the ground. A further spectrum analysis of adjacent time regions of each record would also reveal that the frequency/amplitude contents change during the passages of earthquakes. The reason for this is that there is a time difference between the arrivals of the P and S waves, and that the ground tends to filter out some of the higher frequency components. In order to take the nonstationarity of earthquake histories into account, the duration of the underlying stochastic process needs to be divided into separate contiguous time regions, each having a unique time variable frequency/amplitude content, whose amplitudes can be varied by using a deterministic time envelope or shaping function $\zeta(t)$.

An acceleration history for the $i$th time region, with zero mean and variance $\sigma_{\ddot{x}}{}^2$, may be generated by use of eq. (14.4). Thus

$$\ddot{x}_i(t) = \sum_{s=1}^{p} \phi_{is}\ddot{x}_i(t - s\Delta t) + \sigma_{N\ddot{x}_i}N(t) \tag{14.34}$$

or

$$\ddot{x}_i(t) = \sum_{s=1}^{p} \phi_{is}B_{is}\ddot{x}_i(t) + \sigma_{N\ddot{x}_i}N(t) \tag{14.35}$$

where $B_{is}$ is the *back shift operator* for the $i$th time region, whose meaning is defined by the equality

$$B_{is}\ddot{x}_i(t) = \ddot{x}_i(t + s\Delta t) \tag{14.36}$$

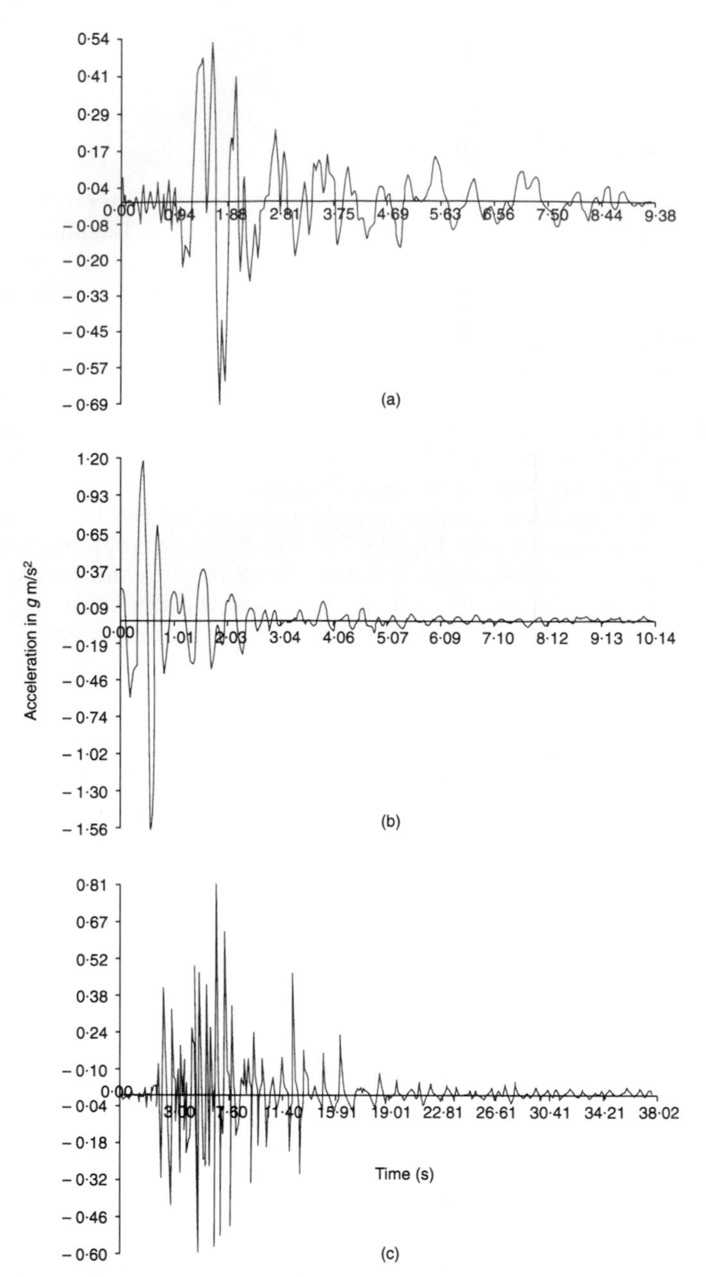


*Fig. 14.3. Accelerograms of main horizontal earthquake components: (a) N–S component of the San Salvador earthquake, 10 November 1986, duration 9·38 s, peak acceleration 0·69g; (b) E–W component of the Friuli 1 earthquake, Italy, 6 May 1976, duration 41·5 s, peak acceleration 0·16g; (c) L component of the Imperial Valley earthquake, USA, 15 May 1979, duration 42·1 s, peak acceleration 0·81g*

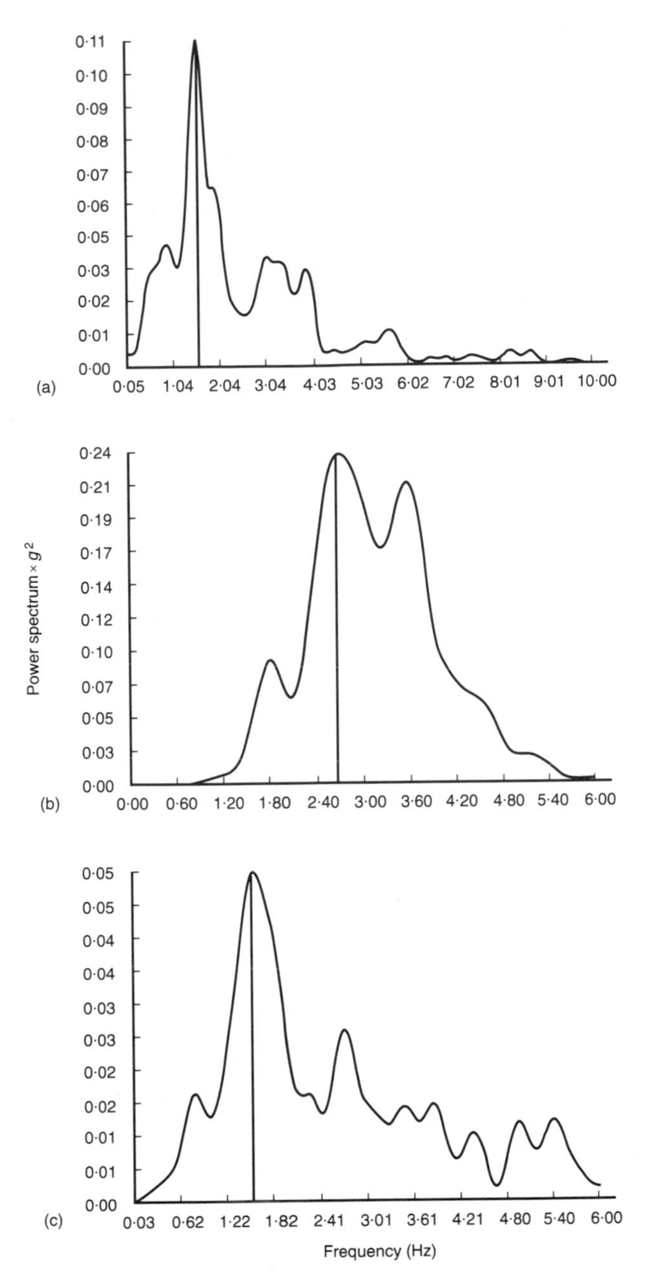


*Fig. 14.4. Power spectral density functions of accelerograms of earthquake components in Fig. 14.3: (a) N–S component of the San Salvador earthquake, 10 November 1986; (b) E–W component of the Friuli 1 earthquake, Italy, 6 May 1976; (c) L component of the Imperial Valley earthquake, USA, 15 May 1979*

Solving eq. (14.36) with respect to $\ddot{x}_i(t)$ yields

$$\ddot{x}_i(t) = \frac{1}{1 - \sum_{s=1}^{p} \phi_{is} B_{is}} \times \sigma_{N\ddot{x}_i} N(t) \tag{14.37}$$

where the parameters $\phi_{is}$ and the variance $\sigma_{N\ddot{x}_i}{}^2$ are determined in the same manner as for wind. Generation of time histories for each time region with different autocovariance functions, but with the same variance $\sigma_{\ddot{x}}{}^2$, leads to

$$c_{\ddot{x}}(k\Delta t) = \sum_{s=1}^{p} \phi_{is} c_{i\ddot{x}}[(k - s)\Delta t] \quad k = 1, 2, \ldots p \tag{14.38}$$

$$\sigma_{N\ddot{x}_i}{}^2 = \sigma_{\ddot{x}}{}^2 \left\{ 1 - \sum_{s=1}^{p} \phi_{is} x_{i\ddot{x}}(s\Delta t) \right\} \tag{14.39}$$

where eq. (14.39) can be written in matrix form (see eq. 14.16). Thus the shape function for the $i$th region $\zeta_i(t)$ can be expressed as

$$\zeta(t) = \sqrt{[\sigma_{i\ddot{x}}{}^2(t)/\sigma_{\ddot{x}}{}^2]} \tag{14.40}$$

where $\sigma_{x_i}{}^2$ is the variance for the $i$th time region. Hence

$$\ddot{x}_i(t) = \zeta_i(t) \times \ddot{x}_t(t) \tag{14.41}$$

Ashmawy (1991) investigated the above method and found that when generating earthquakes it is better to use 10 $\phi$ parameters rather than 3–5, as indicated by Fig. 14.1 in the case of wind, together with a time step of 0·02 s. Using the autocovariance functions of a number of different earthquakes, he showed that it is possible to generate families of earthquake histories with power spectra similar to those of the parent ones.

## Cross-correlation of earthquake histories

The motion of earthquakes is usually recorded in the form of accelerograms along three mutually perpendicular axes—two horizontal axes and one vertical axis. Thus, in order to generate families of earthquakes with the same statistical properties as the parent one it is necessary to generate three accelerograms not only with similar power spectra, but also with similar cross-covariance. Because the motion is nonstationary, each of the three recorded accelerograms and the corresponding underlying stochastic processes needs to be divided into the same number of separate contiguous time regions, with the latter being correlated region by region. This may be achieved as follows. Let the cross-covariance matrix for the $r$th time region of a recorded quake at time lag $\tau = 0$ be

$$\mathbf{C}_{\ddot{x},r}(0) = [\sigma_{i,j}]_r = \mathbf{Z}_r \boldsymbol{\lambda}_r \mathbf{Z}_r{}^{\mathrm{T}} \quad i = 1, \ldots 2, 3 \tag{14.42}$$

where $\boldsymbol{\lambda}_r = \text{diag.}[\lambda_1, \lambda_2, \lambda_3]_r$ is the eigenvalue matrix, and $\mathbf{Z}_r = [\mathbf{Z}_1, \mathbf{Z}_2, \mathbf{Z}_3]_r$ is the normalized eigenvector matrix of $\mathbf{C}_{\ddot{x},r}(0)$. If the uncorrelated acceleration histories are denoted by the vector $\ddot{\boldsymbol{x}}_r(\boldsymbol{t})$ and the correlated one by $\ddot{\boldsymbol{\chi}}_r(\boldsymbol{t})$, then it follows from eq. (14.43) that the correlated acceleration during the $r$th time region is given by

$$\ddot{\boldsymbol{\chi}}_r(\boldsymbol{t}) = \mathbf{Z}_r\ddot{\boldsymbol{x}}_r(t) \tag{14.43}$$

## Design earthquakes

The number of actual strong earthquake records available is limited, and even if they were available it is unlikely that they would form a basis for believing that future earthquakes occurring at the same site would be similar to those previously recorded. Thus there is a need for a method that enables the simulation of realistic earthquakes with different but defined statistical characteristics. Ashmawy (1991) found that: (*a*) it is possible to generate realistic time histories by assuming rectilinear autocovariance functions as shown in Fig. 14.5, and (*b*) the magnitudes of the dominant frequencies of the simulated quakes varied with the slope of the assumed autocovariance function. Thus he found that when the value of the time lag $\tau$ increased in steps from 0·2 to 5 s, the dominant frequency decreased from 3·6 to 0·48 Hz, while the frequency spectra of the simulated histories for $\tau = 0{\cdot}2$ s ranged from 3·6 to 15·5 Hz, and for $\tau = 5$ s from 0·05 to 0·38 Hz. Ashmawy's results are summarized in Figs 14.6 and 14.7. Design earthquakes simulated with the same slope of

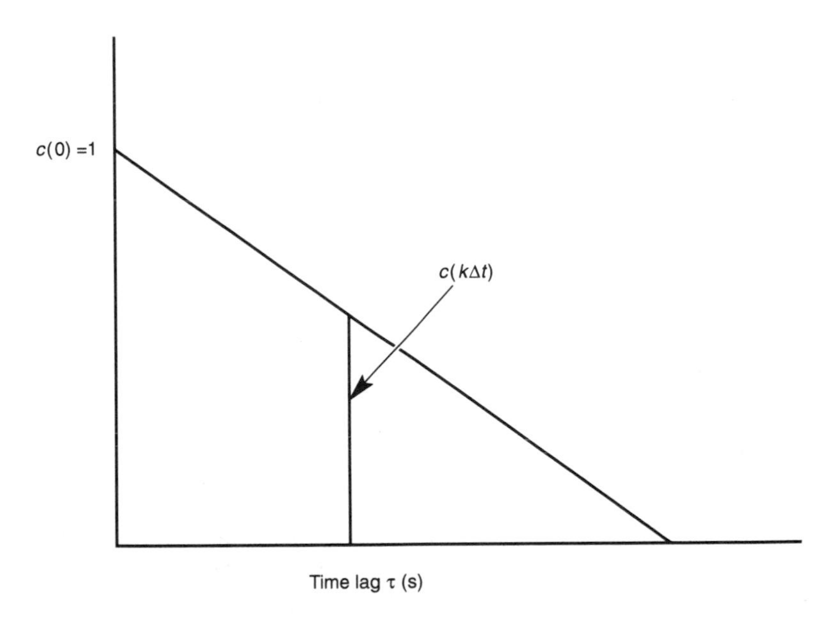


*Fig. 14.5. Linearized autocovariance function for the simulation of design earthquakes*

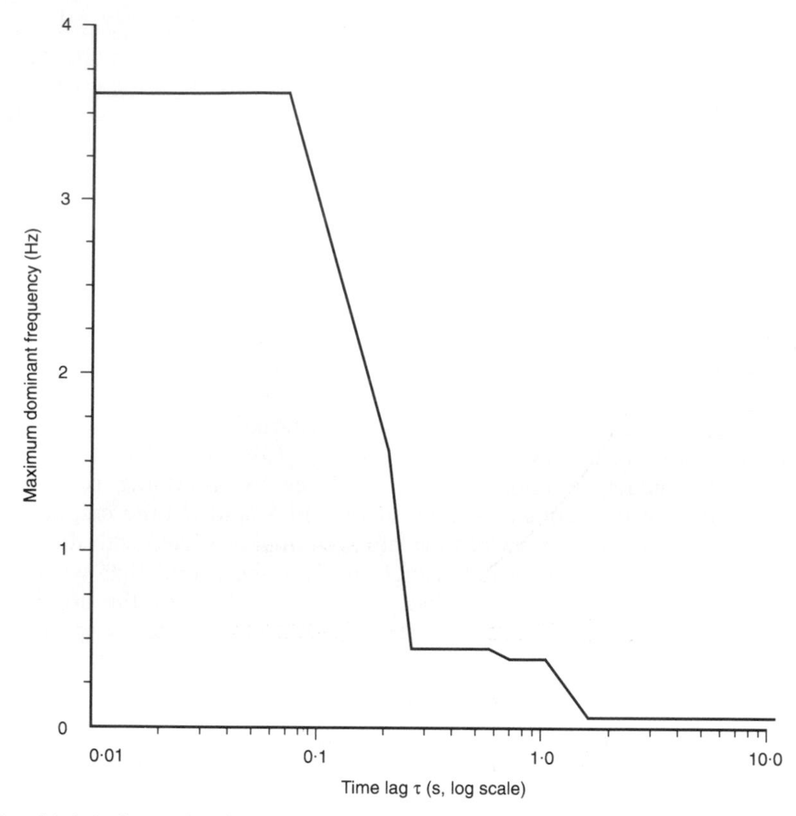


*Fig. 14.6. Relationship between total time lag τ of the autocovariance function and the maximum dominant frequencies of the resulting design earthquakes (after Ashmawy, 1991)*

the assumed autocovariance function but with different series of random numbers will have different power spectra. The curves in Fig. 14.7 should therefore by taken as indicating trends and not exact relationships. The smooth curves show median values derived from scattered points on a graph, with the degree of scatter being a function of the underlying series of random numbers used in generating with histories.

It is therefore advisable to calculate the response to a family of design earthquakes and not to only a single one. Ashmawy studied the validity of design earthquakes, simulated as described above, by comparing the calculated responses of a 238·6 m tall guyed mast to recorded earthquakes and to design earthquakes with the same peak acceleration and similar power spectra, and found that the two responses were very similar.

The worst design scenario for sites where the dominant ground frequency and the cross-correlation of the three acceleration components are unknown is when, given an assumed peak acceleration, the dominant frequency of the simulated histories coincides with the first natural

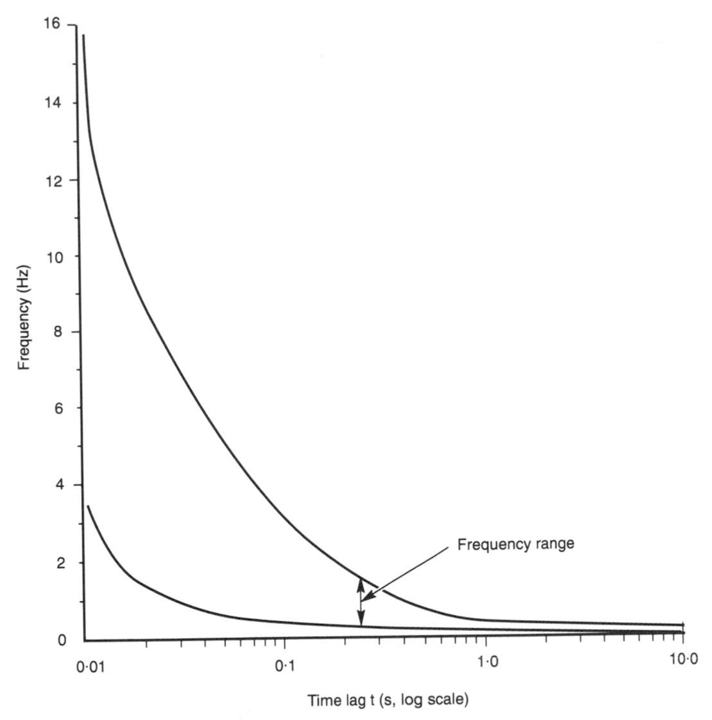


*Fig. 14.7. Relationship between the total time lag τ of the autocovariance function and the frequency range of the resulting design earthquakes (after Ashmawy, 1991)*

frequency of the structure and the cross-correlations of all three ground acceleration components are unity, The latter will be the case when these components are generated with the same set of random numbers.

## References

Ashmawy, M. A. *Nonlinear dynamic analysis of guyed masts for wind and earthquake loading*. PhD thesis, Polytechnic of Central London, 1991.

Box, G. E. P. and Jenkins, C. M. *Time series analysis: forecasting and control.* Holden Day, San Francisco, 1977.

Buchholdt, H. A. *Introduction to cable roof structures.* Cambridge University Press, Cambridge, 1985.

Buchholdt, H. A., Moossevinejad, S. & Iannuzzi, A. Non-linear dynamic analysis of guyed masts subjected to wind and guy ruptures. *Proc. Instn Civ. Engrs*, 1986, Part 2, Sept., 353–359.

Iannuzzi, A. *Response of guyed masts to simulated wind.* PhD thesis, Polytechnic of Central London, 1987.

Iwatani, Y. Simulation of multidimensional wind fluctuations having any arbitrary power spectra and cross spectra. *J. Wind Engng,* 1982, No. 1, Jan.

Shinozuka, M. and Jan, C. M. Digital simulation of random processes and its applications. *J. Aeronaut. Sci.* 1952, **19**, No. 12, Dec., 793–800.